AKADEMIE DER WISSENSCHAFTEN DER DDR

Zentralinstitut für Physik der Erde (ZIPE)

Seismological Bulletin 1972 Station Moxa (MOX)

By

Johannes Stelzner, Dorothea Güth, and Joachim Weyrauch

With 1 Figure

AKADEMIE-VERLAG · BERLIN
1977

Erschienen im Akademie-Verlag, 108 Berlin, Leipziger Straße 3—4

Lizenznummer: 202 · 100/460/77
Gesamtherstellung:
VEB Druckerei „Thomas Müntzer", DDR-582 Bad Langensalza
Bestellnummer: 762 494 5 (2004/B/1972) · LSV 1437
Printed in GDR
DDR 42,— M

TABLE OF CONTENTS

P R E F A C E

The annual Seismological Bulletin 1972 for the Station Moxa (MOX) continues the series of publications about seismological records at the Central Station Moxa edited by the Central Earth Physics Institute of the Academy of Sciences of the German Democratic Republic.

The provisional analysis of the records of station Moxa was performed in the Seismological Service under the direction of JOHANNES STELZNER by JOACHIM WEYRAUCH and BRIGITTE HÄNSCH.

The annual Bulletin 1972 was prepared by JOHANNES STELZNER and DOROTHEA GÜTH with the technical assistance of URSULA DÖRING.

Control of the instruments of the station Moxa was carried out under the direction of CHRISTIAN TEUPSER.

H. KAUTZLEBEN
Director

Preliminary notes for the interpretation of seismograms

In the Bulletin the international code is used:

1. Phase interpretation

Pg — direct longitudinal wave in near epicentral distances (D < 10°)

Pb, Pn — guided longitudinal head waves along the CONRAD- or MOHOROVIČIĆ-discontinuity (D < 10°)

P — direct longitudinal wave travelled through the earth mantle

P diff — direct longitudinal wave diffracted around the core boundary

PKIKP — direct longitudinal wave travelled through the inner core (traved-time branch DF)

PKHKP — direct longitudinal wave refracted in the intermediary zone between inner and outer core. Phase symbol according to BOLT [1] (travel-time branch GH)

PKP2 — direct longitudinal wave travelled through the outer core only (travel-time branch AB)

PKP — first noticeable onset of longitudinal core phase not identified

PP, PPP — waves reflected on the earth surface with permanent longitudinal character

PKKP — core phase reflected once within the core at the outer core boundary

PKPPKP — longitudinal core phase reflected at the earth surface

Sg — direct transversal wave in near epicentral distances (D < 10°)

Sb, Sn — guided transversal head waves along the CONRAD- or MOHOROVIČIĆ-discontinuity (D < 10°)

S — direct transversal wave travelled through the earth mantle

SKS —	direct wave travelled transversal through the mantle and longitudinal through the core
SS, SSS —	waves reflected on the earth surface with permanent transversal character
SKKS —	wave travelled transversal through the mantle, longitudinal through the core and reflected within the core at the outer core boundary
PcP, ScS, PcS, ScP —	longitudinal and transversal waves with steady or changing character reflected at the outer core
PS, SP, PPS —	longitudinal and transversal waves with changing character reflected at the surface of the earth
pP, sP, pPP, sPP, pPKIKP, sPKP2, pS —	phases of deep-focus earthquakes of longitudinal or transversal waves with steady or changing character. p ;s — reflected near the epicentre
pPKP, sPKP —	phases of deep focus earthquakes of longitudinal core waves not exactly to be coordinated
SKP, PKS —	core phases with different character before and after the direct transit of the core
SKSP —	SKS wave with longitudinal character after the reflection at the surface of the earth
P1, P2, P3,...S1, S2,... —	multiple onsets of body waves
Pn, Sn —	teleseismic Pn and Sn waves in the epicentral distances $23° < D < 40°$ after Båth [2]
Pa, Sa —	waves probably guided in the astenosphere channel or higher modes of surface waves
PL —	leaking modes, normal dispersed train of waves of periods greater than about 10 s, beginning at or near the time of initial P-wave
X, Y, Z —	remarkable phases of body waves, not to be identified
LmV, LmH —	maximum of the vertical and horizontal component respectively of longperiodical surface waves. If there are several maxima with comparable proportions in A/T, the numeration was carried out in a temporal sequence e.g. Lm1H, Lm2H

The phase symbol is followed by the designation of the type of seismometer from which the time of onsets is taken.

A —	seismograph with amplitude characteristic of type A (short-period)

B — seismograph with amplitude characteristic of type B (middle-period)
C — seismograph with amplitude characteristic of type C (long-period)

2. Measurements of amplitudes and calculation of magnitudes

All data of amplitudes and periods printed in the column "remarks" are always taken from the records of the same instruments, from which are taken the onset-times of the corresponding phases. The symbol of phase and component is followed by the symbol of the type of instruments e. g.: PV A, PV B, LmHB, LmV C.

Data of amplitudes obtained from records of instruments of type A are given in units of length of nm (1 nm = 1 nanometre = 10^{-6} millimetre). Data of amplitudes obtained from instruments of type B and such obtained from instruments of type C are given in units of length μm (1μm = 1 mikrometre = 10^{-3} millimetre) e. g.: PVA 1.3 s 38.6 nm, SHB 10 s 3.2 μm, LmH B 22 s 15 μm.

Magnitudes are determined from all those phases, for which calibrating functions are known and internationally used, i. e.

for maxima of body waves P(PH, PV), PP(PPH, PPV), and S(SH)-Q-functions from GUTENBERG and RICHTER [3] — and

for maxima of surface waves (h $<$ 100 km) LmH, LmV — calibrating functions from Prague σ [4] —.

The station correction S was not yet taken into consideration.

MB — magnitude of vertical component V of the first onset of P-waves given by NEIC
MS — magnitude of horizontal component H of the maximum surface wave given by NEIC
M — magnitude calculated from given data of station Moxa. Notice the wave type and the type of instruments written on the same line

3. Direction of body-wave onsets

If the direction of motion at the beginning of a wave onset is clearly to be recognized, the sign + or — is placed before the phase symbol. It means:

in the Z component	+ ground motion upwards, compression
	— ground motion downwards, dilatation
in the N component	+ ground motion to the north
	— ground motion to the south
in the E component	+ ground motion to the east
	— ground motion to the west

4. Further abbreviations

i — sharp beginning of phase motion (impetus)
e — gradual beginning of phase motion (emersio)

D — epicentral distances in degree (°), calculated according to geocentric coordinates, the maximum error of the own calculations amounts to ± 0,1°

Az — azimuth: clockwise measured angle between north direction in epicentre and the connecting line from epicentre to station Moxa

h — depth of focus in km, our data for depth of focus are based on travel-time curves foı deep focus earthquakes after GUTENBERG and RICHTER [5]

H — origin time in GMT (Greenwich Mean Time)

NEIC — National Earthquake Information Center, Boulder, Colorado, USA

BCIS — Bureau Central International de Séismologie Strasbourg France,

ANUSSR — Academia Nauk USSR, Moscow

USAEC — United States Atomic Energy Commission, Washington, D.C.

ISC — International Seismological Centre, Edinburgh

HAN — Niedersächsisches Landesamt für Bodenforschung Hannover-Buchholz, FRG

For abbreviations of seismological stations and other agencies in the international three letter code see the introductions to the Regional Catalogue of Earthquakes, Edinburgh and the Bulletins of the International Seismological Centre, Edinburgh.

Round brackets indicate uncertainties in interpretation of phase, time, depth of focus or epicentral distances respectively.

[1] BOLT, A., The velocity of seismic waves near the earths center. Bull. Seism. Soc. Am. 54 (1964) 1, 191—208.

[2] BÅTH, M., Propagation of Sn and Pn teleseismic distances. Pure and Applied Geophysics 64 (1966/II) 19—30.

[3] GUTENBERG, B. and RICHTER, C. F., Magnitude and energy of earthquakes. Annali di Geofisica 9 (1956) 1, 1—15.

[4] KÁRNÍK, V., KONDORSKAJA, N. V. u. a., Standardization of the earthquake magnitude scale. Stud. Geophys. et Geodet., Prague 6 (1962) 41—48.

[5] GUTENBERG, B. and RICHTER, C. F., Materials for the study of deep-focus earthquakes. Bull. Seism. Soc. Am. 26 (1936) 4, 341—390.

Seismological Station Moxa (MOX)
of the Central Earth Physics Institute

Elevation above mean sea level:	455 m
Bedrock:	clay slate of the lower carboniferous formation
Geographic coordinates:	$\varphi = 50°38'46''$N $\lambda = 11°36'58''$ E
Address:	Central Earth Physics Institute Seismological Service GDR-69, Jena, Burgweg 11 German Democratic Republic Telex: 05886275 seis dd

Seismographs and their parameters 1972

T_s — seismometer free period
T_g — galvanometer free period
D_s — seismograph damping
D_g — galvanometer damping
V_0 — magnification factor
N — north-south component
E — east-west component
Z — vertical component
σ^2 — coupling coefficient

SKM — Seismograph Kirnos modified
SSJ — Seismic Station Apparatus Type Jena
VSJ — Vertical Seismograph Type Jena

	Type of Seismograph	Comp.	T_s/[s]	T_g/[s]	D_s	D_g	V_o	σ^2
	VSJ II	Z	0.23	0.065	0.33	1.2	300 000	0.048
	VSJ II	Z	1.0	1.0	0.5	0.5	47 200	0.56
A	SKM III	N	1.64	0.39	0.52	1.98	25 700	0.047
		E	1.64	0.39	0.51	1.93	25 400	0.046
		Z	1.64	0.39	0.51	1.99	24 600	0.050
B	SSJ I	N	20	1.13	0.50	8.87	109	0.077
			20	1.14	0.50	8.79	1 110	0.077
		E	20	1.13	0.49	8.85	103	0.070
			20	1.16	0.49	8.61	1 050	0.070
		Z	20	1.13	0.48	8.82	108	0.057
			20	1.24	0.48	8.05	1 070	0.057
C	SSJ I/L	N	30	89.5	1.49	0.5	1 000	0.085
		E	30	74.4	1.24	0.5	1 010	0.051
		Z	30	91.0	1.52	0.5	1 060	0.108
	Strain/L	N		85.4		0.70	65*	
		E		86.2		0.70	67*	
	(coupled)	N + E		86.2		0.70	42*	

*) for wave velocity 5 km · s^{-1}

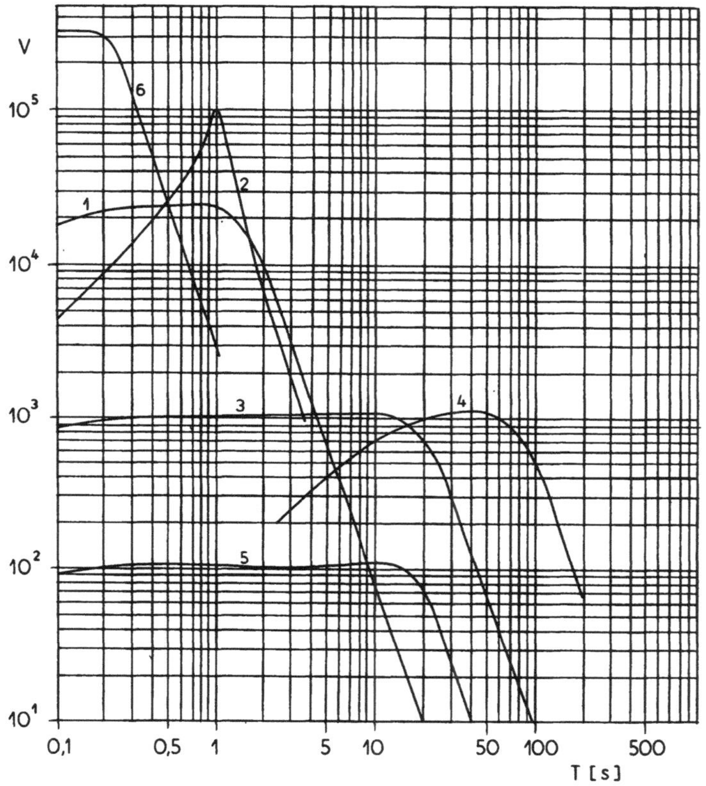

Fig. 1. Station Moxa, mean amplitude characteristics 1972

1 — Seismograph Kirnos Modified-III (SKM-III) (NS-, EW- and Z-component)

2 — Seismograph Type Jena II (Z-component)

3 — Seismic Station Apparatus Type Jena I/1000 (SSJ-I/1000) (NS-, EW- and Z-component)

4 — Seismic Station Apparatus Type Jena I/L (SSJ-I/L) (NS-, EW- and Z-component)

5 — Seismic Station Apparatus Type Jena I/100 (SSJ-I/100) (NS-, EW- and Z-component)

6 — Seismograph Type Jena II (Z-component)

Seismological Recordings at Station Moxa 1972

Day	Phase		h m s	Remarks
1.	ePKIKP	A	10 33 52	Solomon Islands 4.62 S 155.24 E
				H = 10 15 48.0 h = 506 km MB = 5.2
				D = 124.93 Az = 332.6 (NEIC)
				PKIKPV A 2.0s 51.3nm
1.	eP	A	13 06 15	Iceland 64.00 N 22.35 W
	e(PcP)	A	10 30	H = 13 01 18.9 h = normal MB = 4.5
	LmH	C	17.0	D = 22.34 Az = 110.6 (NEIC)
	LmV	C	22.0	PV A 1.2s 16.3nm M = 4.4
				LmV C 13 1.12/um 4.6
1.	eP	A	14 46 51	Iceland 64.16 N 22.37 W
				H = 14 41 53.3 h = normal MB = 4.3
				D = 22.41 Az = 111.0 (NEIC)
1.	+ePKP	AB	22 25 34	Fiji Islands Region 16.74 S 174.79 E
	e	C	26 24	H = 22 05 59.0 h = 56 km MB = 5.8
	e	B	27 02	D = 143.57 Az = 341.9 (NEIC)
	ePP	B	28 28	PKPV A 1.7s 69.7nm
	ei	B	29 40	PPV B 9 3.6/um M = 6.6
	eSS	B	47 30	LmH B 20 15.7/um
	eSSP	B	48 10	LmV B 19 15.6/um
	eSSS	B	52 44	
	LmH	B	23 29.7	
	LmV	B	37.8	
1.	ePKP	A	23 04 33	Fiji Islands Region 16.97 S 174.21 E
	epPKP	A	04 44	H = 22 44 58.9 h = 36 km MB = 5.2
				D = 143.62 Az = 341.3 (NEIC)
				PKPV A 1.9s 30.3nm
				pPKPV A 1.5 25.1nm
2.	ePKIKP	A	04 27 58	South of Fiji Islands 24.92 S 179.99 W
	ePKHKP	A	28 05.5	H = 04 09 03.8 h = 497 km MB = 4.5
	ePKP2	A	28 18.5	D = 152.79 Az = 343.7 (NEIC)
				PKIKP traces

Day	Phase		h m s	Remarks
2.	eP	A	10 36 22.5	Southern Sinkiang Prov., China
	eX	A	36 44	41.80 N 84.47 E
	eY	A	36 51.5	H = 10 27 34.9 h = normal MB = 5.2
	LmH	B	58.0	D = 49.28 Az = 306.6 (NEIC)
	LmV	B	58.0	PV A 1.2s 28.5nm M = 5.2
				XV A 1.3 19.6nm
				YV A 1.1 14.1nm
				LmH B 12 0.4/um 4.6
				LmV B 12 0.4/um 4.7
2.	ePKP2	A	18 57 26	Kermadec Islands 30.0 S 177.4 W
				H = 18 37 00 h = 66 km
				D = 158.26 Az = 344 (ISC)
2.	LmH	C	20 30.0	Kyushu, Japan 30.56 N 130.66 E
	LmV	C	30.4	H = 19 36 00.5 h = 62 km MB = 4.7 (NEIC)
				D = 82.90
				LmV C 15s 0.47/um
2.	eP	AB	22 10 00	Near Coast of Guerrero, Mexico
	ePP	B	13 30	16.15 N 98.36 W
	ePPP	B	15 28	H = 21 57 05.0 h = 54 km MB = 5.4
	eSKS	B	20 30	D = 89.76 Az = 36.8 (NEIC)
	eS	B	20 59	PV B 12s 0.9/um M = 6.0
	e	B	21 18	PPV B 10 1.3/um 6.3
	eSS	B	26 45	PPPV B 8 0.5/um
	LmH	B	52.8	LmH B 20 2.0/um
	LmV	B	53.7	LmV B 18 2.8/um
2.	eP	A	22 22 35.5	Bonin Islands Region 27.43 N 141.31 E
				H = 22 09 38.4 h = 54 km MB = 5.3
				D = 90.46 Az = 330.7 (NEIC)
3.	eP	A	06 48 15.5	Off East Coast of Kamchatka
				51.56 N 159.37 E
				H = 06 36 37.9 h = normal MB = 4.8
				D = 74.56 Az = 339.4 (NEIC)
				PV A 0.6s 13.4nm M = 5.1

Day	Phase		h m s	Remarks
3.	eP	A	17 18 17	Rat Islands, Aleutian Is.
	e	A	19 11	51.14 N 178.90 E
	ePP	C	21 20	H = 17 06 22.3 h = 46 km MB = 5.5
	eS	B	28 30	D = 78.02 Az = 351.8 (NEIC)
	eSS	B	33 25	PV A 2.0s 47.0nm M = 5.2
	LmH	B	18 02.9	LmH B 15.5 1.6 µm 5.5
	LmV	B	03.3	LmV B 15.5 1.4 µm 5.4
4.	eP	A	00 32 20	Near East Coast of Honshu, Japan
	ePP	A	35 33	35.82 N 140.45 E
				H = 00 19 59.7 h = 43 km MB = 5.1
				D = 82.83 Az = 330.0 (NEIC)
				PV A 1.6s 22.0nm M = 5.0
4.	+iP	AB	03 29 27.8	Taiwan Region 22.55 N 122.11 E
	PmV	A	29 48	H = 03 16 54.5 h = normal MB = 6.1
	ePP	A	32 46	D = 84.95 Az = 323.2 (NEIC)
	e(SKS)	C	39 46	PV A 1.5s 106.0nm M = 5.9
	iS	B	39 52	PV B 10 4.64 µm 6.7
	iSS	C	45 27	PmV A 2.6 1000.0nm 6.6
	ePKKP	A	47 38	PPV A 1.9 193.0nm 6.2
	eP'P'	A	55 56	SH B 15 9.3 µm 6.6
	LmH	B	04 05.2	P'P'V A 2.4 82.9nm
	LmV	B	13.2	LmH B 17.5 196.0 µm 7.6
				LmV B 17 157.0 µm 7.5
4.	e	A	12 28 25.5	Taiwan Region 22.41 N 122.19 E
	LmH	C	13 07.8	H = 12 15 16.6 h = normal MB = 4.8 (NEIC)
	LmV	C	10.1	D = 85.11
				LmH C 18s 0.4 µm M = 4.9
				LmV C 14 1.0 µm 5.4
4.	eP	A	18 23 51	Molucca Sea 0.05 S 124.81 E
	LmV	C	19 10.3	H = 18 09 46.1 h = normal MB = 5.6
	LmH	C	10.6	D = 104.56 Az = 322.8 (NEIC)
				LmV C 20s 0.8 µm M = 5.2

Day	Phase		h m s	Remarks
4.	ePKHKP	A	20 26 29.5	South of Fiji Islands 23.84 S 179.95 E
	ePKP2	A	26 41	H = 20 07 35.8 h = 558 km MB = 4.9
				D = 151.75 Az = 344.2 (NEIC)
				PKP2V A 1.2s 12.2nm
5.	eP	A	01 10 50.5	Rat Islands, Aleutian Is.
				51.15 N 178.87 E
				H = 00 58 54.9 h = 49 km MB = 5.1
				D = 78.00 Az = 351.7 (NEIC)
5.	eP	A	02 28 08.5	Kurile Islands 43.81 N 147.18 E
				H = 02 16 10.1 h = normal MB = 4.5
				D = 78.30 Az = 332.9 (NEIC)
5.	ePn	A	04 58 45.3	Austria 47.82 N 16.20 E
	iPn	A	58 47	H = 04 57 40.8 h = 11 km MB = 4.0
	iPb	A	58 54	D = 4.12 Az = 315.0 (NEIC)
	iPg	A	59 00.5	LmH B 5.5s 4.2/um M = 4.1
	i	A	59 24	LmV B 6 2.3/um
	iSn	A	59 32	
	iSg	AB	59 49	
	LmH	B	05 00.0	
	LmV	B	03.0	
5.	eP	A	06 07 28	Central Mid-Atlantic Ridge
	e	A	07 38	3.33 N 31.34 W
				H = 05 57 26.6 h = normal MB = 4.9
				D = 59.36 Az = 30.3 (NEIC)
				PV A 2.2s 38.2nm M = 5.1
5.	LmH	C	12 41.0	Nicaragua 12.29 N 86.24 W
	LmV	C	41.3	H = 11 52 26.9 h = normal MB = 4.2 (NEIC)
				D = 85.53
				LmV C 18s 0.44/um M = 4.9
6.	+iPKIKP	AB	00 48 56.8	New Britain Region 4.58 S 151.84 E
	epP	A	49 37.5	H = 00 30 17.2 h = 160 km MB = 5.9
	+ePP	AB	50 40	D = 123.29 Az = 330.8 (NEIC)

Day	Phase		h m s	Remarks
cont. 6.	eX	B	00 51 32	PKIKPV A 1.3s 78.6nm
	e	A	51 35	pPV A 1.2 20.4nm
	e(SKP)	A	52 17	PPV A 1.6 121.0nm M = 5.8
	ePS	C	01 00 20	PPV B 8 0.76/um 5.9
	e	C	01 24	XV B 8 0.98/um
	eSS	C	07 20	SKPV A 1.7 66.0nm
	ePSPS	C	08 20	
	LmH	C	34.0	
	LmV	C	44.4	
6.	eP	A	04 26 54	North Atlantic Ocean 55.40 N 35.18 W H = 04 21 04.3 h = normal MB = 4.5 D = 28.12 Az = 80.0 (NEIC) traces
6.	e	A	05 47 57.5	North Atlantic Ocean 55.51 N 35.11 W
	eX	A	48 30.5	H = 05 41 53.9 h = normal MB = 4.3
	eY	A	48 45	D = 28.06 Az = 80.2 (NEIC)
	LmH	B	06 01.6	XV A 1.4s 16.3nm
	LmV	B	01.6	YV A 2.0 29.9nm
				LmH B 13 1.4/um M = 4.7
				LmV B 12 1.5/um 4.9
6.	eP	A	09 48 26	Iran 30.30 N 50.52 E H = 09 41 33.2 h = 41 km MB = 5.2 D = 35.35 Az = 316.3 (NEIC) PV A 1.0s 9.8nm M = 4.7
6.	ePKP	A	12 33 13.5	Fiji Islands Region 17.06 S 175.92 E
	LmV	B	13 40.9	H = 12 13 38.9 h = normal MB = 5.1
	LmH	B	41.3	D = 144.21 Az = 342.9 (NEIC)
				LmH B 24s 1.5/um M = 5.6
				LmV B 20 1.2/um 5.6
6.	eSg	A	13 08 48.5	Poland 50.28 N 18.92 E H = 13 06 14.5 M = 2.8 (WAR) D = 4.7

Day	Phase		h m s	Remarks
6.	LmH	B	18 04.5	<u>Rat Islands, Aleutian Is.</u>
	LmV	B	04.7	51.0 N 178.6 E
				H = 17 26 21.4 h = 26 km MB = 4.0 (NEIC)
				D = 78.14
				LmV B 18s 0.8 µm M = 5.1
6.	e	A	20 36 07.5	<u>Poland</u> 50.30 N 18.92 E
	eSg	A	36 17.5	H = 20 33 37.6 M = 2.6 (WAR)
				D = 4.7
6.	ePg	A	23 05 18	<u>Federal Rep. of Germany</u>
	eiSg	A	05 50	51°38' N 7°07' E
				Rockburst in the Ruhr mining district
				H = 23 04 20 (BCIS)
				D = 3.0 Az = 107 (ISC)
				PgV A 1.0s 19.7nm
7.	eSg	A	04 27 48	<u>Poland</u> 50.27 N 18.99 E
				H = 04 25 17.4 M = 2.9 (WAR)
				D = 4.7
7.	ePKIKP	A	06 44 28	<u>Near N. Coast of West New Guinea</u>
	eX	A	44 34	2.13 S 138.98 E
	ePP	A	45 27	H = 06 25 48.4 h = normal MB = 5.9
	eY	A	45 40	D = 114.49 Az = 326.2 (NEIC)
	LmH	B	07 26.5	PKIKP traces
	LmV	B	35.3	XV A 1.8s 27.0nm
				PPV A 2.0 77.0nm M = 6.1
				YV A 2.0 59.9nm
				LmH B 23 5.5 µm 6.1
				LmV B 20 4.9 µm 6.1
7.	eSg	A	16 32 30	<u>Poland</u> 50.2 N 18.92 E
				H = 16 29 52.7 M = 2.5 (WAR)
				D = 4.7

Day	Phase		h m s	Remarks
7.	LmV	C	18 25.0	Near N. Coast of West New Guinea
				1.87 S 139.13 E
				H = 17 14 37.4 h = normal MB = 4.8 (NEIC)
				D = 114.36
7.	+iPKHKP	A	23 52 28.5	Fiji Islands Region 20.74 S 178.62 W
				H = 23 33 46.1 h = 600 km MB = 4.4
				D = 149.08 Az = 347.3 (NEIC)
				PKHKPV A 1.2s 18.3nm
8.	ePKIKP	A	03 27 37.5	Tonga Islands 18.78 S 173.31 W
	ePKHKP	A	27 40	H = 03 07 56.2 h = normal MB = 5.3
	LmH	C	04 37.2	D = 147.97 Az = 354.1 (NEIC)
	LmV	C	40.3	PKIKPV A 1.9s 37.9nm
				PKHKPV A 2.0 128.0nm
				LmH C 20 0.88 μm M = 5.5
				LmV C 18 1.32 μm 5.8
8.	ePn	A	04 46 55	Austria 47.6 N 16.1 E
	iPn	A	46 56	H = 04 45 48 h = 0 km
	ePg	A	47 08	D = 4.21 Az = 318 (ISC)
	iSn	A	47 42	
	eiSg	A	47 58	
8.	+iP	A	05 40 27.8	Philippine Islands Region
	-iP	B	40 29	20.87 N 120.21 E
	ePP	B	43 40	H = 05 27 52.9 h = normal MB = 6.2
	ePP	A	43 45	D = 85.22 Az = 322.7 (NEIC)
	eiS	B	50 55	PV A 1.9s 855.0nm M = 6.7
	ePS	B	51 25	PV B 9.5 9.9 μm 7.0
	i	B	51 40	PPV A 2.2 311.0nm 6.4
	eSS	C	56 30	PPV B 14 3.3 μm 6.6
	ei	B	57 15	SH B 11 6.1 μm 6.6
	eSSS	C	06 00 00	P'P'V A 1.9 34.1nm
	e	C	00 55	LmH B 14 58.0 μm 7.1
	e	A	06 32	LmV B 13.5 69.3 μm 7.2
	eP'P'	A	06 39	
	LmH	B	20.4	
	LmV	B	24.2	

Day	Phase		h m s	Remarks
8.	iP	A	05 48 07.5	<u>Taiwan Region</u> 21.31 N 120.22 E
				H = 05 35 34.8 h = normal MB = 5.1
				D = 84.88 Az = 322.7 (NEIC)
				PV A 1.0s 29.5nm M = 5.4
8.	eP	A	06 22 24	<u>Philippine Islands Region</u>
	eX	A	22 33.5	20.91 N 120.46 E
	eY	A	23 07.5	H = 06 09 48.7 h = normal MB = 4.8
				D = 85.34 Az = 322.8 (NEIC)
				PV A 1.6s 30.2nm M = 5.3
				XV A 1.6 27.5nm
				YV A 1.3 19.7nm
8.	eP	A	06 37 20	<u>Philippine Islands Region</u>
				20.91 N 120.26 E
				H = 06 24 43.5 h = 20 km MB = 4.2
				D = 85.22 Az = 322.7 (NEIC)
8.	eP1	A	08 26 55.5	<u>Philippine Islands Region</u>
	+iP2	A	27 03.5	20.87 N 120.26 E
	LmH	B	09 08.6	H = 08 14 20.0 h = normal MB = 5.3
	LmV	B	10.8	D = 85.25 Az = 322.7 (NEIC)
				P1V A 1.5s 30.2nm M = 5.3
				P2V A 1.4 69.8nm 5.7
				LmH B 14.5 2.2/um 5.7
				LmV B 14 1.8/um 5.6
8.	ePP	A	11 53 52	<u>South Sandwich Islands Region</u>
	epPP	A	53 56.5	55.76 S 28.72 W
	(SP)	B	12 03 38	H = 11 34 49.2 h = 60 km MB = 6.2 (NEIC)
	eSS	C	09 22	D = 111.21
	LmH	B	34.2	PPV A 1.7s 42.4nm M = 5.8
	LmV	B	34.2	pPPV A 1.7 72.7nm
				LmH B 22 3.1/um
				LmV B 20 2.8/um
8.	eP	A	14 44 50	<u>Philippine Islands Region</u>
				20.92 N 120.28 E

Day	Phase		h m s	Remarks
cont. 8.	e	A	14 45 00	H = 14 32 14.8 h = normal MB = 5.0 D = 85.23 Az = 322.7 (NEIC) PV A 1.5s 22.6nm M = 5.2
8.	ePKP2	A	21 53 55	North Island, New Zealand 37.76 S 175.56 E H = 21 33 04.8 h = 37 km (NEIC) D = 162.77
9.	e eX	A A	07 18 44.5 18 48	Loyalty Islands Region 22.50 S 171.32 E H = 06 58 52.3 h = 59 km MB = 4.6 D = 147.74 Az = 335.6 (NEIC) XV A 1.6s 16.5nm
9.	eSn eSg	A A	12 54 54 55 06	Federal Rep. of Germany 48.3 N 9.0 E H = 12 53 33 (BCIS) D = 2.88
9.	e	A	14 12 30.5	Off East Coast of Kamchatka 55.69 N 163.62 E H = 14 00 59.3 h = normal MB = 4.3 (NEIC) D = 71.50
10.	+iP e	A A	05 36 28 36 36	Philippine Islands Region 20.90 N 120.37 E H = 05 23 52.5 h = 32 km MB = 5.0 D = 85.29 Az = 322.8 (NEIC) PV A 1.8s 40.6nm M = 5.3
10.	eP	A	15 16 39.5	Dodecanese Islands 35.08 N 27.75 E H = 15 12 14.4 h = 42 km MB = 4.1 D = 19.48 Az = 328.0 (NEIC) PV A 1.7s 18.2nm M = 4.1
10.	eP	A	19 33 06.5	South of Honshu, Japan 30.99 N 137.04 E H = 19 21 20.5 h = 489 km MB = 5.1 D = 85.52 Az = 328.6 (NEIC) PV A 1.6s 24.7nm M = 4.7

Day	Phase		h m s	Remarks
11.	eP	A	03 14 11.5	North Atlantic Ridge 15.16 N 46.61 W
				H = 03 04 16.2 h = normal MB = 4.6
				D = 58.39 Az = 39.5 (NEIC)
12.	eP	A	08 21 39.5	United Arab Republic 27.53 N 33.75 E
	e	A	21 47	H = 08 15 46.1 h = 54 km MB = 5.1
				D = 28.57 Az = 329.9 (NEIC)
				PV A 2.0s 64.1nm M = 4.9
12.	-eiP	AB	10 11 15	Western Brazil 6.86 S 71.82 W
	iSKS	B	20 53	H = 09 59 10.3 h = 580 km MB = 5.9
	eS	B	21 22	D = 91.10 Az = 39.2 (NEIC)
	eSS	B	27 56	PV A 1.7s 270.0nm M = 6.0
	ePKKP	A	28 36	PV B 8 2.07/um 6.2
	esSS	B	31 00	LmV B 16 1.64/um
	LmH	B	50.6	
	LmV	B	50.6	
12.	+iP	A	13 55 26.5	Crete 34.99 N 23.47 E
	(LmH)	B	14 03.8	H = 13 51 20.2 h = 48 km MB = 4.9
	(LmV)	B	03.9	D = 17.87 Az = 334.8 (NEIC)
				PV A 0.9s 44.7nm M = 4.6
12.	+iP	A	18 45 37.5	Tadzhik-Sinkiang Border Region
	+e(pP)	A	46 12.5	37.69 N 75.07 E
	eiPcP	A	47 28	H = 18 37 20.6 h = 84 km MB = 5.6
				D = 45.96 Az = 307.6 (NEIC)
				PV A 0.8s 96.1nm M = 5.7
				pPV A 1.3 122.2nm
				PcPV A 2.0 119.8nm
12.	eP	A	20 31 35	Off East Coast of Kamchatka
	e	A	31 41	55.57 N 163.92 E
				H = 20 20 14.6 h = normal MB = 4.8
				D = 71.67 Az = 341.8 (NEIC)

Day	Phase		h m s	Remarks
13.	eP	A	14 06 05.5	Kurile Islands 46.80 N 152.46 E H = 13 54 16.3 h = 60 km MB = 5.2 D = 77.22 Az = 335.7 (NEIC) PV A 1.0s 15.8nm M = 5.0
13.	+eP	A	17 34 43.5	Eastern Siberia 61.93 N 147.06 E
	e	A	34 48.5	H = 17 24 22.6 h = normal MB = 5.3
	ePcP	A	35 23	D = 62.30 Az = 329.7 (NEIC)
	LmV	B	18 03.7	PV A 1.4s 60.5nm M = 5.6
	LmH	B	04.1	LmH B 14 4.5 µm 5.8 LmV B 16 2.1 µm 5.5
13.	LmV	B	22 52.2	Off Coast of Central Mexico
	LmH	B	52.3	22.69 N 107.90 W H = 21 58 49.1 h = normal MB = 5.0 (NEIC) D = 89.65 LmH B 17.5s 1.3 µm M = 5.4 LmV B 16 1.4 µm 5.5
14.	e	A	22 16(37)	Iran-Iraq Border Region 32.82 N 46.86 E H = 22 10 03.7 h = normal MB = 5.1 D = 31.37 Az = 315.1 (NEIC) traces
15.	ePKIKP	A	03 58 46	Tonga Islands 18.30 S 174.62 W
	ePKHKP	A	58 48	H = 03 39 17.0 h = 130 km MB = 5.6
	ePKP2	A	58 51.5	D = 147.34 Az = 352.6 (NEIC)
	epPKP	A	59 23	h = 130 km PKHKPV A 1.3s 87.4nm pPKPV A 1.9 75.7nm
15.	eP	A	19 33 57	Mexico-Guatemala Border Region 15.96 N 90.99 W H = 19 21 20.8 h = normal MB = 4.7 D = 85.58 Az = 38.5 (NEIC) PV A 1.2s 20.3nm M = 5.2

Day	Phase		h m s	Remarks
15.	+eP1	A	20 30 20.2	Southern Sinkiang Prov., China
	eiP2	A	30 23.5	40.33 N 78.99 E
	ePP	CE	32 15	H = 20 21 50.1 h = normal MB = 5.4
	eS	B	37 10	D = 46.82 Az = 306.3 (NEIC)
	eSS	B	40 30	P1V A 1.6s 60.4nm M = 5.3
	LmH	B	50.1	P2V A 1.5 100.5nm 5.6
	LmV	B	53.9	LmH B 15 23.1/um 6.3
				LmV B 15 31.1/um 6.5
15.	ePKHKP	A	21 06 02.5	Fiji Islands Region 20.57 S 176.18 W
	ePKP2	A	06 09.5	H = 20 46 41.6 h = 245 km MB = 4.4
				D = 149.36 Az = 350.2 (NEIC)
				PKHKPV A 0.7s 17.2nm
15.	eP	A	23 54 25.5	Southern Sinkiang Prov., China
				40.18 N 78.73 E
				H = 23 45 59.3 h = normal MB = 5.0
				D = 46.74 Az = 306.4 (NEIC)
16.	ePn	A	11 01 18.5	Czechoslovakia 49.15 N 18.84 E
	ePg	A	01 38.5	explosion
	e(Sg)	A	02 29.5	H = 11 00 00 yield 140 t (PRU)
				D = 4.90 Az = 291 (ISC)
16.	ePKHKP	A	24 18 07	Tonga Islands 21.16 S 173.51 W
	e	A	18 24	H = 23 58 17.4 h = normal MB = 5.2
				D = 150.30 Az = 353.4 (NEIC)
				PKHKPV A 1.3s 35.0nm
17.	e	A	00 26 14	Northern Italy 44.22 N 7.9 E
	eSg	A	26 49	H = 00 23 02.6 h = 0 km (ISC)
				D = 7.0
17.	eSn	A	04 07 14	Northern Italy 44.21 N 8.0 E
	eSg	A	08 01	H = 04 04 12 h = 0 km (ISC)
				D = 7.0

Day	Phase		h m s	Remarks
17.	eP	A	04 30 45.5	<u>Kurile Islands</u> 45.96 N 149.25 E H = 04 18 59.2 h = 125 km MB = 5.0 D = 77.04 Az = 333.9 (NEIC) PV A 1.6s 49.5nm M = 5.0
17.	e	A	05 59 06.5	<u>Crete</u> 34.86 N 27.0 E H = 15 54 20 h = 25 km MB = 4.3 (ISC) D = 19.3
17.	ePn	A	12 09 46	<u>Northern Italy</u> 44.15 N 8.18 E
	ePg	A	10 21	H = 12 08 06.0 h = 0 km
	iSn	A	11 03	D = 6.90 Az = 19 (ISC)
	eSg	A	11 57	
17.	eP	A	21 55 31.5	<u>Southern Sumatra</u> 1.94 S 99.56 E
	eX	A	55 39	H = 21 42 32.9 h = normal MB = 5.2 MS=5.1 D = 90.17 Az = 320.5 (NEIC) PV A 1.0s 23.6nm M = 5.4 XV A 1.9 60.6nm
18.	+iP	A	00 29 10	<u>Kodiak Island Region</u> 56.51 N 153.50 W
	epP	A	29 15.5	H = 00 17 43.0 h = 23 km MB = 5.1
	eX	A	29 23	D = 72.51 Az = 9.9 (NEIC) h = 20 km PV A 1.0s 43.3nm M = 5.4 pPV A 1.1 28.2nm XV A 1.4 46.5nm
18.	+iP	AB	12 54 51	<u>Near East Coast of Kamchatka</u>
	LmH	B	13 25.0	54.68 N 161.43 E
	LmV	B	32.0	H = 12 43 27.8 h = 33 km MB = 5.6 D = 72.05 Az = 340.3 (NEIC) PV A 1.4s 205.0nm M = 6.0
18.	eP	A	14 13 56.5	<u>Kurile Islands</u> 44.58 N 149.11 E H = 14 02 01.1 h = 52 km MB = 4.8 D = 78.23 Az = 333.9 (NEIC) PV A 1.0s 15.8nm M = 4.9

Day	Phase		h m s	Remarks
18.	eP	A	21 18 04.5	Northwestern Iran 37.55 N 48.68 E
				H = 21 12 01.7 h = normal MB = 4.9
				D = 29.34 Az = 308.4 (NEIC)
18.	ePKIKP	A	22 14 07.5	Near North Coast of New Guinea
	e	A	14 16	4.78 S 145.04 E
	e	A	14 44	H = 21 55 15.5 h = normal MB = 5.7
	ePP	A	15 45	D = 120.00 Az = 327.7 (NEIC)
	ePKKP	A	24 27	PPV A 1.9s 91.0nm M = 6.1
	eSS	B	32 00	PKKPV A 1.7 42.4nm
	LmH	B	57.4	LmH B 25 33.4 µm 6.9
	LmV	B	23 05.7	LmV B 22 19.5 µm 6.7
18.	ePKIKP	A	22 27 03.5	Near North Coast of New Guinea
	e	A	27 14	4.70 S 145.03 E
	ePP	A	28 44.5	H = 22 08 12.9 h = normal MB = 5.5
	ePKKP	A	37 19	D = 119.92 Az = 327.7 (NEIC)
				PPV A 2.2s 107.0nm M = 6.1
18.	ePn	A	23 27 52	Northern Italy 44.24 N 8.25 E
	e	A	27 58	H = 23 26 11.9 h = 25 km MB = 4.1
	eiPg	A	28 24.5	D = 6.80 Az = 18.4 (NEIC)
	iSn	A	29 07	
	eSg	A	29 57	
19.	eP	A	00 42 50	Madeira Islands Region
	e	A	43 05.5	31.21 N 13.87 W
				H = 00 37 06.7 h = normal MB = 4.9
				D = 27.12 Az = 36.9 (NEIC)
19.	ePKIKP	A	15 19 47	Near North Coast of New Guinea
	iPP	B	21 10	4.70 S 144.97 E
	ePPP	B	23 44	H = 15 00 54.2 h = normal MB = 5.8
	iPS	B	30 55	D = 119.90 Az = 327.7 (NEIC)
	eSS	C	37 45	PKIKPV A 1.4s 23.3nm
	eSSS	C	41 50	PPV B 9 3.80 µm M = 7.0
	LmV	B	16 10.6	LmH B 17 7.3 µm 6.4
	LmH	B	18.7	LmV B 19 10.9 µm 6.5

Day	Phase		h m s	Remarks
19.	ePKHKP	A	20 41 55	West of Macquarie Island
	ePKP2	A	42 10	60.15 S 150.37 E
	e	A	42 20.5	H = 20 21 53.8 h = normal MB = 5.4
				D = 155.12 Az = 266.1 (NEIC)
20.	eP	A	00 56 29	Dodecanese Islands 36.65 N 27.08 E
	e	A	56 39	H = 00 52 21.4 h = normal MB = 4.6
	LmH	C	01 03.5	D = 17.86 Az = 326.4 (NEIC)
	LmV	C	05.5	PV A 1.7s 66.6nm M = 4.5
20.	eP	A	02 19 16.5	Dodecanese Islands 36.55 N 27.14 E
				H = 02 15 06.7 h = normal MB = 4.8
				D = 17.97 Az = 326.5 (NEIC)
				PV A 1.7s 66.6nm M = 4.5
20.	+iP	AB	11 44 16.5	Hindu Kush Region 36.38 N 70.71 E
	iPP	A	45 57	H = 11 36 28.6 h = 213 km MB = 6.0
	e	B	47 14	D = 44.01 Az = 308.2 (NEIC)
	eScP	A	49 27	PV A 2.0s 769.0nm M = 5.8
	e	B	51 51	PPV A 1.3 131.0nm 5.5
	e	B	54 27	ScPV A 1.2 52.8nm
20.	ePg	A	18 51 28.5	Northern Italy 44.18 N 8.19 E
	eSg	A	52 55	H = 18 49 13.4 h = 0 km (ISC)
				D = 6.8
21.	e	A	01 12 16	Western Mediterranean Sea
	eSn	A	12 55.5	23.42 N 6.9 E
				H = 01 09 20 h = 0 km
				D = 8.39 Az = 21 (ISC)
				V A 0.5s 11.5nm
21.	i	A	11 00 50.5	Czechoslovakia 51.00 N 14.41 E
	iSg	A	01 15.5	explosion
				H = 11 00 00 yield 17.8 t (PRU)
				D = 1.81 Az = 260 (ISC)

Day	Phase		h m s	Remarks
21.	eP	A	11 13 20	Southwestern Ryukyu Islands
				24.66 N 124.99 E
				H = 11 00 48.4 h = 62 km MB = 4.8
				D = 84.83 Az = 324.1 (NEIC)
21.	-eP	AB	19 31 02.5	Western Brazil 6.71 S 71.89 W
	epP	A	32 50.5	H = 19 18 57.2 h = 562 km MB = 5.6
	eSKS	B	40 38	D = 91.03 Az = 39.2 (NEIC)
	e	C	42 20	h = 511 km
	e	C	43 40	PV A 1.8s 121.8nm M = 5.6
	eSSS	C	50 35	
21.	LmH	B	23 04.0	Chile-Argentina Border Region
	LmV	B	04.0	35.02 S 70.46 W
				H = 21 56 42.8 h = 20 km
				MB = 5.1 MS = 4.9 (NEIC)
				D = 111.62
				LmH B 18s 1.0 µm M = 5.5
				LmV B 18 1.28 µm 5.6
22.	eP	A	13 21 26	Guatemala 14.03 N 91.01 W
	epP	A	21 43.5	H = 13 08 50.3 h = 102 km MB = 5.5
	isP	A	21 54	D = 87.10 Az = 38.5 (NEIC)
	eSKS	B	31 47	h = 70 km
	ePS	B	33 21	PV A 1.4s 44.2nm M = 5.3
	eSS	B	37 48	pPV A 1.4 76.7nm
	LmV	B	58.7	sPV A 1.8 115.0nm
	LmH	B	58.8	LmH B 19.5 2.7 µm
				LmV B 18 2.5 µm
22.	ePKP	AB	22 10 17.5	New Hebrides Islands Region
	eX	A	10 38.5	20.48 S 172.85 E
	eY	AB	11 12	H = 21 50 38.9 h = normal MB = 5.3
	eSS	B	32 49	D = 146.45 Az = 338.2 (NEIC)
	eSSS	B	38 05	PKPV A 3.0s 500.0nm

Day	Phase		h m s	Remarks
cont. 22.	LmV	E	23 11.0	PKPV B 6s 1.74 μm
	LmH	B	11.3	XV A 1.9 254.0nm
				YV A 1.9 170.0nm
				LmH B 19 3.3 μm M = 6.1
				LmV B 20 3.5 μm 6.1
22.	ePKP	A	22 57 10	Tonga Islands 18.11 S 174.59 W
	epPKP	A	57 44.5	H = 22 37 41.1 h = 135 km MB = 5.2
				D = 147.16 Az = 352.7 (NEIC)
				h = 129 km
				PKPV A 1.1s 60.5nm
23.	eP	A	02 17 30	Yunnan Province, China 23.65 N 102.70 E
				H = 02 06 01.2 h = normal MB = 5.2
				D = 72.75 Az = 318.2 (NEIC)
				PV A 1.3s 21.8nm M = 5.0
23.	eSg	A	05 40 26	Poland 50.27 N 19.07 E
				H = 05 37 54.4 M = 3.1 (WAR)
				D = 4.8
23.	eP	A	11 47 44	Andreanof Islands, Aleutian Is.
				52.03 N 178.67 W
				H = 11 35 59.7 h = 102 km MB = 4.9
				D = 77.33 Az = 353.3 (NEIC)
23.	ePg	A	11 48 32	Northern Italy 44.1 N 8.3 E
	eSn	A	49 18	H = 11 46 20 h = 56 km (ISC)
	eSg	A	50 08	D = 7.0
23.	ePKHKP	A	13 12 58.5	Fiji Islands Region 20.90 S 178.64 W
	ePKP2	A	13 05.5	H = 12 54 16.1 h = 603 km MB = 4.8
				D = 149.23 Az = 347.2 (NEIC)
				PKHKPV A 1.0s 17.7nm
23.	LmH	C	18 40.1	New Hebrides Islands 13.11 S 166.34 E
				H = 17 18 39.5 h = normal MB=5.3 MS=5.6
				(NEIC) D = 137.28
				LmH C 21s 1.43 μm M = 5.6

Day	Phase		h m s	Remarks
23.	LmH	C	19 26.0	New Hebrides Islands 13.19 S 166.56 E
	LmV	B	26.2	H = 18 04 00.2 h = 33 km
				MB = 5.1 MS = 5.7 (NEIC)
				D = 137.44
				LmH C 20s 1.56/um M = 5.7
				LmV B 20 2.22/um 6.0
23.	ePKIKP1	A	21 37 07	New Hebrides Islands 13.17 S 166.38 E
	iPKIKP2	B	37 26	H = 21 17 52.1 h = normal MB = 5.9
	ePP1	A	40 00	D = 137.35 Az = 336.4 (NEIC)
	ePP2	B	40 08	PKIKP1V A 3.0s 263.0nm
	ePS	B	50 08	PKIKP2V B 14 6.38/um
	ePPS	C	52 05	PP2V B 16 14.13/um M = 6.9
	LmH	B	22 42.8	LmH B 19 35.4/um 7.1
	LmV	B	42.8	LmV B 19 31.8/um 7.1
23.	ePKIKP	A	21 54 06	New Hebrides Islands 13.02 S 166.19 E
				H = 21 34 37.7 h = normal MB = 5.1
				D = 137.14 Az = 336.3 (NEIC)
24.	ePKP	A	04 07 28	New Hebrides Islands Region
	eX	A	07 40	20.41 S 172.91 E
				H = 03 47 50.1 h = normal MB = 5.2
				D = 146.40 Az = 338.3 (NEIC)
				PKPV A 2.5s 107.0nm
				XV A 2.2 109.0nm
24.	ePKIKP	A	04 15 05	New Hebrides Islands 13.04 S 166.40 E
	e(PP)	A	17 51	H = 03 55 42.5 h = 28 km MB = 5.6
				D = 137.24 Az = 336.5 (NEIC)
				PKIKPV A 2.2s 98.1nm
				PPV A 3.0 211.0nm M = 5.8
24.	eP	A	04 43 08.5	Celebes Sea 3.52 N 123.81 E
	ePP	A	47 27	H = 04 30 02.6 h = 417 km MB = 5.3
				D = 101.13 Az = 323.1 (NEIC)
				PPV A 1.5s 30.2nm M = 5.1

Day	Phase		h m s	Remarks
24.	eP	A	09 51 21.5	North Atlantic Ridge 15.36 N 45.84 W
				H = 09 41 31.9 h = normal MB = 4.9
				D = 57.76 Az = 39.4 (NEIC)
				PV A 2.0s 51.3nm M = 5.2
24.	eP	A	17 36 49	Alaska Peninsula 55.32 N 159.52 W
	epP	A	36 57.5	H = 17 25 13.1 h = 28 km MB = 4.8
				D = 74.15 Az = 5.9 (NEIC)
				h = 31 km
				PV A 0.8s 15.4nm M = 5.1
24.	ePKIKP	A	19 57 49.5	South of Fiji Islands 26.15 S 177.24 W
	ePKHKP	A	57 58.5	H = 19 38 07.1 h = 92 km MB = 5.5
	e	A	58 06.5	D = 154.61 Az = 346.8 (NEIC)
	ePKP2	A	58 13.5	PKIKPV A 2.0s 29.9nm
24.	eP	A	23 05 16.5	Hokkaido, Japan Region 43.16 N 143.77 E
				H = 22 53 34.2 h = 142 km MB = 5.1
				D = 77.70 Az = 331.1 (NEIC)
25.	eP1	A	02 18 57	Taiwan Region 22.46 N 122.26 E
	eiP2	A	19 04	H = 02 06 23.3 h = normal MB = 6.3
	eiP3	AB	19 17	D = 25.11 Az = 323.3 (NEIC)
	ePP	B	22 25	P1V A 1.5s 286.0nm M = 6.3
	eSKS	B	29 18	P2V A 1.8 1385.0nm 6.9
	eS	B	29 35	P3V A 1.9 1895.0nm 7.0
	eSS	B	35 00	P'P'V A 1.6 44.0nm
	ePKKP	A	37 06	LmH B 20 1047.0/um 8.2
	eP'P'	A	45 17.5	LmV B 17 968.0/um 8.3
	LmH	B	54.7	
	LmV	B	03 02.5	
25.	eiP	A	03 53 56	Taiwan Region 23.03 N 122.15 E
	i	A	53 59.3	H = 03 41 23.7 h = normal MB = 5.9
	eS	B	04 04 18	D = 84.59 Az = 323.2 (NEIC)
	LmV	B	37.2	PV A 1.6s 280.0nm M = 6.4
	LmH	B	37.3	LmH B 16 177.0/um 7.5
				LmV B 16 174.8/um 7.6

Day	Phase		h m s	Remarks
25.	+iP	A	05 49 41.3	Hindu Kush Region 35.60 N 69.84 E
	+ipP	A	50 06.3	H = 05 41 39.7 h = 74 km MB = 5.4
				D = 43.94 Az = 308.7 (NEIC)
				h = 117 km
				PV A 1.5s 60.3nm M = 5.2
				pPV A 1.1 38.3nm
25.	eP	A	10 14 07	Near East Coast of Kamchatka
				53.86 N 160.87 E
				H = 10 02 40.3 h = normal MB = 4.6
				D = 72.71 Az = 340.1 (NEIC)
25.	eP	A	15 17 27	Volcano Islands Region 22.54 N 141.93 E
	ePP	A	21 21	H = 15 04 28.4 h = 220 km MB = 5.2
				D = 94.99 Az = 330.8 (NEIC)
				PV A 1.2s 20.3nm M = 5.2
				PPV A 2.0 47.0nm 5.2
25.	eiPn	A	20 26 20.5	Central Italy 43.76 N 13.42 E
	iSn	A	27 39.5	H = 20 24 38.9 h = normal MB = 4.5
	eSg	A	28 29	D = 7.00 Az = 350.5 (NEIC)
				PnV A 0.8s 36.5nm M = 5.4
25.	ePn	A	21 18 32	Central Italy 43.83 N 12.99 E
	ePg	A	19 11	H = 21 16 45.8 h = 0 km
	eSg	A	19 46	D = 6.88 Az = 3.53 (ISC)
25.	eiPn	A	23 23 59	Central Italy 43.79 N 13.44 E
	eiSg	A	25 16.5	H = 23 22 17.1 h = normal MB = 4.8
				D = 6.97 Az = 350.4 (NEIC)
				PnV A 0.7s 21.1nm M = 5.2
26.	eP	A	03 54 50.5	Southern Sumatra 2.60 S 102.49 E
				H = 03 41 57.1 h = 174 km MB = 5.5
				D = 92.55 Az = 320.4 (NEIC)
26.	eP	A	04 43 33	Sea of Okhotsk 52.16 N 150.92 E
				H = 04 33 04.6 h = 561 km MB = 4.0
				D = 71.93 Az = 334 (ISC)

Day	Phase		h m s	Remarks
26.	ePn	A	10 51 50	Central Italy 43.86 N 13.30 E
	ePg	A	52 29	H = 10 50 09.2 h = normal MB = 4.3
	eSn	A	53 06.5	D = 6.89 Az = 351.0 (NEIC)
	eSg	A	53 58.5	PV A 0.6s 23.0nm M = 5.3
26.	epP	A	19 04 38	Hokkaido, Japan Region 42.76 N 145.52 E
				H = 18 52 23.1 h = 48 km MB = 4.6
				D = 78.67 Az = 332.1 (NEIC)
26.	ePKIKP	A	23 18 54.5	Fiji Islands Region 20.18 S 178.95 W
	iPKIKP	AB	18 56.6	H = 23 00 24.4 h = 668 km MB = 5.7
	PKIKPm	A	19 02	D = 148.47 Az = 347.1 (NEIC)
	ipPKP	AB	21 28	h = 680 km
	eisPKP	B	22 30	iPKIKPV A 2.0s 530.0nm
				iPKIKPV B 3.5 5.5/um
				PKIKPmV A 2.4 4060.0nm
26.	eiPKHKP	A	23 27 36.5	Fiji Islands Region 20.32 S 178.69 W
	ePKP2	A	27 42.5	H = 23 09 01.7 h = 686 km MB = 4.9
				D = 148.66 Az = 347.3 (NEIC)
				PKHKPV A 1.1s 56.4nm
				PKP2V A 1.4 50.2nm
26.	iPKIKP	A	23 51 05.1	Fiji Islands Region 20.18 S 178.99 W
	iPKHKP	A	51 09.3	H = 23 32 36.5 h = 689 km MB = 5.1
	iPKP2	A	51 15.8	D = 148.46 Az = 347.1 (NEIC)
				PKIKPV A 1.2s 28.5nm
				PKHKPV A 1.3 131.0nm
				PKP2V A 1.2 87.4nm
28.	-iPKP	AB	01 35 42	New Hebrides Islands 19.38 S 169.11 E
	e	B	36 22	H = 01 16 21.7 h = 124 km MB = 6.1
	LmH	B	02 32.0	D = 144.06 Az = 335.5 (NEIC)
	LmV	B	33.8	PKPV A 1.4s 860.0nm
				PKPV B 12 9.75/um
				LmH B 24 3.4/um
				LmV B 21 2.3/um

Day	Phase		h m s	Remarks
28.	eP	A	10 35 30	West Pakistan 26.64 N 66.32 E
	ePP	A	37 16.5	H = 10 26 53.6 h = normal MB = 5.9
				D = 47.69 Az = 315.4 (NEIC)
				PV A 1.5s 25.1nm M = 5.0
				PPV A 1.4 32.6nm 5.1
28.	iPn	A	11 54 51	Czechoslovakia 50.58 N 14.01 E
	iPg	A	54 53.5	explosion
	eSn	A	55 11	H = 11 54 00 yield 9.6 t (PRU)
	iSg	A	55 12.5	H = 11 54 24 50.7 N 13.8 E
				D = 1.53 Az = 273 (ISC)
28.	-eiP	A	16 31 01	Philippine Islands Region
	LmH	B	17 07.0	20.92 N 120.19 E
	LmV	B	16.0	H = 16 18 26.0 h = normal MB = 5.4
				D = 85.17 Az = 322.7 (NEIC)
				PV A 1.5s 67.9nm M = 5.6
				LmH B 16 1.1/um 5.3
				LmV B 14 0.7/um 5.2
29.	ePKHKP	A	06 37 01	South Pacific Cordillera
	ePKP	A	37 19	55.60 S 129.39 W
	LmH	B	07 42.5	H = 06 17 01.0 h = normal MB = 4.6
	LmV	B	42.6	D = 156.32 Az = 85.7 (NEIC)
				PKHKPV A 1.2s 16.3nm
				LmH B 20 0.9/um
				LmV B 19 0.8/um
29.	ePKHKP	A	08 05 22	Tonga Islands 19.12 S 174.59 W
				H = 07 45 43.4 h = 84 km MB = 5.0
				D = 148.16 Az = 352.5 (NEIC)
				PKHKPV A 1.0s 17.7nm
29.	eP	A	11 24 50	Off East Coast of Honshu, Japan
				40.18 N 143.82 E
				H = 11 12 38.0 h = 19 km MB = 4.5
				D = 80.33 Az = 331.4 (NEIC)
				PV A 1.3s 15.3nm M = 4.9

Day	Phase		h m s	Remarks
29.	ePKP	A	17 16 19	Fiji Islands Region 18.07 S 178.60 W H = 16 57 47.3 h = 652 km MB = 4.3 D = 146.49 Az = 348.2 (NEIC)
29.	LmH	C	23 27.5	LmH C 16s 0.53 /um
	LmV	C	28.4	LmV C 8 0.66 /um
30.	LmH	C	04 27.5	LmH C 18s 0.60 /um
	LmV	C	41.7	LmV C 15 0.70 /um
30.	LmH	B	20 00.7	Near Coast of Nicaragua
	LmV	B	01.3	11.26 N 86.77 W H = 19 10 07.4 h = 48 km MB = 4.6 MS = 4.4 (NEIC) D = 86.65 LmH B 16s 0.8 /um M = 5.2 LmV B 18 0.8 /um 5.2
30.	eP	A	20 34 13.5	North of Svalbard 84.03 N 0.36 W H = 20 27 32.1 h = normal MB = 4.5 D = 33.68 Az = 166.2 (NEIC)
30.	ePKP	A	20 37 22.5	Fiji Islands Region 17.82 S 178.78 W H = 20 18 51.1 h = 649 km MB = 4.8 D = 146.21 Az = 348.1 (NEIC) PKPV A 1.1s 10.1nm
31.	+iPKP	A	01 00 48	Fiji Islands Region 16.46 S 178.16 W H = 00 42 03.6 h = 477 km MB = 4.9 D = 145.00 Az = 349.1 (NEIC) PKPV A 1.3s 48.0s
31.	eP	A	06 02 45	Fox Islands, Aleutian Is.
	e	A	03 25.5	52.48 N 170.60 W H = 05 50 54.7 h = 52 km MB = 4.7 D = 77.23 Az = 358.6 (NEIC)

Day	Phase		h m s	Remarks
31.	e(Pg)	A	14 27 51.5	Poland 51.3 N 16.2 E
	i	A	27 30	explosion in mine Polkowice
	iSg	A	27 32	H = 14 26 54.5 (WAR)
				D = 2.97 Az = 259 (ISC)
31.	ePKIKP	A	19 25 12.5	South of Fiji Islands 23.44 S 179.97 W
	ePKHKP	A	25 19.5	H = 19 06 27.0 h = 550 km MB = 5.0
	epPKP	A	27 30	D = 151.38 Az = 344.5 (NEIC)
	e	A	27 44	

Day	Phase		h m s	Remarks
1.	-iP	A	00 36 20	Rat Islands, Aleutian Is. 51.77 N 177.66 E H = 00 24 30.7 h = 57 km MB = 5.2 D = 77.28 Az = 350.9 (NEIC) PV A 0.8s 26.9nm M = 5.3
1.	ePKIKP	A	02 56 31	Solomon Islands 9.82 S 160.20 E
	ePP	A	58 53	H = 02 37 26.8 h = 57 km MB = 5.4
	LmH	C	03 56.2	D = 131.77 Az = 334 (ISC)
	LmV	B	57.3	LmH C 20s 1.17 µm
1.	e	A	15 35 03.5	Loyalty Islands Region 22.58 S 171.57 E H = 15 15 02.04 h = 48 km MB = 5.2 D = 147.91 Az = 335.7 (NEIC)
2.	eP1	A	21 23 00.5	Greece 38.90 N 21.19 E
	eP2	A	23 07.5	H = 21 19 49.2 h = 44 km MB = 4.6 D = 13.56 Az = 333.1 (NEIC) P1V A 1.1s 12.1nm M = 4.7 P2V A 0.9 19.5nm
3.	e	A	02 35 28.5	Eastern Caucasus 40.72 N 48.40 E
	e	A	35 49	H = 02 29 21.9 h = 39 km MB = 5.1
	LmH	C	45.3	D = 27.30 Az = 303.8 (NEIC)
	LmV	C	46.8	LmH C 26s 3.24 µm M = 4.8 LmV C 20 3.7 µm 5.1
3.	eP	A	05 43 22	North Atlantic Ocean 57.88 N 32.18 W
	LmH	C	53.8	H = 05 37 48.0 h = normal MB = 4.6
	LmV	C	53.8	D = 26.18 Az = 87.0 (NEIC) PV A 1.2s 26.4nm M = 4.7
3.	eP	A	07 20 51	North Atlantic Ocean 58.13 N 32.04 W
	LmH	B	31.6	H = 07 15 18.0 h = normal MB = 4.5
	LmV	B	31.8	D = 26.10 Az = 87.6 (NEIC) PV A 2.0s 77.0nm M = 5.0 LmH B 14 4.7 µm 5.2 LmV B 14 3.7 µm 5.2

Day	Phase		h m s	Remarks
3.	ePKIKP	A	08 54 55	New Britain Region 4.56 S 151.80 E
	ePP	A	56 39.5	H = 08 36 16.6 h = 168 km MB = 5.4
				D = 123.26 Az = 330.8 (NEIC)
				PPV A 1.5s 50.2nm M = 5.5
4.	iPn	A	02 43 59.5	Central Italy 43.82 N 13.31 E
	iPg	A	44 39	H = 02 42 18.9 h = 25 km MB = 4.8
	iSn	A	45 16.5	D = 6.93 Az = 351.0 (NEIC)
	eSg	A	46 07.5	LmH B 14s 5.7/um M = 4.3
	LmH	B	46.7	LmV B 7 2.9/um
	LmV	B	47.7	
4.	eSn	A	04 44 04	Central Italy 43.92 N 13.02 E
				H = 04 40 56.2 h = 33 km (ISC)
				D = 6.9
				SnV A 0.8s 11.5nm
4.	LmH	C	06 52.0	LmV C 18s 2.31/um
	LmV	C	52.0	
4.	ePKP	A	07 21 31	Fiji Islands Region 17.81 S 178.60 W
				H = 07 02 57.6 h = 611 km MB = 4.7
				D = 146.24 Az = 348.3 (NEIC)
				PKPV A 2.0s 77.0nm
4.	iPn	A	09 20 12.5	Central Italy 43.93 N 13.21 E
	ePg	A	20 53	H = 09 18 31.5 h = 23 km MB = 4.4
	iSn	A	21 29	D = 6.81 Az = 351.4 (NEIC)
	eSg	A	22 20	LmH B 14s 2.58/um M = 3.8
	LmH	B	23.0	
4.	LmH	B	11 26.9	Volcano Islands Region
	LmV	B	30.7	24.10 N 143.29 E
				H = 10 30 49.2 h = normal
				MB = 4.9 MS = 4.8 (NEIC)
				D = 94.23

Day	Phase		h m s	Remarks
4.	eP	A	14 18 05.5	Tibet 30.38 N 84.61 E
				H = 14 08 21.7 h = 18 km MB = 5.2
				D = 56.71 Az = 313.2 (NEIC)
				PV A 1.2s 22.4nm M = 5.1
4.	eP	A	15 32 03.5	Nicobar Islands Region 7.95 N 94.03 E
				H = 15 20 04.1 h = 55 km MB = 5.4
				D = 79.09 Az = 320.0 (NEIC)
4.	iPn	A	17 21 34.5	Central Italy 43.78 N 13.31 E
	ePg	A	22 12.5	H = 17 19 52.4 h = 23 km MB = 4.4
	iSn	A	22 50	D = 6.97 Az = 351.1 (NEIC)
	eSg	A	23 41	PnV A 0.8s 50.0nm
	LmH	B	24.3	LmH B 14 2.5 μm M = 4.0
	LmV	B	25.4	LmV B 7.5 1.6 μm
4.	ePn	A	18 19 10.5	Central Italy 43.82 N 13.42 E
	ePg	A	19 49	H = 18 17 29.5 h = normal MB = 4.8
	eSn	A	20 30.5	D = 6.94 Az = 350.5 (NEIC)
	eSg	A	21 18.5	PnV A 1.4s 16.3nm M = 4.8
4.	ePn	A	18 36 23	Northern Italy 44.8 N 12.6 E
	e(Sn)	A	37 37	H = 18 34 52 h = 0 km
	e	A	38 31	D = 5.93 Az = 354 (ISC)
4.	ePn	A	19 04 37	Central Italy 43.83 N 13.35 E
	ePg	A	05 13.5	H = 19 02 56.0 h = normal MB = 4.8
	eSn	A	05 55.5	D = 6.92 Az = 350.8 (NEIC)
				PnV A 0.8s 30.8nm M = 5.3
4.	ePn	A	19 31 34.5	Central Italy 43.82 N 12.93 E
	ePg	A	32 09	H = 19 29 48.3 h = 0 km (ISC)
	eSn	A	32 46.5	D = 6.88 Az = 353
4.	ePn	A	19 55 37	Central Italy 43.7 N 13.4 E
	ePg	A	56 16.5	H = 19 53 59 (BCIS)
	eSn	A	57 02	D = 7.05 Az = 351 (ISC)
	Sg	A	57 57	

Day	Phase		h m s	Remarks
5.	ePKIKP	A	00 35 43.5	South Pacific Cordillera
	ePKHKP	A	35 53	55.43 S 128.65 W
	LmH	B	01 44.0	H = 00 15 51.3 h = normal MB = 5.9
	LmV	B	44.0	D = 155.88 Az = 84.8 (NEIC)
				PKIKPV A 1.8s 47.3nm
				PKHKPV A 1.8 67.6nm
				LmH B 20 1.6 /um M = 5.7
				LmV B 18 1.9 /um 6.0
5.	iPn	A	01 28 12.5	Central Italy 43.83 N 13.29 E
	iPg	A	28 49.5	H = 01 26 32.3 h = normal MB = 4.8
	eiSn	A	29 28	D = 6.92 Az = 351.1 (NEIC)
	eSg	A	30 18	PnV A 0.8s 61.5nm M = 5.6
	LmH	B	30.9	
5.	ePn	A	02 15 24	Central Italy 43.66 N 13.1 E
	ePg	A	16 00	H = 02 13 39 h = 0 km
	eSn	A	16 46	D = 7.06 Az = 353 (ISC)
5.	ePn	A	03 51 34.5	Central Italy 43.24 N 13.73 E
	ePg	A	52 13	H = 03 49 45.1 h = normal MB = 4.4
	eSn	A	52 57.5	D = 7.55 Az = 349.7 (NEIC)
	e(Sg)	A	53 38.5	PnV A o.7s 13.4nm M = 5.1
5.	eP	A	04 28 33	North Atlantic Ridge 14.58 N 45.09 W
	e	A	28 39.5	H = 04 18 42.4 h = normal MB = 5.0
				D = 57.91 Az = 38.9 (NEIC)
				PV A 1.2s 20.3nm M = 5.1
5.	ePn	A	05 07 33	Central Italy 43.74 N 13.45 E
	ePg	A	08 09.5	H = 05 05 50.7 h = normal MB = 4.6
	eSn	A	08 48	D = 7.02 Az = 350.4 (NEIC)
	eSg	A	09 42	PnV A 0.9s 21.4nm M = 5.1
	LmH	B	10.2	
5.	ePn	A	07 09 53	Central Italy 43.85 N 13.26 E
	e	A	10 06.5	H = 07 08 12.9 h = normal MB = 4.7
	eiSn	A	11 07	D = 6.89 Az = 351.3 (NEIC)

Day	Phase		h m s	Remarks
cont. 5.	eSg	A	07 11 56	PnV A 1.2s 30.5nm M = 5.1
	LmH	B	12.7	LmH B 15 1.7 µm 3.7
5.	ePn	A	12 35 31.5	Central Italy 43.88 N 13.12 E
	e	A	35 40	H = 12 33 51.1 h = normal
				D = 6.85 Az = 352.0 (NEIC)
5.	ePn	A	15 16 29.5	Central Italy 43.75 N 13.37 E
	ePg	A	17 03.5	H = 15 14 48.0 h = 35 km MB = 4.7
	eiSn	A	17 45	D = 7.01 Az = 350.8 (NEIC)
	e	A	18 02.5	PnV A 1.0s 37.4nm M = 5.3
	e	A	18 19	LmH B 15 2.7 µm 4.0
	eSg	A	18 38	LmV B 7.5 1.6 µm
	LmH	B	19.2	
	LmV	B	20.4	
5.	ePn	A	18 10 47.5	Central Italy 43.76 N 13.02 E
	iSn	A	12 02	H = 18 09 02.3 h = 0 km
				D = 6.95 Az = 353 (ISC)
5.	eSg	A	20 22 13	Poland 50.35 N 18.91 E
				H = 20 19 37.7 M = 2.8 (WAR)
				D = 4.65
5.	ePn	A	22 41 52	Central Italy 43.78 N 13.17 E
				H = 22 40 02.9 h = 0 km
				D = 6.95 Az = 352 (ISC)
6.	eP	A	00 42 41	Andreanof Islands, Aleutian Is.
				51.19 N 179.33 W
				H = 00 30 44.7 h = 45 km MB = 4.8
				D = 78.12 Az = 352.9 (NEIC)
				PV A 1.2s 12.2nm M = 4.8
6.	iPn	A	01 36 00.5	Adriatic Sea 44.01 N 13.19 E
	ePg	A	36 38	H = 01 34 22.4 h = normal MB = 4.9
	iSn	A	37 18	D = 6.72 Az = 351.4 (NEIC)
	eSg	A	38 08	PnV A 0.8s 73.0nm M = 5.6

Day	Phase		h m s	Remarks
cont. 6.	LmH	B	01 38.7	LmH B 14s 1.9/um M = 3.8
6.	ePKP	A	12 14 23	New Hebrides Islands 17.83 S 168.37 E
	e	A	16 23	H = 11 55 09.0 h = 135 km MB = 5.2
				D = 142.36 Az = 335.7 (NEIC)
6.	ePn	A	21 46 11	Central Italy 43.81 N 13.24 E
	eSn	A	47 25.5	H = 21 44 28.8 h = normal MB = 4.4
	eSg	A	48 20	D = 6.93 Az = 351.4 (NEIC)
6.	eP	A	22 17 28	South of Mariana Islands
	ePP	A	21 46	12.54 N 141.91 E
	LmH	C	59.6	H = 22 03 30.4 h = 50 km MB = 5.5
	LmV	C	23 01.5	D = 103.63 Az = 330.0 (NEIC)
7.	ePKHKP	A	04 37 25.5	South of Fiji Islands 24.02 S 179.91 E
	ePKP2	A	37 36.5	H = 04 18 31.0 h = 550 km MB = 4.2
				D = 151.90 Az = 344.1 (NEIC)
				PKHKPV A 1.3s 21.8nm
				PKP2V A 1.2 16.3nm
7.	eP	A	05 20 05	Off East Coast of Honshu, Japan
	epP	A	20 17.5	39.55 N 143.38 E
	esP	A	20 26.5	H = 05 07 53.7 h = 41 km MB = 5.0
	LmH	B	54.7	D = 80.73 Az = 331.2 (NEIC)
	LmV	B	59.3	h = 50 km
				PV A 1.5s 25.1nm M = 5.0
				pPV A 1.8 37.2nm
				sPV A 1.4 20.9nm
				LmH B 18 2.6/um 5.6
				LmV B 16 1.5/um 5.5
7.	eP	A	08 01 21	Off East Coast of Kamchatka
	e	A	01 30.5	52.26 N 160.05 E
	LmH	B	30.4	H = 07 49 48.2 h = 60 km MB = 4.8
	LmV	B	37.0	D = 74.05 Az = 339.7 (NEIC)

Day	Phase		h m s	Remarks
7.	e	A	11 09 43	Atlantic-Indian Rise 27.94 S 64.13 E
				H = 10 56 32.9 h = normal MB = 5.0
				D = 90.95 Az = 329.7 (NEIC)
7.	eP	A	14 33 46	Kurile Islands 49.73 N 154.64 E
				H = 14 22 17.8 h = 130 km MB = 4.4
				D = 75.12 Az = 336.7 (NEIC)
7.	eP1	A	19 27 34.5	Costa Rica 8.54 N 83.88 W
	eP2	AB	27 40	H = 19 14 47.6 h = 14 km MB = 5.5
	eS	C	38 10	D = 86.93 Az = 39.4 (NEIC)
	eS	C	44 05	P1V A 1.0s 15.7nm M = 5.2
	LmH	B	20 05.0	P2V A 2.1 273.0nm 6.1
	LmV	B	05.2	P2V B 4 0.66 μm 6.2
				LmH B 18.5 2.4 μm 5.6
				LmV B 19 2.4 μm 5.6
7.	e	A	20 16 18	Costa Rica 8.60 N 83.80 W
				H = 20 03 35.4 h = normal MB = 4.4
				D = 86.84 Az = 39.4 (NEIC)
				traces
8.	eP	AB	03 50 36.5	Philippine Islands Region
	ePP	AB	54 03	19.35 N 122.05 E
	eS	B	04 01 08	H = 03 37 52.4 h = 50 km MB = 5.7
	ePS	B	02 10	D = 87.47 Az = 323.3 (NEIC)
	LmV	B	34.9	PV A 2.0s 162.0nm M = 5.9
	LmH	B	35.1	PPV A 2.5 169.0nm 6.0
				LmH B 13.5 2.6 μm 5.8
				LmV B 14.5 3.3 μm 5.9
8.	iPn	A	12 20 55.5	Central Italy 43.84 N 13.28 E
	iPg	A	21 35	H = 12 19 14.9 h = normal MB = 4.6
	iSn	A	22 09	D = 6.90 Az = 351.2 (NEIC)
	e(Sg)	A	22 59	PnV A 0.7s 23.0nm M = 5.2
	LmH	C	23.7	
8.	eP	A	19 08 53	Near East Coast of Honshu, Japan
	epP	A	09 06	39.99 N 142.61 E

Day	Phase		h m s	Remarks
cont. 8.	LmH	B	19 43.0	H = 18 56 47.5 h = 55 km MB = 5.1
	LmV	B	48.6	D = 80.05 Az = 330.8 (NEIC)
				h = 46 km
				PV A 1.0s 21.6nm M = 5.0
				LmH B 17 0.4/um
				LmV B 16 0.4/um
8.	LmH	C	21 37.7	Peru 14.06 S 71.52 W
	LmV	C	37.8	H = 20 44 08.5 h = 48 km MB = 4.8 (NEIC)
				D = 96.45
9.	eP	A	01 46 48	Taiwan Region 24.08 N 122.39 E
	LmH	B	02 28.5	H = 01 34 21.6 h = 50 km MB = 4.8
	LmV	B	29.7	D = 83.89 Az = 323.2 (NEIC)
9.	ePKIKP	A	14 53 04	South of Fiji Islands 23.72 S 178.72 E
	ePKHKP	A	53 11	H = 14 34 18.8 h = 549 km MB = 4.4
	ePKP2	A	53 21	D = 151.31 Az = 342.8 (NEIC)
	epPKP	A	55 23.5	PKHKPV A 1.3s 10.9nm
				pPKPV A 1.2 16.2nm
9.	eP	A	16 00 36	South of Panama 4.98 N 82.44 W
	e	A	00 39	H = 15 47 43.1 h = normal MB = 4.9
				D = 88.77 Az = 39.4 (NEIC)
9.	LmH	C	16 33.0	LmH C 23s 0.69/um
	LmV	C	33.0	LmV C 25 1.01/um
9.	eiPKIKP	A	21 03 34	Near Coast of Southern Chile
	LmH	B	56.2	51.80 S 73.99 W
	LmV	B	56.3	H = 20 44 36.4 h = normal MB = 5.5
				D = 125.04 Az = 50.8 (NEIC)
				PKIKPV A 1.6s 60.4nm
				LmH B 20 2.0/um M = 5.8
				LmV B 20 2.7/um 5.9

Day	Phase		h m s	Remarks
10.	ePKIKP	A	03 23 50	Fiji Islands Region 20.82 S 178.77 W
	eiPKHKP	A	23 54	H = 03 05 09.2 h = 575 km MB = 4.9
	ePKP2	A	24 01.5	D = 149.12 Az = 347.1 (NEIC)
				PKHKPV A 1.3s 41.5nm
10.	+iP	A	05 10 49	Eastern Kazakh SSR 49.99 N 78.89 E
	ePn	A	12 25	H = 05 02 57.3 h = 0 km MB = 5.5
				D = 41.60 Az = 297.8 (NEIC)
				PV A 1.0s 66.9nm M = 5.3
				underground explosion
10.	LmH	C	07 44.0	Gulf of California 26.57 N 109.99 W
	LmV	C	44.0	H = 06 53 11.7 h = normal MB = 4.7 (NEIC)
				D = 87.46
				LmH C 16s 0.44/um M = 5.0
				LmV C 16 0.69/um 5.2
11.	e	A	06 04 30.5	Southern Sinkiang Prov., China
				39.87 N 77.36 E
				H = 05 55 46.4 h = 23 km MB = 4.9
				D = 46.08 Az = 306.3 (NEIC)
11.	eP	A	17 24 00	Central Mid-Atlantic Ridge
	e	A	24 20	0.94 S 21.55 W
	eiS	C	32 19	H = 17 14 03.7 h = normal MB = 4.9
	LmH	C	40.3	D = 58.64 Az = 24.1 (NEIC)
	LmV	C	43.3	PV A 1.0s 11.8nm M = 4.9
				LmH C 40 2.46/um 5.0
				LmV C 40 2.34/um 5.1
11.	eP	A	21 47 33	Near East Coast of Kamchatka
				56.11 N 162.91 E
				H = 21 36 17.0 h = 44 km MB = 4.6
				D = 70.98 Az = 341.1 (NEIC)
				PV A 1.1s 12.1nm M = 4.8
12.	iPn	A	05 15 31.5	Federal Republic of Germany
	i(Sg)	A	16 07.5	49.3 N 8.3 E
				H = 05 14 47 (BCIS) D = 2.55

Day	Phase		h m s	Remarks
12.	ePKP	A	12 14 28	<u>New Hebrides Islands</u> 19.15 S 168.78 E
				H = 11 55 00.5 h = 53 km MB = 5.0
				D = 143.72 Az = 335.3 (NEIC)
				PKPV A 0.8s 23.1nm
12.	+eiPKP	AB	19 11 35.5	<u>Tonga Islands</u> 15.31 S 173.37 W
	LmH	B	20 19.0	H = 18 51 57.0 h = 5 km MB = 5.9
	LmV	B	19.0	D = 144.53 Az = 354.5 (NEIC)
				PKPV A 1.6s 143.0nm
				LmH B 20 1.7/um M = 5.8
				LmV B 20 2.5/um 6.0
13.	eP	A	10 12 35.5	<u>Tanzania</u> 4.78 S 34.81 E
	LmH	C	39.0	H = 10 02 40.5 h = normal MB = 5.3
	LmV	C	39.8	D = 58.71 Az = 342.9 (NEIC)
				PV A 1.6s 38.4nm M = 5.3
13.	eP	A	11 31 35	<u>Southern Greece</u> 36.04 N 23.99 E
				H = 11 27 39.2 h = 79 km MB = 4.5
				D = 17.11 Az = 332.4 (NEIC)
13.	eP	A	13 11 03.5	<u>Southern Greece</u> 37.09 N 23.98 E
	LmH	C	16.6	H = 13 07 10.8 h = 27 km MB = 4.5
	LmV	C	17.8	D = 16.19 Az = 330.7 (NEIC)
				PV A 0.8s 13.5nm M = 4.1
13.	ePKP	A	13 21 30.5	<u>South of Fiji Islands</u> 19.22 S 175.13 E
	LmH	C	14 30.0	H = 13 01 52.8 h = 73 km MB = 5.1 (NEIC)
	LmV	C	30.5	D = 146.02
				PKPV A 2.0s 47.0nm
				LmH C 22 0.66/um
				LmV C 22 1.56/um
13.	eP	A	18 07 31	<u>South of Alaska</u> 52.84 N 161.51 W
	epP	A	07 41	H = 17 55 41.5 h = normal MB = 5.0
				D = 76.73 Az = 4.5 (NEIC)
				h = 35 km
				PV A 1.3s 26.2nm M = 5.1
				pPV A 1.2 20.3nm

Day	Phase		h m s	Remarks
13.	+eiP	AB	21 33 28.5	Central Mid-Atlantic Ridge
	eS	B	41 48	0.97 N 28.41 W
	LmH	B	56.2	H = 21 23 22.4 h = normal MB = 5.4
	LmV	B	57.6	D = 59.97 Az = 28.2 (NEIC)
				PV A 1.7s 109.0nm M = 5.7
				PV B 5 0.77/um 6.1
				SH B 13 2.34/um 5.9
				LmH B 13.5 2.7/um 5.6
				LmV B 18 2.7/um 5.5
13.	eP	A	22 51 08	Southern Alaska 59.94 N 154.20 W
	epP	A	51 47	H = 22 40 16.2 h = 153 km MB = 4.9
				D = 69.17 Az = 9.6 (NEIC)
				h = 163 km
				PV A 1.0s 19.7nm M = 4.9
14.	eP diff	BC	23 46 09	Santa Cruz 11.36 S 166.34 E
	+ePKIKP	C	48(36)	H = 23 29 51.7 h = 102 km MB = 6.2
	-ePKIKP	B	48 45	D = 135.68 Az = 337.1 (NEIC)
	ePKIKP	A	48 52	XV A 3.4s 1044.0nm
	eiX	A	49 02	YV A 2.0 496.0nm
	eiY	A	49 09	PPV B 12 22.4/um M = 7.2
	ePP	B	51 36	LmH B 20 49.2/um
	iSKP	B	52 28	LmV B 20 60.8/um
	e	C	24 03 09	
	iSS	C	09 42	
	iSSS	C	14 21	
	LmH	B	53.1	
	LmV	B	53.2	
15.	+ePKP	A	04 28 31	Tonga Islands 15.81 S 173.71 W
	epPKP	A	29 03.5	H = 04 09 03.9 h = 92 km MB = 4.8
				D = 144.99 Az = 354.1 (NEIC)
				h = 119 km
				PKPV A 1.5s 45.2nm
				pPV A 1.9 41.7nm

Day	Phase		h m s	Remarks
15.	e(Pg)	A	19 05 09	Central Italy 43.97 N 11.25 E
	e	A	05 36	H = 19 02 48.2 h = 0 km (ISC)
				D = 6.55
15.	eP	A	21 31 55	Near Coast of Peru 17.51 S 71.19 W
	e(pP)	A	32 18	H = 21 18 21.2 h = 71 km MB = 5.7
	e(sP)	A	32 33	D = 98.89 Az = 39.7 (NEIC)
	LmH	C	22 09.0	h = 92 km
	LmV	C	17.0	PV A 1.4s 27.9nm M = 5.6
				sPV A 1.5 25.1nm
16.	eP	A	00 46 15.5	Southern Greece 36.92 N 24.18 E
	e	A	46 19	H = 00 42 23.6 h = normal MB = 4.5
				D = 16.42 Az = 330.6 (NEIC)
16.	ePKIKP	A	04 09 03	Santa Cruz Islands 11.33 S 166.26 E
				H = 03 49 50.8 h = 100 km MB = 5.2
				D = 135.63 Az = 337.1 (NEIC)
				PKIKPV A 1.0s 13.8nm
16.	eP	A	21 15 59	Fox Islands, Aleutian Is.
				52.24 N 171.13 W
				H = 21 04 07.4 h = 48 km MB = 4.8
				D = 77.46 Az = 358.2 (NEIC)
				PV A 1.4s 14.0nm M = 4.7
16.	eiP	A	23 27 51	Southern Sinkiang Prov., China
	epP	A	27 58.5	41.66 N 80.69 E
				H = 23 19 19.7 h = 29 km MB = 4.8
				D = 47.07 Az = 305.7 (NEIC)
				h = 34 km
				PV A 1.0s 15.7nm M = 5.0
17.	eP	A	00 24 44	Jamaica Region 17.61 N 76.05 W
	epP	A	24 50	H = 00 13 01.5 h = 21 km MB = 5.1
	LmH	C	47.2	D = 75.13 Az = 41.2 (NEIC)

Day	Phase		h m s	Remarks
cont. 17.	LmV	C	00 54.5	h = 22 km
				PV A 1.5s 17.6nm M = 4.9
				pPV A 1.6 41.2nm
				LmH C 22 0.78 µm 5.0
				LmV C 20 0.64 µm 5.0
17.	ePKIKP	A	00 58 07	New Britain Region 6.12 S 151.79 E
				H = 00 39 09.9 h = 40 km MB = 5.5
				D = 124.61 Az = 330.3 (NEIC)
				PKIKPV A 1.2s 18.3nm
17.	eP	A	02 14 45	Fox Islands, Aleutian Is.
	e	A	15 08	52.19 N 171.00 W
				H = 02 02 53.0 h = 44 km MB = 4.8
				D = 77.51 Az = 358.3 (NEIC)
				PV A 1.2s 16.3nm M = 4.9
17.	ePn	A	04 04 26	Belgium 50.6 N 6.0 E
	ePg	A	04 42	H = 04 03 31 (BCIS, Bensberg)
	e	A	05 20	50°32' N, 5°57' E (Brussels)
	eSg	A	05 25	D = 3.58
17.	eP	A	16 09 46.5	South of Honshu, Japan 29.15 N 139.67 E
	ePP	A	13 18	H = 15 57 37.1 h = 383 km MB = 5.1
				D = 88.26 Az = 329.9 (NEIC)
				PV A 1.7s 39.4nm M = 5.0
19.	ePn	A	00 42 19	Belgium 50.47 N 5.86 E
	eSn	A	42 59.5	H = 00 41 22.6 h = 26 km
	eSg	A	43 15.5	D = 3.67 Az = 85 (ISC)
19.	e	A	08 20 20	Mindoro, Philippine Islands
	e	A	21 08	12.89 N 121.31 E
	LmV	B	09 04.4	H = 08 07 02.1 h = 14 km MB = 4.9
	LmH	B	07.0	D = 92.20 Az = 323.1 (NEIC)
				LmH B 20s 1.5 µm M = 5.4
				LmV B 18 1.8 µm 5.6

Day	Phase		h m s	Remarks
19.	eiP	A	13 31 23.5	<u>Kurile Islands</u> 44.45 N 149.15 E
				H = 13 19 25.2 h = normal MB = 5.2
				D = 78.36 Az = 334.0 (NEIC)
				PV A 1.1s 32.3nm M = 5.3
19.	eiP	A	14 06 41.5	<u>Kurile Islands</u> 44.58 N 149.09 E
	eipP	A	06 54.5	H = 13 54 45.8 h = 50 km MB = 5.4
	LmH	B	39.5	D = 78.22 Az = 333.9 (NEIC)
	LmV	B	45.0	h = 48 km
				PV A 1.6s 107.0nm M = 5.6
				pPV A 0.8 26.9nm
				LmH B 18 0.8 µm 5.1
				LmV B 16 0.9 µm 5.3
19.	eP	A	22 52 15	<u>Kurile Islands</u> 44.37 N 149.14 E
				H = 22 40 17.9 h = 48 km MB = 5.0
				D = 78.43 Az = 334.0 (NEIC)
				PV A 1.0s 21.6nm M = 5.1
20.	eP	A	03 11 17	<u>Tibet</u> 34.57 N 80.32 E
				H = 03 02 14.0 h = normal MB = 4.8
				D = 51.22 Az = 310.4 (NEIC)
20.	ePKIKP	A	04 57 39.5	<u>South of Fiji Islands</u> 26.73 S 178.43 W
	ePKHKP	A	57 50	H = 04 38 19.2 h = 275 km MB = 5.1
	iPKP2	A	58 05.8	D = 154.91 Az = 344.8 (NEIC)
				PKIKPV A 1.3s 19.6nm
				PKHKPV A 1.5 25.1nm
				PKP2V A 1.2 40.6nm
20.	eP	A	06 20 58	<u>Gulf of California</u> 29.90 N 113.53 W
	LmV	B	07 02.7	H = 06 08 17.9 h = normal MB = 5.4
	LmH	B	02.8	D = 86.32 Az = 31.4 (NEIC)
				PV A 1.2s 16.3nm M = 5.1
				LmH B 13.5 2.2 µm 5.7
				LmV B 15.5 2.1 µm 5.7

Day	Phase		h m s	Remarks
20.	e	A	13 47 54	New Hebrides Islands 14.62 S 167.26 E
				H = 13 28 35.1 h = 137 km MB = 4.9
				D = 139.01 Az = 336.4 (NEIC)
20.	ePKP	A	14 29 36	Loyalty Islands Region 21.57 S 169.52 E
	eX	A	29 45	H = 14 09 57.3 h = 18 km MB = 4.8
				D = 146.19 Az = 334.5 (NEIC)
				PKPV A 1.6s 11.0nm
				XV A 3.0 184.0nm
20.	eSn	A	20 41 53	Poland 50.35 N 18.92 E
	eSg	A	42 27	H = 20 39 49.2 M = 3.0 (WAR)
				D = 4.67 Az = 276 (ISC)
21.	ePKIKP	A	06 22 08	New Hebrides Islands 17.58 S 167.85 E
				H = 06 02 37.7 h = 27 km MB = 5.1
				D = 141.93 Az = 335.4 (NEIC)
21.	LmH	B	15 48.6	Kyushu, Japan 31.42 N 131.75 E
	LmV	B	49.3	H = 14 55 24.4 h = 40 km MB = 4.6 (NEIC)
				D = 82.72
				LmH B 16s 0.7 μm M = 5.1
				LmV B 18 0.9 μm 5.2
21.	eP	AB	19 46 17.5	Alaska Peninsula 55.90 N 158.27 W
	eipP	A	46 32.5	H = 19 34 50.9 h = 60 km MB = 5.7
	i(PcP)	AB	46 34.5	D = 73.49 Az = 6.7 (NEIC)
	eS	B	55 42	h = 54 km
	LmH	B	20 15.8	PV A 1.4s 188.0nm M = 5.8
	LmV	B	20.8	PV B 6 1.63 μm 6.1
				pPV A 1.0 114.2nm
				PcPV A 1.0 189.0nm
				LmH B 26 4.1 μm
				LmV B 20 2.0 μm
21.	eP	A	22 12 22	Near East Coast of Kamchatka
				54.44 N 161.31 E
				H = 22 00 58.8 h = normal MB = 4.8

Day	Phase		h m s	Remarks
cont. 21.				D = 72.26 Az = 340.3 (NEIC)
				PV A 1.1s 24.2nm M = 5.1
22.	eP	A	01 22 34	Hindu Kush Region 36.44 N 70.57 E
	ipP	A	23 21.3	H = 01 14 47.5 h = 212 km MB = 5.3
	e	A	23 36	D = 43.88 Az = 308.1 (NEIC)
	e	A	23 45	h = 224 km
				PV A 1.8s 81.0nm M = 4.9
				pPV A 1.5 45.3nm
22.	LmH	C	10 08.5	South of Sumbawa Islands
				10.23 S 116.83 E
				H = 09 07 56.85 h = normal
				MB = 5.5 MS = 4.8 (NEIC)
				D = 107.49
22.	eiP	A	18 55 28.5	Andaman Islands Region 10.42 N 92.46 E
	e	A	55 45	H = 18 43 42.0 h = normal MB = 5.4
	e	A	57 00.5	D = 76.22 Az = 319.7 (NEIC)
				PV A 1.3s 39.3nm M = 5.3
22.	+iP	A	20 11 53.9	Hokkaido, Japan Region 41.76 N 142.80 E
	LmH	C	49.3	H = 19 59 56.8 h = 56 km MB = 5.3
	LmV	C	49.3	D = 78.58 Az = 330.7 (NEIC)
				PV A 1.0s 25.6nm M = 5.1
22.	iPKP2	A	22 11 46.4	Ballenly Islands Region 62.88 S 166.60 E
	LmH	C	23 31.5	H = 21 51 02.0 h = normal MB = 5.1 (NEIC)
	LmV	C	40.1	D = 161.77
23.	eP	A	03 19 03	Kurile Islands Region 43.72 N 148.35 E
	epP	A	19 16	H = 03 07 03.6 h = 41 km MB = 4.8
				D = 78.76 Az = 333.6 (NEIC)
				h = 45 km
				PV A 1.0s 9.8nm M = 4.7
				pPV A 1.3 19.7nm

Day	Phase		h m s	Remarks
23.	eP	A	03 33 28.5	<u>Kurile Islands</u> 44.16 N 148.43 E
	epP	A	33 40	H = 03 21 30.7 h = 40 km MB = 4.7
				D = 78.39 Az = 333.6 (NEIC)
				h = 43 km
				pPV A 1.4s 14.0nm
23.	eP	A	03 54 40.5	<u>Kurile Islands Region</u> 43.92 N 148.27 E
				H = 03 42 41.1 h = 39 km MB = 4.9
				D = 78.55 Az = 333.5 (NEIC)
23.	e	A	11 36 09.5	
23.	eiPKP	AC	18 38 [illegible]6	<u>Tonga Islands</u> 15.11 S 172.97 W
	epPKP	A	38 54	H = 18 19 04.6 h = 51 km MB = 5.7
	LmV	B	19 36.7	D = 144.37 Az = 355.0 (NEIC)
	LmH	B	37.3	h = 79 km
				PKPV A 2.0s 85.5nm
				pPKPV A 1.4 55.8nm
				LmH B 22.5 0.9 /um
				LmV B 26 0.9 /um
24.	eiP	A	01 54 31.5	<u>Alaska Peninsula</u> 55.83 N 158.25 W
	LmH	C	02 24.0	H = 01 43 04.5 h = 66 km MB = 5.3
	LmV	C	24.0	D = 73.56 Az = 6.7 (NEIC)
				PV A 1.6s 104.3nm M = 5.5
24.	iP	A	10 31 23.2	<u>Kurile Islands</u> 48.81 N 155.73 E
	LmH	B	11 09.0	H = 10 19 36.6 h = normal MB = 5.0
	LmV	C	09.6	D = 76.26 Az = 337.4 (NEIC)
				PV A 1.0s 27.6nm M = 5.2
				LmH B 20 0.9 /um 5.1
				LmV C 18.5 1.2 /um 5.3
24.	LmH	B	12 27.9	<u>Loyalty Islands Region</u>
	LmV	C	29.8	22.42 S 171.81 E
				H = 15 51 11.9 h = normal MS = 5.2 (NEIC)
				D = 86.63
				LmH B 13.5s 0.5 /um M = 5.1
				LmV C 20 0.6 /um 4.9

Day	Phase		h m s	Remarks
24.	ePKHKP	A	16 10 54	Loyalty Islands Region
	ePKP2	A	10 58	22.42 S 171.81 E
	LmH	C	17 15.7	H = 15 51 11.9 h = normal MB = 5.5
	LmV	C	20.6	D = 147.86 Az = 336.1 (NEIC)
				LmH C 22s 0.8 µm M = 5.4
				LmV C 20 0.7 µm 5.5
24.	LmH	B	16 44.8	Central California 36.58 N 121.22 W
	LmV	C	51.5	H = 15 56 51.1 h = 5 km MB = 4.9 MS = 4.0 (NEIC)
				D = 83.73
				LmH B 17s 1.2 µm M = 5.4
				LmV C 20 0.6 µm 5.0
25.	ePKIKP	A	01 35 54	South Sandwich Islands Region
	eiPP	AC	36 55	60.90 S 25.67 W
	ePKKP	A	46 35	H = 01 17 12.5 h = normal MB=6.0 MS=6.1
	ePS	B	46 38	D = 114.88 Az = 25.2 (NEIC)
	eSS	C	53 00	PPV A 1.7s 78.8nm M = 6.2
	ePcSPKP1	A	54 19	PcSPKP2V A 1.7 30.3nm
	ePcSPKP2	A	54 24.5	LmH B 18.5 4.1 µm 6.1
	LmH	B	02 21.1	LmV C 18 8.9 µm 6.4
	LmV	C	22.3	
25.	ePn	A	02 37(33)	Austria 47.4 N 10.0 E
	ePg	A	37(47)	H = 02 36 46 (BCIS)
	eSn	A	38 14	D = 3.39
	iSg	A	38 31.5	
25.	ePKIKP	A	09 03 10.5	South of Fiji Islands 25.09 S 179.74 W
	ePKHKP	A	03 18.5	H = 08 44 12.0 h = 450 km MB = 5.2
	ePKP2	A	03 31.5	D = 153.02 Az = 344.0 (NEIC)
	epPKP	A	05 20	PKP2V A 1.0s 19.7nm
25.	LmV	C	15 10.4	LmH B 22s 0.9 µm
	LmH	B	15.5	LmV C 23 1.4 µm

Day	Phase		h m s	Remarks
26.	-iP	A	02 24 42.5	Kurile Islands 49.17 N 156.17 E
	LmH	B	03 04.4	H = 02 12 57.3 h = normal MB = 4.9
	LmV	C	04.5	D = 76.03 Az = 337.6 (NEIC)
				PV A 1.2s 34.6nm M = 5.2
				LmH B 18 0.8 µm 5.1
				LmV C 16 1.3 µm 5.4
26.	eP	A	06 10 14.5	Kurile Islands 46.81 N 152.62 E
				H = 05 58 22.1 h = normal MB = 4.9
				D = 77.26 Az = 335.7 (NEIC)
				PV A 0.8s 11.5nm M = 5.0
26.	ePKIKP	A	08 30 55	New Britain Region 5.19 S 152.70 E
				H = 08 12 01.1 h = 58 km MB = 5.2
				D = 124.24 Az = 331.1 (NEIC)
26.	eP	A	15 38 10	Mexico-Guatemala Border Region
				15.99 N 91.02 W
				H = 15 25 33.2 h = normal MB = 4.6
				D = 85.58 Az = 38.5 (NEIC)
26.	ePKHK̇P	A	21 05 49	Fiji Islands Region 19.55 S 177.75 W
				H = 20 47 04.4 h = 556 km MB = 4.2
				D = 148.10 Az = 348.7 (NEIC)
				traces
26.	eP1	A	23 40 11	USSR-Mongolia Border Region
	eiP2	AB	40 14	50.61 N 97.31 E
	e(pP2)	A	40 22	H = 23 31 09.6 h = normal MB=5.3 MS=4.6
	ePcP2	A	41 25.5	D = 51.33 Az = 305.6 (NEIC)
	ePP2	A	42 10	P1V A 1.5s 22.6nm M = 4.9
	LmH	B	24 04.5	P2V A 1.6 165.0nm 5.8
	LmV	B	04.5	(pP2)V A 1.6 82.4nm
				PcP2V A 1.2 24.4nm
				PP2V A 2.0 94.0nm 5.4
				LmH B 12 4.5 µm 5.7
				LmV B 12 4.0 µm 5.7

Day	Phase		h m s	Remarks
27.	eP	A	10 10 13.5	North of Franz Josef Land
	e(pP)	A	10 20	87.05 N 53.54 E
	eX	A	16 39	H = 10 03 02.6 h = normal MB = 4.9
	LmH	C	33.0	D = 37.38 Az = 224.5 (NEIC)
	LmV	C	33.0	(h = 31 km)
				PV A 2.5s 92.0nm M = 5.2
				(pPV) A 1.7 60.6nm
				XV A 1.0 11.8nm
27.	eP	A	11 19 23.5	South Indian Ocean 1.47 S 65.05 E
				H = 11 08 19.9 h = normal MB = 5.0
				D = 68.93 Az = 326.8 (NEIC)
				PV A 1.6s 33.0nm M = 5.2
27.	eP	A	12 18 55.5	West of Gibraltar 34.80 N 9.14 W
	ePP	A	19 11	H = 12 14 03.7 h = normal MB = 4.7
	LmH	C	26.0	D = 21.87 Az = 37.3 (NEIC)
				PV A 1.4s 23.3nm M = 4.4
				PPV A 1.4 32.6nm
28.	eP1	A	10 56 36	Southern Greece 36.98 N 24.16 E
	iP2	A	56 39.3	H = 10 52 43.7 h = 11 km MB = 4.8
	LmH	C	11 02.1	D = 16.35 Az = 330.6 (NEIC)
	LmV	C	04.4	P2V A 1.0s 35.4nm M = 4.5
				LmH C 18.5 2.4 μm 4.4
				LmV C 14 2.9 μm 4.8
28.	ePn	A	15 41 46	Federal Republic of Germany
	ePg	A	41 54.5	49.50 N 8.21 E
	eiSn	A	42 15	H = 15 41 05.3 h = 10 km
	eiSg	A	42 26	D = 2.48 Az = 61.2 (NEIC)
28.	eP	A	17 43 13.5	Near East Coast of Eastern Russia
				43.43 N 132.25 E
				H = 17 32 29.4 h = 457 km MB = 4.5
				D = 73.03 Az = 325.1 (NEIC)

Day	Phase		h m s	Remarks
28.	LmH	C	19 19.5	Northern Easter I. Cordillera
	LmV	C	20.5	4.6 S 105.2 W
				H = 18 19 25.57 h = normal
				MB = 4.8 MS = 5.6 (NEIC)
				D = 110.29
28.	e	A	20 26 22	Poland 50.37 N 18.83 E
	eSg	A	26 33.5	H = 20 24 00.8 M = 2.8 (WAR)
				D = 4.61 Az = 276 (ISC)
29.	+iP1	AB	09 35 31	South of Honshu, Japan 33.34 N 140.78 E
	iP2	AB	35 45	H = 09 22 59.8 h = 56 km MB = 6.3 (NEIC)
	ePP	B	38 58	D = 83.73
	ePPP	B	40 44	P1V A 1.3s 430.0nm M = 6.3
	eS	B	45 52	P2V A 2.0 4060.0nm 7.1
	ePS	B	46 48	PPV B 12 11.22 µm 7.1
	e(SS)	B	51 49	SH B 15 56.8 µm 7.5
	e	B	53 00	PcPPKPV A 2.1 105.2nm
	eSSS	B	55 00	P'P'V A 1.6 38.5nm
	eSSSS	B	57 40	LmH B 16 185.0 µm
	ePcPPKP	A	58 25.5	LmV B 18 132.0 µm
	P'P'	A	10 01 36.5	
	LmH	B	14.8	
	eP'P'P'	A	22 26.5	
	LmV	B	22.7	
29.	+iP	A	09 55 12.0	South of Honshu, Japan 33.46 N 140.86 E
				H = 09 42 41.9 h = 63 km MB = 5.8 (NEIC)
				D = 85.04
				PV A 1.7s 182.0nm M = 5.8
29.	e(P)	A	11 20 33	Off East Coast of Honshu, Japan
	eX	A	20 38	33.50 N 140.96 E
				H = 11 07 57.3 h = 51 km MB = 5.0 (NEIC)
				D = 85.04
				XV A 1.6s 19.2nm

Day	Phase		h m s	Remarks
29.	eP	A	12 13 58.5	South of Honshu, Japan 33.57 N 140.86 E H = 12 01 25.5 h = 53 km MB = 4.7 (NEIC) D = 84.95 PV A 1.6s 19.2nm M = 5.0
29.	eP	A	13 13 31	Luzon, Philippine Islands 18.19 N 120.42 E H = 13 00 45.0 h = 34 km MB = 5.1 D = 87.47 Az = 322.9 (NEIC) traces
29.	ePn	A	20 56 31	Adriatic Sea 42.04 N 15.38 E
	eiSn	A	58 07	H = 20 54 21.2 h = normal MB = 4.7 (NEIC)
	eSg	A	59 19.5	D = 8.99
				PnV A 1.0s 23.6nm
29.	ePn	A	21 34 30	Adriatic Sea 42.02 N 15.20 E
	iSn	A	36 03.5	H = 21 32 18.6 h = 57 km MB = 4.1
	eSg	A	37 17	D = 8.98 Az = 345 (ISC)

Day	Phase		h m s	Remarks
1.	ePKIKP	A	09 24 22.5	Tonga Islands 18.92 S 173.83 W
	+iPKHKP	A	24 26.3	H = 09 04 41.7 h = normal MB = 5.8
	LmH	B	10 19.4	D = 148.05 Az = 353.4 (NEIC)
	LmV	B	22.0	PKIKPV A 1.3s 21.8nm
				PKHKPV A 1.6 181.0nm
				LmH B 16 7.8 μm M = 6.5
				LmV B 15 6.4 μm 6.4
1.	eP	A	09 38 08	South of Honshu, Japan 33.48 N 140.95 E
	epP	A	38 21	H = 09 25 36.5 h = 62 km MB = 5.0
				D = 85.06 Az = 330.4 (NEIC)
				h = 47 km
1.	eP1	A	09 41 15	Off Coast of Northern California
	eP2	A	41 17	40.51 N 125.24 W
	eP3	A	41 21.5	H = 09 28 56.7 h = normal MB = 5.4
				D = 81.68 Az = 26.1 (NEIC)
				P1V A 1.0s 11.8nm M = 4.9
				P2V A 1.5 40.2nm 5.2
				P3V A 1.4 51.2nm 5.4
1.	epP	A	13 21 58.5	South of Honshu, Japan 33.55 N 140.78 E
				H = 13 09 15.8 h = 59 km MB = 4.5
				D = 84.93 Az = 330.3 (NEIC)
				pPV A 1.4s 18.6nm
2.	ePKP	A	09 49 34.5	New Hebrides Islands 18.21 S 168.12 E
				H = 09 30 08.7 h = 37 km MB = 4.9
				D = 142.61 Az = 335.3 (NEIC)
				PKPV A 1.2s 10.2nm
2.	eP1	A	10 56 50	South of Panama 4.79 N 78.28 W
	eP2	A	56 53	H = 10 44 09.3 h = normal MB = 5.3
				D = 86.27 Az = 39.6 (NEIC)
				P2V A 1.4s 34.9nm M = 5.4

Day	Phase		h m s	Remarks
2.	eP	A	12 23 26.5	Philippine Islands Region
				20.21 N 123.00 E
				H = 12 10 41.5 h = 33 km MB = 4.9
				D = 87.31 Az = 323.6 (NEIC)
				PV A 1.6s 22.0nm M = 5.2
2.	eP	A	12 53 46	Norwegian Sea 72.43 N 3.28 E
				H = 12 48 48.5 h = normal MB = 4.5
				D = 22.18 Az = 165.8 (NEIC)
				PV A 1.8s 33.8nm M = 4.5
2.	+iP	A	20 22 39.0	South of Honshu, Japan 33.43 N 140.82 E
	ePP	A	25 53	H = 20 10 08.2 h = 57 km MB = 5.7
	LmH	C	57.5	D = 85.05 Az = 330.3 (NEIC)
	LmV	C	21 08.5	PV A 1.3s 74.3nm M = 5.6
3.	e	A	07 39 32.5	Western Gulf of Aden 11.7 N 45.7 E
				H = 07 30 53 h = 16 km
				D = 47.81 Az = 331 (ISC)
3.	ePn	A	21 28 42	Yugoslavia 44.69 N 18.42 E
	iX	A	28 45	H = 21 26 51.3 h = 32 km MB = 4.9
	iSn	A	30 07	D = 7.52 Az = 324.8 (NEIC)
	eSg	A	31 02	PnV A 0.7s 15.3nm M = 5.1
	LmH	B	31.3	XV A 0.7 51.7nm
3.	eSg	A	21 36 55	Yugoslavia 44.6 N 18.5 E
				H = 21 32 58 h = 0 km (ISC)
				D = 7.6
4.	ePKIKP	A	03 17 08.5	Fiji Islands Region 21.77 S 179.52 W
	-iPKHKP	A	17 14.5	H = 02 58 28.0 h = 570 km MB = 5.4
	iPKP2	A	17 22.5	D = 149.88 Az = 345.8 (NEIC)
	epPKP	A	19 38	PKIKPV A 1.4s 23.3nm
				PKHKPV A 1.2 142.2nm
				PKP2V A 1.3 122.2nm
				pPKPV A 2.0 81.2nm

Day	Phase		h m s	Remarks
4.	ePKIKP	A	04 38 08.5	New Hebrides Islands 14.70 S 166.62 E
	eX	A	38 16.5	H = 04 18 44.9 h = 32 km MB = 5.0
				D = 138.84 Az = 335.9 (NEIC)
				XV A 1.4s 23.2nm
4.	e	A	11 00 51	
	e	A	00 57	
	e	A	01 35	
4.	eP	A	18 32 15	Tadzhik-Sinkiang Border Region
				38.28 N 73.98 E
				H = 18 24 11.1 h = 130 km MB = 5.1
				D = 44.91 Az = 307.0 (NEIC)
				PV A 0.8s 19.2nm M = 4.9
4.	eiP	A	19 17 18.5	Jan Mayen Islands Region 71.34 N 5.28 W
	LmH	B	26.5	H = 19 12 24.6 h = normal MB = 5.0
				D = 22.15 Az = 150.6 (NEIC)
				PV A 1.7s 248.0nm M = 5.4
5.	eP	AB	03 15 36.5	Fox Islands, Aleutian Is.
	e	A	15 49.5	51.32 N 170.91 W
				H = 03 03 37.0 h = normal MB = 5.2
				D = 78.39 Az = 358.4 (NEIC)
				PV A 1.1s 52.4nm M = 5.5
5.	eP	A	10 19 23	Off East Coast of Honshu, Japan
				33.68 N 141.15 E
				H = 10 06 49.9 h = 42 km MB = 4.8
				D = 84.97 Az = 330.5 (NEIC)
6.	ePKP	A	00 45 19.5	Kermadec Islands Region 18.2 S 174.4 W
				H = 00 25 18 h = 0 km MB = 4.8
				D = 157.14 Az = 350 (ISC)
				PV A 1.0s 13.8nm
6.	-iP	AB	19 00 50.5	Sea of Okhotsk 50.17 N 148.77 E
	epP	A	02 53.5	H = 18 50 18.2 h = 592 km MB = 5.4

Day	Phase		h m s	Remarks
cont. 6.	e(sP)	A	19 03 46	D = 73.12 Az = 333.1 (NEIC) h = 627 km PV A 1.6s 302.0nm M = 5.6 pPV A 1.6 110.0nm
7.	eP	A	06 32 24.5	Taiwan Region 22.33 N 122.18 E H = 06 19 50.0 h = normal MB = 5.0 D = 85.17 Az = 323.3 (NEIC)
7.	-iPKIKP eiPKHKP eiPKP2 eiPP LmH	AB AB AB AB C	08 04 53.5 05 02.5 05 23 09 00 09 01.0	Kermadec Islands Region 28.23 S 178.35 W H = 07 45 22.0 h = 192 km MB = 6.2 (NEIC) D = 156.37 PKIKPV A 2.0s 512.0nm PKHKPV A 2.1 540.0nm PKP2V A 1.7 673.0nm PPV A 2.4 676.0nm
7.	eP e	A A	14 10 02.5 10 09	Near Coast of Chiapas, Mexico 14.63 N 93.80 W H = 13 57 10.6 h = normal MB = 4.8 D = 88.30 Az = 37.9 (NEIC) PV A 2.0s 59.8nm M = 5.6
7.	ePKP	A	18 42 22	Loyalty Islands Region 22.1 S 169.9 E H = 18 22 40.5 h = 33 km MB = 3.4 D = 146.82 Az = 335 (ISC)
8.	eP LmH LmV	A B B	04 04 01 48.0 05 03.2	West New Guinea Region 3.72 S 131.39 E H = 03 45 25.2 h = normal MB = 5.9 D = 111.42 Az = 323.6 (NEIC) LmH B 20s 3.2/um M = 5.9 LmV B 16 1.1/um 5.5
8.	eP	A	06 27 28	Fox Islands, Aleutian Is. 52.71 N 166.18 W H = 06 15 38.9 h = 47 km MB = 4.8 D = 77.01 Az = 1.4 (NEIC) PV A 0.8s 17.3nm M = 5.1

Day	Phase		h m s	Remarks
8.	ePKP2	A	06 56 19	Loyalty Islands Region 22.46 S 171.55 E
	LmH	C	07 50.0	H = 06 36 27.2 h = 30 km MB = 5.0
	LmV	C	53.0	D = 147.79 Az = 335.8 (NEIC)
8.	eP	A	15 34 41.5	Alaska Peninsula 54.50 N 162.57 W
				H = 15 22 58.8 h = 20 km MB = 4.9
				D = 75.12 Az = 3.8 (NEIC)
				PV A 1.1s 18.1nm M = 5.0
8.	eP	A	19 56 42.5	South of Honshu, Japan 33.33 N 140.58 E
				H = 19 44 12.7 h = 67 km MB = 4.9
				D = 85.04 Az = 330.2 (NEIC)
				PV A 0.7s 15.3nm M = 5.2
9.	e	A	05 09 04	Fiji Islands Region 14.89 S 176.44 W
	LmV	B	06 11.8	H = 04 49 26.8 h = 143 km MB = 4.9
	LmH	B	14.3	D = 143.75 Az = 351.3 (NEIC)
				LmH B 18s 0.8 µm
				LmV B 20 0.7 µm
9.	eP	A	15 41 21	Fox Islands, Aleutian Is.
				52.67 N 169.39 W
				H = 15 29 31.8 h = 50 km MB = 4.7
				D = 77.06 Az = 359.3 (NEIC)
				PV A 1.3s 15.3nm M = 4.9
10.	epP	A	02 35 26.5	South of Honshu, Japan 33.45 N 140.82 E
				H = 02 22 40.3 h = 69 km MB = 4.6 (NEIC)
				D = 85.03
				pPV A 1.5s 25.1nm
10.	+iP	A	05 04 46.3	Eastern Kazakh SSR 49.76 N 78.18 E
	ePn	A	06 19	H = 04 56 57.4 h = 0 km MB = 5.5
				D = 41.31 Az = 297.8 (NEIC)
				Underground explosion MB = 6.1 (UPP)
				PV A 0.9s 97.3nm M = 5.5
				PnV A 1.1 22.2nm

Day	Phase		h m s	Remarks
10.	eP	A	10 36 14.5	Near Coast of Venezuela 10.85 N 62.94 W
				H = 10 25 03.4 h = 128 km MB = 5.2
				D = 71.89 Az = 40.2 (NEIC)
				PV A 0.8s 11.5nm M = 4.7
10.	eP	A	14 44 44	West Pakistan 33.76 N 72.71 E
	LmH	C	15 08.4	H = 14 36 16.5 h = 45 km MB = 4.9
	LmV	C	08.6	D = 46.93 Az = 310.3 (NEIC)
11.	eP	AB	04 03 22.5	Near Coast of Guatemala 13.37 N 90.04 W
	eX	A	03 35	H = 03 50 45.9 h = 85 km MB = 5.1
	eS	B	14 00	D = 87.02 Az = 38.6 (NEIC)
	ePS	B	15 10	PV A 1.1s 20.2nm M = 5.1
	eSS	B	20 08	XV A 1.6 41.2nm
	eSSS	B	23 45	LmH B 20 2.8 μm
	LmH	B	39.7	LmV B 18 5.6 μm
	LmV	B	43.5	
11.	eP	A	04 45 04.5	Southeast of Taiwan 22.54 N 123.95 E
				H = 04 32 27.3 h = 31 km MB = 4.9 (NEIC)
				D = 85.97
				PV A 1.8s 27.0nm M = 5.3
11.	ePKP	A	16 14 04.5	Tonga Islands 15.21 S 174.91 W
				H = 15 54 58.7 h = 252 km MB = 4.6 (NEIC)
				D = 144.26
				PKPV A 1.2s 20.3nm
13.	ePKHKP	A	05 59 46	South of Fiji Islands 23.86 S 179.92 E
				H = 05 40 50.3 h = 523 km MB = 4.9
				D = 151.76 Az = 344.2 (NEIC)
14.	eP	A	00 59 49	South of Honshu, Japan 33.26 N 140.88 E
	e	A	01 00 29	H = 00 47 14.0 h = 38 km MB = 5.4
	ePP	A	03 02.5	D = 85.22 Az = 330.4 (NEIC)
	LmH	C	40.8	PV A 2.0s 81.2nm M = 5.6
	LmV	C	47.0	

Day	Phase		h m s	Remarks
14.	ePKHKP	A	07 35 55.5	Tonga Islands Region 22.66 S 175.95 W
				H = 07 16 16.6 h = 135 km MB = 4.7
				D = 151.45 Az = 349.9 (NEIC)
14.	eP	A	11 19 39.5	Near East Coast of Honshu, Japan
				38.55 N 141.71 E
				H = 11 07 30.4 h = 64 km MB = 4.6
				D = 80.96 Az = 330.4 (NEIC)
14.	eP1	AB	14 09 44	Turkey 39.28 N 29.42 E
	eiP2	AB	09 47.5	H = 14 05 45.8 h = 33 km MB = 5.4
	e	A	10 12	D = 16.92 Az = 318.0 (NEIC)
	eS	B	12 55	P1V A 1.3s 56.7nm M = 4.5
	LmH	B	15.6	P2V A 1.8 541.0nm 5.4
	LmV	B	17.5	P2V B 6 2.6 μm 5.5
				LmH B 16 10.2 μm 5.2
				LmV B 12 6.7 μm 5.3
14.	ePn	A	23 05 29	Central Italy 43.54 N 12.34 E
	ePg	A	06 07	H = 23 03 53.1 h = normal
	eSg	A	07 44	D = 7.13 Az = 356.3 (NEIC)
15.	eP	A	06 10 14	Tibet 30.43 N 84.50 E
				H = 06 00 32.4 h = normal MB = 5.3
				D = 56.61 Az = 313.2 (NEIC)
				PV A 1.4s 27.9nm M = 5.1
15.	+iP	AC	11 35 00.8	Rat Islands, Aleutian Is.
	e	AB	35 14	51.62 N 176.77 E
	ePS	B	45 38	H = 11 23 08.7 h = 39 km MB = 5.4
	eSS	C	50 00	D = 77.33 Az = 350.4 (NEIC)
	LmH	B	12 16.3	PV A 1.4s 51.1nm M = 5.3
	LmV	B	16.3	LmH B 16 0.8 μm 5.1
				LmV B 16 0.4 μm 4.8
15.	e(PKP2)	A	15 18 45	South of Kermadec Islands
				32.75 S 178.80 W
				H = 14 58 00.9 h = 28 km MB = 5.0 (NEIC)
				D = 160.54

Day	Phase		h m s	Remarks
15.	eP	A	19 54 58	Mindanao, Philippine Islands
				6.67 N 126.40 E
				H = 19 41 07.7 h = 42 km MB = 4.9
				D = 149.24 Az = 151.2 (NEIC)
16.	eP	A	03 39 08	Greece 37.95 N 23.40 E
				H = 03 35 36.6 h = 146 km MB = 4.2
				D = 15.22 Az = 330.3 (NEIC)
				PV A 1.0s 19.7nm M = 4.4
16.	eP	A	05 22 02	Luzon, Philippine Islands
	epP	A	22 13.5	15.70 N 121.81 E
	LmH	B	06 06.3	H = 05 09 06.2 h = 53 km MB = 5.1
	LmV	B	10.9	D = 90.25 Az = 323.3 (NEIC)
				h = 41 km
				PV traces
				LmH B 17s 1.0 µm
				LmV B 16 0.5 µm
16.	eP	A	21 35 30	Spain 37.45 N 2.22 W
	LmH	C	41.0	H = 21 31 33.4 h = normal MB = 4.8
	LmV	B	43.0	D = 16.49 Az = 32.5 (NEIC)
				PV A 1.6s 38.5nm M = 4.3
17.	+ePKIKP	AB	00 40 27.5	South of Fiji Islands 24.70 S 179.62 W
	+iPKHKP	AB	40 35.8	H = 00 21 25.8 h = 407 km MB = 5.6
	eiPKP2	AB	40 48	D = 152.68 Az = 344.3 (NEIC)
	e	A	43 21	PKIKPV A 1.4s 58.2nm
	ePP	A	44 21	PKHKPV A 1.3 142.0nm
	LmH	C	01 25.0	PKP2V A 1.1 298.0nm
				PPV A 2.0 120.0nm M = 5.7
17.	ePg	A	05 00 49	France 44.70 N 6.76 E
	eSg	A	02 18	H = 04 58 37.6 h = normal
				D = 6.79 Az = 27.1 (NEIC)

Day	Phase		h m s	Remarks
17.	eP	A	07 32 37	Ryukyu Islands 27.05 N 127.70 E
	LmH	B	08 07.8	H = 07 20 07.6 h = 38 km MB = 4.9
				D = 84.31 Az = 324.9 (NEIC)
				PV traces
				LmH B 20s 0.85 μm M = 5.1
17.	eiP	A	08 00 48.5	Kurile Islands 49.02 N 156.23 E
				H = 07 49 02.3 h = normal MB = 5.2
				D = 76.19 Az = 337.7 (NEIC)
				PV A 1.0s 43.3nm M = 5.4
17.	eP	A	09 24 55	Tadzhik SSR 40.08 N 69.74 E
	ePP	A	26 30	H = 09 17 10.5 h = 26 km MB = 5.2
	eX	A	26 36	D = 41.19 Az = 304.8 (NEIC)
	LmH	B	45.5	PV A 1.9s 37.9nm M = 4.8
	LmV	B	45.5	PPV A 1.6 35.7nm 4.9
				LmH B 16 1.0 μm 4.8
				LmV B 10 1.2 μm 5.2
17.	iP	A	12 40 32	Fox Islands, Aleutian Is.
				52.96 N 165.63 W
				H = 12 28 41.5 h = 26 km MB = 5.1
				D = 76.74 Az = 1.8 (NEIC)
				PV A 0.8s 15.4nm M = 5.1
18.	eiP	A	00 52 44	Sakhalin Islands 46.94 N 143.69 E
	epP	A	54 15	H = 00 41 48.3 h = 405 km MB = 5.0
	ePP	A	55 28	D = 74.37 Az = 330.6 (NEIC)
				h = 409 km
				PV A 1.5s 35.2nm M = 4.8
				pPV A 1.8 37.2nm
18.	eiPKP	A	03 53 12	New Hebrides Islands 19.05 S 169.33 E
				H = 03 34 06.2 h = 241 km MB = 4.7
				D = 143.85 Az = 335.8 (NEIC)
				PKPV A 1.2s 28.5nm
18.	+iP1	AB	15 06 02	Iceland Region 68.83 N 17.33 W

Day	Phase		h m s	Remarks
cont. 18.	eiP2	A	15 06 03.5	H = 15 00 59.7 h = normal MB = 4.9
	eS	B	10 14	D = 22.95 Az = 127.8 (NEIC)
	LmV	B	16.6	P1V A 1.3s 70.0nm M = 5.0
	LmH	B	16.8	P1V B 5 0.9/um 5.5
				P2V A 1.2 110.0nm 5.2
				LmH B 14.5 2.2/um 4.8
				LmV B 14 1.7/um 4.8
18.	eP	A	23 21 01	Off East Coast of Honshu, Japan
				33.47 N 141.20 E
				H = 23 08 29.2 h = 60 km MB = 5.0
				D = 85.17 Az = 330.5 (NEIC)
18.	-eiP	AB	23 30 13	Off East Coast of Honshu, Japan
	ePP	A	33 28	33.54 N 141.21 E
	eS	B	40 40	H = 23 17 40.9 h = 47 km MB = 5.9
	LmH	B	24 05.5	D = 85.12 Az = 330.5 (NEIC)
	LmV	B	13.4	PV A 1.2s 179.0nm M = 6.1
				LmH B 18 3.2/um 5.8
				LmV B 14 1.7/um 5.6
19.	ePKHKP	A	05 14 30	South of Fiji Islands 23.72 S 176.13 W
				H = 04 54 50.3 h = 140 km MB = 4.4
				D = 152.45 Az = 349.3 (NEIC)
19.	-eiP	AB	16 09 47.5	Near East Coast of Honshu, Japan
	ePP	B	12 42	40.83 N 141.90 E
	eS	B	19 40	H = 15 57 50.4 h = 76 km MB = 6.0
	ePS	B	20 10	D = 79.05 Az = 330.4 (NEIC)
	eSS	C	24 40	PV A 1.5s 176.0nm M = 5.8
	LmH	B	43.2	PV B 6 1.2/um 6.0
	LmV	B	47.6	PPV B 9 0.9/um 6.1
				SH B 10 1.7/um 6.1
				LmH B 18.5 4.9/um 5.9
				LmV B 17 1.7/um 5.5
20.	eiP1	AB	07 47 03.5	Northern Peru 6.77 S 76.79 W
	eiP2	A	47 06	H = 07 33 49.7 h = 64 km MB = 6.1
	e	A	47 18.5	D = 94.17 Az = 39.7 (NEIC)

Day	Phase		h m s	Remarks
cont. 20.	ePP	AB	07 50 51	P1V A 0.8s 23.5nm M = 5.7
	e	B	57 18	P2V A 2.3 548.0nm 6.6
	ei	B	58 00	PPV A 2.5 353.0nm 6.4
	eS	C	58 12	PPV B 14 3.1 μm 6.6
	ePKKP	A	08 04 07	P'P'V A 2.5 138.2nm
	e	A	04 44	LmH B 17 8.6 μm 6.3
	eSS	B	04 50	LmV B 16 5.9 μm 6.2
	eP'P'	A	12 16.5	
	LmH	B	30.2	
	LmV	B	30.2	
20.	eP	A	13 24 21	Southern Sumatra 1.08 S 100.28 E
	epP	A	24 40	H = 13 11 24.3 h = 55 km MB = 4.8
				D = 89.98 Az = 320.5 (NEIC)
				h = 71 km
20.	eP	A	17 03 45	Northern Peru 6.81 S 76.85 W
				H = 16 50 27.1 h = normal MB = 5.4
				D = 94.24 Az = 39.7 (NEIC)
				PV A 1.5s 15.1nm M = 5.2
20.	LmH	C	18 16.0	Probably to:
	LmV	C	16.0	Easter Island Cordillera
				23.79 S 115.29 W
				H = 17 02 46.5 h = normal MB = 5.4 (NEIC)
				D = 131.27
				LmH C 20s 0.3 μm M = 5.0
				LmV C 24 0.5 μm 5.1
20.	+iP	AB	23 43 43.5	Andreanof Islands, Aleutian Is.
	ePP	A	46 51	51.29 N 179.22 W
	eS	B	53 35	H = 23 31 48.8 h = 46 km MB = 6.0
	eSS	C	58 55	D = 78.03 Az = 353.0 (NEIC)
	LmH	B	24 20.0	PV A 1.0s 181.0nm M = 6.4
	LmV	B	26.2	PPV A 2.0 68.4nm 5.4
				LmH B 20 2.5 μm 5.5
				LmV B 16 1.4 μm 5.4

Day	Phase		h m s	Remarks
20.	ePKIKP	A	24 01 49.5	South of Fiji Islands 24.63 S 179.73 W
	ePKHKP	A	01 57	H = 23 42 56.8 h = 516 km MB = 5.0
	ePKP2	A	02 09.5	D = 152.58 Az = 344.2 (NEIC)
	epPKIKP	A	03 52	h = 538 km
	epPKHKP	A	03 57.5	PKIKPV A 1.6s 27.5nm
	epPKP2	A	04 09	PKHKPV A 1.5 50.2nm
				PKP2V A 1.4 60.5nm
				pPKIKPV A 2.0 34.2nm
				pPKHKPV A 1.8 40.5nm
				pPKP2V A 1.7 54.5nm
21.	eP	A	01 09 11	Near East Coast of Kamchatka
				53.05 N 159.67 E
				H = 00 57 42.5 h = 46 km MB = 4.9
				D = 73.23 Az = 339.4 (NEIC)
				PV A 1.0s 23.6nm M = 5.1
21.	+iP	A	09 59 44	Andreanof Islands, Aleutian Is.
	e	A	59 46.5	50.01 N 176.17 W
	e	A	59 53.5	H = 09 47 38.3 h = 33 km MB = 5.4
				D = 79.50 Az = 355.0 (NEIC)
				PV A 1.6s 55.0nm M = 5.3
21.	e	A	18 22 11	Turkey 40.7 N 42.4 E
	e	A	22 22	H = 18 16 54 h = 22 km MB = 3.9
				D = 23.54 Az = 305
21.	eP	A	23 10 20.5	Mediterranean Sea 35.83 N 14.84 E
	ei	A	10 23.5	H = 23 06 48.5 h = normal MB = 4.6
	eS	B	13 10	D = 15.00 Az = 352.1 (NEIC)
	LmH	B	16.1	LmH B 18s 6.0 μm M = 4.8
	LmV	B	17.2	LmV B 15 1.3 μm 4.4
22.	eP	A	00 57 00	Turkey 40.28 N 42.13 E
	e	A	57 07.5	H = 00 51 51.6 h = 34 km MB = 4.8
	e	A	57 12	D = 23.62 Az = 306.2 (NEIC)
	e	A	57 16	LmH B 12s 0.5 μm M = 4.2
	LmH	B	01 10.2	LmV B 12 0.4 μm 4.3
	LmV	B	10.9	

Day	Phase		h m s	Remarks
22.	+eP	AB	03 10 38	Near East Coast of Kamchatka
	e	A	10 57	53.00 N 159.65 E
	LmV	C	41.6	H = 02 59 09.0 h = 41 km MB = 5.2
	LmH	C	41.7	D = 73.27 Az = 339.4 (NEIC)
				PV A 1.5s 65.5nm M = 5.4
				LmH C 24 0.65 /um 4.8
				LmV C 26 1.0 /um 5.0
22.	ePKHKP	A	08 49 09	North of New Zealand 21.91 S 174.78 W
				H = 08 29 28 h = 117 km MB = 5.0 (NEIC)
				D = 150.88 Az = 352 (ISC)
22.	ePKHKP	A	09 09 20	Tonga Islands Region 22.19 S 174.37 W
				H = 08 49 28.8 h = normal MB = 5.0
				D = 151.21 Az = 352.1 (NEIC)
				PKHKPV A 1.4s 11.6nm
22.	e(Pn)	A	10 21 40	France 49°14' N 5°58' E
	e	A	21 44.5	H = 10 20 33 (BCIS)
	eSg	A	22 38	D = 3.94
22.	-iP	AB	10 39 12	Kurile Islands 49.06 N 153.57 E
	+ipP	B	39 47	H = 10 27 41.9 h = 134 km MB = 6.3
	+isP	B	39 58	D = 75.46 Az = 336.1 (NEIC)
	+iPP	B	42 00	h = 136 km
	+isPP	B	42 48	PV A 1.7s 2361.0nm M = 6.7
	e(PPP)	C	43 48	PV B 6.5 13.5 /um 6.8
	isPPP	B	44 28	pPV B 5.5 4.1 /um
	eS	B	48 35	sPV B 13 11.6 /um
	eiSPP	B	49 28	PPV B 12 6.1 /um 6.8
	eiSS	B	54 28	sPPV B 11 10.0 /um
	eSSS	C	57 40	SH B 16 6.1 /um 6.2
	eSKPPKP	A	11 09 49	SKPPKPV A 2.0 34.2nm
	LmH	B	19.3	LmH B 15.5 17.9 /um
	LmV	B	20.2	LmV B 14.5 15.2 /um
22.	ePKIKP	A	16 13 16	Tonga Islands 21.22 S 174.38 W
	+iPKHKP	AB	13 20.8	H = 15 53 30.9 h = normal MB = 5.6

Day	Phase		h m s	Remarks
cont. 22.	e(PKP2)	A	16 13 28	D = 150.26 Az = 352.3 (NEIC)
	ei	B	13 32	PKHKPV A 1.7s 163.8nm
	e	A	13 37	
22.	eP1	AB	16 38 25.5	Jan Mayen Island Region
	eiP2	A	38 28.5	70.97 N 6.81 W
	PmV	A	38 54	H = 16 33 32.4 h = normal MB = 5.3
	eS	B	42 25	D = 22.09 Az = 147.6 (NEIC)
	LmH	B	47.9	P1V A 0.8s 38.4nm M = 4.9
	LmV	B	50.1	P1V B 16 1.3 µm 5.1
				P2V A 1.5 110.8nm 5.1
				PmV A 1.9 183.0nm 5.2
				SH B 14.5 3.7 µm 5.6
				LmH B 16 6.3 µm 5.1
				LmV B 13 4.9 µm 5.3
23.	ePKP	A	07 36 27	Samoa Islands Region 15.43 S 172.93 W
	e	A	36 37	H = 07 16 53.3 h = normal MB = 5.5
	eX	A	36 46.5	D = 144.68 Az = 355.0 (NEIC)
				PKPV A 1.5s 12.6nm
				XV A 1.8 67.6nm
23.	LmH	C	22 36.0	Molucca Passage 1.50 N 126.24 E
	LmV	C	37.0	H = 21 48 12.0 h = normal MB = 5.1 (NEIC)
				D = 104.19
23.	eX	A	23 30 46.5	Probably Off West Coast of S. Island, N.Z.
				45.29 S 166.40 E
				H = 23 09 59.3 h = normal MB = 5.5 (NEIC)
				D = 162.32
				XV A 0.8s 13.5nm
24.	+iP	AB	03 49 52.0	Alaska Peninsula 56.14 N 157.18 W
	e	A	50 03.5	H = 03 38 27.1 h = 69 km MB = 6.0
	ei	A	50 09.5	D = 73.18 Az = 7.4 (NEIC)
	eiX	AB	50 18.5	PV A 1.4s 530.0nm M = 6.3
	eiY	A	50 20.5	PV B 6 1.8 µm 6.2
	eZ	A	59 14	XV A 0.8 192.4nm

Day	Phase		h m s	Remarks
cont. 24.	eS	B	03 59 16	YV A 1.2s 366.0nm
	e	B	59 40	ZV A 2.5 115.2nm
	e	A	59 50	LmH C 50 2.5 µm
	LmH	C	04 11.4	LmV C 25 1.6 µm
	LmV	C	21.3	
24.	ePKIKP	A	06 26 15	<u>Fiji Islands Region</u> 19.96 S 177.67 W
	-eiPKHKP	A	26 19	H = 06 07 35.2 h = 570 km MB = 5.0
	ePKP2	A	26 24	D = 148.51 Az = 348.7 (NEIC)
				PKHKPV A 1.0s 39.0nm
24.	eP	A	08 20 49	<u>Northern Sinkiang Prov., China</u>
	eipP	A	20 56.3	42.94 N 87.44 E
				H = 08 11 52.8 h = normal MB = 5.0
				D = 50.36 Az = 306.7 (NEIC)
				h = 30 km
				PV A 1.1s 12.1nm M = 4.8
				pPV A 1.2 28.5nm
24.	iPg	A	14 45 29	<u>Adelebsen, Fed. Rep. of Germany</u>
	iSg	A	45 50	51°36.5' N 9°44.8' E
				explosion
				H = 14 45 00.8 yield 8.5 t (Hannover)
				D = 1.5
24.	eP	A	17 31 15.5	<u>Mindanao, Philippine Islands</u>
	LmV	C	18 18.2	7.15 N 123.75 E
	LmH	B	18.7	H = 17 17 40.7 h = normal MB = 5.5
				D = 98.20 Az = 323.4 (NEIC)
				PV A 1.8s 20.3nm M = 5.4
				LmH B 18 0.4 µm 5.0
				LmV C 18 0.8 µm 5.3
24.	eP	AB	23 07 53.5	<u>Near East Coast of Kamchatka</u>
	epP	A	08 09	53.08 N 159.67 E
	esP	A	08 17	H = 22 56 24.7 h = 41 km MB = 5.4
	LmV	C	38.8	D = 73.19 Az = 339.4 (NEIC)

Day	Phase		h m s	Remarks
cont. 24.	LmH	B	23 42.0	h = 56 km
				PV A 2.2s 87.2nm M = 5.3
				pPV A 1.0 25.6nm
				sPV A 1.8 81.0nm
				LmH B 16 0.9 µm 5.2
				LmV C 25 0.8 µm 4.9
25.	+iP	AB	01 07 40	Kurile Islands 48.04 N 153.16 E
	epP	A	08 18	H = 00 56 05.1 h = 134 km MB = 5.8
	eS	B	17 10	D = 76.28 Az = 335.9 (NEIC)
	ePS	B	18 05	h = 117 km
	LmH	B	40.4	PV A 1.3s 209.5nm M = 5.7
	LmV	B	46.8	pP A 1.7 78.8nm
				LmH C 15 1.2 µm
				LmV C 13 0.8 µm
25.	ePKIKP	A	05 46 27	Tonga Islands 21.74 S 174.61 W
	ePKHKP	A	46 31.5	H = 05 26 41.0 h = normal MB = 5.4
	ePKP2	A	46 47	D = 150.74 Az = 351.9 (NEIC)
	LmH	C	06 54.5	PKHKPV A 1.6s 55.0nm
	LmV	C	55.0	LmH C 24 0.4 µm M = 5.0
				LmV C 20 0.4 µm 5.2
25.	eP	A	06 07 48	Mongolia 45.35 N 100.81 E
				H = 05 58 08.2 h = normal MB = 4.8
				D = 56.42 Az = 310.2 (NEIC)
25.	eP	A	06 20 16.5	Dodecanese Islands 36.63 N 27.24 E
	e	A	20 27	H = 06 16 06.6 h = normal MB = 4.3
	LmH	B	29.4	D = 17.96 Az = 326.2 (NEIC)
	LmV	B	29.4	PV A 1.6s 22.0nm M = 4.0
				LmH B 14 0.45 µm 3.9
				LmV B 12 0.65 µm 4.3
25.	ePKIKP	A	12 34 21	New Ireland Region 3.54 S 150.09 E
	eX	A	34 30	H = 12 15 29.6 h = 41 km MB = 5.5
	ePP	B	35 55	D = 121.54 Az = 330.3 (NEIC)
	ePPP	B	38 28	PKIKPV A 1.8s 16.9nm

Day	Phase		h m s	Remarks
cont. 25.	eiSKSP	B	12 45 36	XV A 1.8s 30.4nm
	eiPS	B	45 46	LmH B 18.5 7.3 µm M = 6.4
	iSS	B	52 52	LmV B 20 6.6 µm 6.3
	LmH	B	13 27.2	
	LmV	B	29.1	
25.	eiP	AB	23 11 37	Kurile Islands 43.42 N 146.03 E
	ei(pP)	A	11 54.3	H = 22 59 40.3 h = 43 km MB = 5.8
	ePP	C	14 36	D = 78.26 Az = 332.3 (NEIC)
	iS	B	21 28	PV A 1.2s 150.5nm M = 5.9
	iSKS	B	21 47	PV B 9 3.2 µm 6.3
	eSS	B	26 30	pPV A 1.5 287.0nm
	LmH	B	49.4	SH B 10 3.3 µm 6.4
	LmV	B	52.5	LmH B 18.5 14.3 µm 6.3
				LmV B 19 6.9 µm 6.0
26.	ePKP	A	04 29 15	Fiji Region 16.67 S 178.12 W
				H = 04 09 41.2 h = 0 km MB = 4.8
				D = 145.21 Az = 349 (ISC)
				PV A 1.0s 11.8nm
27.	ePKHKP	A	05 52 54	Fiji Islands Region 19.27 S 177.72 W
				H = 05 34 11.3 h = 576 km MB = 4.3
				D = 147.82 Az = 348.8 (NEIC)
27.	eP	A	08 26 08	South Atlantic Ridge 17.81 S 13.92 W
				H = 08 14 47.7 h = normal MB = 5.1
				D = 71.77 Az = 16.8 (NEIC)
				PV A 1.4s 27.9nm M = 5.1
27.	eP	A	13 38 09.5	Near Coast of Libya 32.27 N 21.64 E
	eX	A	38 17.5	H = 13 33 34.2 h = normal MB = 4.6
				D = 19.81 Az = 340.9 (NEIC)
				PV A 1.6s 49.5nm M = 4.6
				XV A 1.6 52.2nm

Day	Phase		h m s	Remarks
28.	+iP	A	04 29 46.6	Eastern Kazakh SSR 49.73 N 78.19 E
	ePn	A	31 20	H = 04 21 57.3 h = 0 km MB = 5.2
				D = 41.32 Az = 297.8 (NEIC)
				Underground explosion MB = 5.7 (UPP)
				PV A 0.7s 38.3nm M = 5.2
28.	eiPKIKP	AB	14 17 38.5	Kermadec Islands Region
	ePKHKP	A	17 51	30.72 S 179.82 W
	eiPKP2	AB	18 16.5	H = 13 58 21.9 h = 337 km MB = 5.7 (NEIC)
	e	A	20 13.5	D = 158.33
	ePP	A	21 51	PKIKPV A 1.7s 127.3nm
				PKHKPV A 1.6 55.0nm
				PKP2V A 2.0 660.0nm
				PPV A 2.0 85.5nm M = 5.5
28.	ePKP2	A	18 38 34	South of Tonga Islands 24.27 S 175.74 W
	epPKP2	A	38 45.5	H = 18 18 27.7 h = 44 km MB = 5.1
				D = 153.06 Az = 349.6 (NEIC)
				PKP2V A 0.8s 13.5nm
28.	e	A	21 04 57.5	Unimak Island Region 54.04 N 163.44 W
				H = 20 53 06.2 h = 25 km MB = 4.8
				D = 75.61 Az = 3.2 (NEIC)
29.	ePg	A	11 33 56.5	Northern Italy 44.4 N 9.4 E
	eiSn	A	34 35	H = 11 31 54 (BCIS)
	eSg	A	35 11.5	D = 6.48
	e	A	35 18	
29.	ePKIKP	A	16 31 54	New Ireland Region 4.12 S 153.60 E
				H = 16 13 20.3 h = 215 km MB = 5.0
				D = 123.74 Az = 331.9 (NEIC)
				PV : traces
30.	ePKIKP	A	02 51 21	South of Fiji Islands 23.35 S 176.82 W
	ePKHKP	A	51 28	H = 02 31 47.8 h = 125 km MB = 5.2
	ePKP2	A	51 37.5	D = 151.97 Az = 348.5 (NEIC)
				PKHKPV A 1.4s 27.9nm
				PKP2V A 1.5 35.2nm

Day	Phase		h m s	Remarks
30.	+iPKIKP	AB	05 53 45.2	South of Fiji Islands 25.69 S 179.39 E
	iPKHKP	A	53 51.2	H = 05 34 54.7 h = 532 km MB = 6.2
	iPKP2	A	54 00.2	D = 153.37 Az = 342.5 (NEIC)
	eiX	AB	54 15	PKIKPV A 2.0s 329.0nm
	epPKP	B	55 38	PKIKPV B 6.0 2.6 /um
	i	B	56 56	PKHKPV A 2.0 478.0nm
	e	B	06 03 45	PKP2V A 1.8 632.0nm
	ei(PKKS)	B	06 12	XV A 2.0 837.0nm
	eY	A	09 43	YV A 2.5 115.0nm
	eiSS	B	16 38	SH B 16.5 22.0 /um
	e	C	17 40	
	eiSSSS	C	19 38	
	ei	C	22 00	
30.	ePn	A	06 43 28.5	Yugoslavia 44.4 N 17.4 E
				H = 06 41 39 h = 0 km
				D = 7.41 Az = 330 (ISC)
				PnV A 1.4s 9.3nm
30.	ePKIKP	A	07 23 02	South of Fiji Islands 25.83 S 179.70 E
	ePKP2	A	23 24	H = 07 04 06.9 h = 488 km MB = 5.2 (NEIC)
				D = 153.59
30.	ePn	A	07 44 30	Yugoslavia 44.33 N 17.38 E
	e	A	44 54	H = 07 42 35.0 h = 0 km
	eSn	A	45 51.5	D = 7.43 Az = 330 (ISC)
	e	A	46 26	
30.	ePP	AB	11 19 02.5	Near Coast of Central Chile
	epPP	A	19 17	29.83 S 71.39 W
	esPP	A	19 23	H = 11 00 20.1 h = 72 km MB = 5.6
	eSKS	B	25 12	D = 108.32 Az = 41.7 (NEIC)
	ePS	B	28 25	PPV A 2.0s 154.0nm M = 6.4
	LmH	B	12 03.0	pPPV A 2.0 94.0nm
	LmV	B	06.0	sPPV A 2.0 68.3nm
				LmH B 20 1.4 /um
				LmV B 19 1.7 /um

Day	Phase		h m s	Remarks
30.	e	BC	15 40 55	LmH B 18s 0.45/um
	LmH	B	50.5	LmV B 20 0.6/um
	LmV	B	50.5	
30.	ePKHKP	A	16 17 04	South of Fiji Islands 25.73 S 179.36 E
	ePKP2	A	17 19	H = 15 58 02.6 h = 500 km MB = 5.0
				D = 153.40 Az = 342.4 (NEIC)
				PKHKPV A 1.3s 19.7nm
30.	eP	A	16 56 11	North Atlantic Ridge 28.0 N 44.6 W
	eS	BC	17 03 00	H = 16 47 30 h = 0 km MB = 4.6
	LmH	B	13.0	D = 47.66 Az = 46 (ISC)
	LmV	B	13.0	LmH B 17s 0.9/um M = 4.8
				LmV B 18 0.9/um 4.9
30.	eP	A	20 31 03.5	Eastern Siberia 63.8 N 146.0 E
				H = 20 20 50 h = 0 km
				D = 60.47 Az = 328 (ISC)
31.	eP	A	03 01 43	Southern Greece 36.46 N 21.22 E
	eiX	A	01 48.5	H = 02 58 02.4 h = normal MB = 4.7
	LmH	B	07.7	D = 15.78 Az = 337.0 (NEIC)
	LmV	B	09.3	PV A 1.0s 25.6nm M = 4.3
				XV A 1.3 43.3nm
				LmH B 14 2.9/um 4.6
				LmV B 12 1.35/um 4.5
31.	eP	A	14 18 39	South of Honshu, Japan 30.16 N 140.24 E
				H = 14 06 04.1 h = 133 km MB = 5.2
				D = 87.64 Az = 330.1 (NEIC)
				PV: traces
31.	ePP	B	15 56 00	South Sandwich Islands Region
	ePS	B	16 05 30	55.31 S 29.07 W
	ePPS	B	06 38	H = 15 36 53.5 h = normal MB = 5.5 (NEIC)
	eSS	B	11 30	D = 110.89
	LmH	B	37.4	PPV B 8s 0.55/um M = 6.3
	LmV	B	40.5	LmH B 22 3.5/um 5.9
				LmV B 17 2.2/um 5.8

Day	Phase		h m s	Remarks
31.	eP	A	20 08 46	Dodecanese Islands 36.64 N 27.15 E
	LmH	C	16.5	H = 20 04 32.5 h = normal MB = 4.3
				D = 17.90 Az = 326.3 (NEIC)
				LmV: traces
31.	eP	AB	20 36 12	Dodecanese Islands 36.62 N 27.14 E
	eX	A	36 19.5	H = 20 32 00.9 h = 18 km MB = 4.5
	eS	BZ	39 41	D = 17.92 Az = 326.4 (NEIC)
	LmV	B	43.7	PV A 2.0s 68.4nm M = 4.4
	LmH	B	43.8	XV A 1.8 50.7nm
				LmH B 12 0.8 µm 4.2
				LmV B 12 0.9 µm 4.4

Day	Phase		h m s	Remarks
1.	+iPn1	A	05 53 39.7	Yugoslavia 43.64 N 15.98 E
	+iPn2	A	53 40.7	H = 05 51 48.8 h = normal MB = 4.9
	ei	A	54 08.2	D = 7.61 Az = 338.6 (NEIC)
	eiSn	A	54 59.5	Pn1V A 0.6s 26.8nm M = 5.5
	eSg	A	55 44.5	Pn2V A 0.7 138.2nm
	LmV	B	56.4	LmH B 5.5 0.5 μm 3.7
	LmH	B	56.6	LmV B 5 0.5 μm
1.	eP	A	10 27 11.5	Kurile Islands 50.69 N 157.08 E
				H = 10 15 31.7 h = normal MB = 4.8
				D = 74.99 Az = 338.4 (NEIC)
				PV A 1.0s 11.8nm M = 4.8
1.	ePKIKP	AB	11 20 13	Fiji Islands Region 21.42 S 176.76 W
	-iPKHKP	AB	20 18.5	H = 11 00 52.4 h = 216 km MB = 5.2
	pPKP	A	21 13.5	D = 150.10 Az = 349.3 (NEIC)
				PKIKPV A 1.3s 28.4nm
				PKHKPV A 1.1 202.0nm
1.	ePKHKP	A	16 01 50	Tonga Islands 21.49 S 174.18 W
	ePKP2	A	01 55	H = 15 42 00.9 h = normal MB = 4.7
				D = 150.54 Az = 352.5 (NEIC)
1.	ePKIKP	AB	24 11 21	Auckland Islands Region
	ePKHKP	A	11 40	49.55 S 163.73 E
	ePKP2	AB	12 10.5	H = 23 51 23.0 h = normal MB = 6.0
	e	B	14 32	D = 162.12 Az = 284.2 (NEIC)
	e	B	14 50	PKIKPV A 3.4 410.0nm
	ePP	B	15 56	PKIKPV B 9 1.52 μm
	eSKKS	B	22 38	PKHKPV A 2.4 95.0nm
	eSKKKS	B	23 36	PKP2V A 3.8 727.0nm
	e	B	26 16	PKP2V B 9 2.40 μm
	eScSPKP	B	27 02	LmH B 18 4.0 μm M = 6.2
	ePPS	B	29 28	LmV B 20 3.7 μm 6.2
	eSS	B	36 05	
	e(PSS)	B	37 10	
	eSSS	B	42 25	
	ei(PSS2)	B	42 52	

Day	Phase		h m s	Remarks
cont. 1.	LmV	B	25 30.8	
	LmH	B	39.0	
2.	ePKP2	A	00 38 30.5	<u>Auckland Islands Region</u> 49.60 S 164.17 E H = 00 17 43.8 h = normal MB = 5.5 (NEIC) D = 162.40
2.	ePKP2	A	00 59 48	<u>Auckland Islands Region</u> 49.56 S 164.18 E H = 00 38 59.6 h = normal MB = 5.5 (NEIC) D = 162.40
2.	eP	A	03 42 48	<u>Northwestern Kashmir</u> 36.13 N 73.62 E
	e	A	42 55	H = 03 34 27.8 h = 47 km MB = 5.0
	LmH	C	59.4	D = 46.00 Az = 308.6 (NEIC) LmH C 35s 0.74/um M = 4.4
2.	ePKP2	A	05 09 02	<u>Loyalty Islands Region</u> 22.34 S 171.60 E
	LmH	C	57.5	H = 04 49 17.7 h = 53 km MB = 5.3 D = 147.70 Az = 335.9 (NEIC) PKP2V A 2.0s 34.2nm LmH C 40 1.3/um
2.	+iPKP	AB	09 21 00.0	<u>Tonga Islands</u> 16.13 S 173.08 W
	LmH	B	10 27.2	H = 09 01 23.8 h = normal MB = 5.9
	LmV	B	27.2	D = 145.36 Az = 354.7 (NEIC) PKPV A 1.2s 268.0nm LmH B 20 1.09/um LmV B 18 1.27/um
2.	ePKP	AB	14 52 05.5	<u>Samoa Islands Region</u> 15.87 S 172.87 W
	LmH	C	15 50.0	H = 14 32 30.5 h = normal MB = 4.9
	LmV	C	57.5	D = 145.12 Az = 355.0 (NEIC) PKPV A 1.3s 26.2nm LmH C 24 0.6/um M = 5.2

Day	Phase		h m s	Remarks
2.	-eiP	AB	15 56 18.5	<u>Samoa Islands Region</u> 16.25 S 172.88 W
				H = 15 36 46.7 h = 69 km MB = 5.4
				D = 145.50 Az = 354.9 (NEIC)
				PV A 1.3s 83.0nm
2.	ePKIKP	A	21 49 25	<u>New Hebrides Islands</u> 13.29 S 166.13 E
	e	B	49 30	H = 21 30 05.4 h = normal MB = 5.4
	ePP	B	56 15	D = 137.37 Az = 336.1 (NEIC)
	eSS	C	22 10 17	PKIKPV A 1.9s 30.3nm
	LmH	B	56.4	LmH B 19.5 13.8 µm M = 6.7
	LmV	B	58.0	LmV B 18.5 9.8 µm 6.6
3.	eP	A	08 13 21.5	<u>Southern Iran</u> 28.51 N 52.64 E
	eX	A	13 27.5	H = 08 06 06.9 h = 45 km MB = 4.7
				D = 37.92 Az = 317.2 (NEIC)
				PV: traces
				XV A 1.3s 17.5nm
3.	eP	A	09 14 55	<u>Southern Iran</u> 28.12 N 57.19 E
	epP	A	15 05	H = 09 07 16.0 h = 52 km MB = 5.0
	LmH	B	36.2	D = 40.96 Az = 316.1 (NEIC)
				h = 45 km
				PV A 1.5s 32.7nm M = 4.9
				pPV A 1.8 47.3nm
				LmH B traces
3.	eP	A	09 49 13	<u>South of Honshu, Japan</u> 33.46 N 140.82 E
				H = 09 36 40.7 h = 59 km MB = 4.9
				D = 85.02 Az = 330.3 (NEIC)
3.	eP	AB	18 58 51	<u>North Atlantic Ocean</u> 54.26 N 35.10 W
	e(pP)	A	58 55.5	H = 18 52 59.3 h = normal MB = 5.4
	ePa	B	59 30	D = 28 29 Az = 77.9 (NEIC)
	eS	B	19 03 32	PV A 1.6s 71.5nm M = 5.1
	LmH	B	09.6	pPV A 1.5 60.3nm
	LmV	B	12.1	SH B 10 2.2 µm 5.6
				LmH B 16 7.8 µm 5.4
				LmV B 14.5 11.1 µm 5.7
				Pa B : Wave velocity 8.18 km s^{-1}

Day	Phase		h m s	Remarks
3.	eP	AB	20 42 14	North Atlantic Ocean 54.32 N 35.14 W
	e(pP)	A	42 19	H = 20 36 22.2 h = normal MB = 5.2
	ePa	B	42 46	D = 28.30 Az = 78 (NEIC)
	eS	B	46 55	PV A 2.0s 103.0nm M = 5.2
	iSa	B	48 00	(pP)V A 2.2 174.0nm
	LmH	B	53.0	SH B 10 2.6 µm 5.7
	LmV	B	55.4	LmH B 16 10.7 µm 5.5
				LmV B 14.5 14.4 µm 5.8
				Wave velocities:
				Pa B 8.18 km s^{-1}
				Sa B 4.5
3.	eP	A	22 20 47.5	South of Honshu, Japan 33.45 N 138.07 E
				H = 22 08 50.9 h = 309 km MB = 4.5
				D = 83.87 Az = 329.0 (NEIC)
3.	eSg	A	22 23 04	France 44.4 N 7.0 E
				H = 22 19 17 (BCIS)
				D = 7.04
4.	eP	A	01 48 49.5	Taiwan Region 21.63 N 120.01 E
				H = 01 36 25.3 h = 94 km
				D = 84.51 Az = 323 (ISC)
4.	ePn	A	08 28 21.5	Central Italy 43.7 N 13.5 E
	e	A	28 22.5	H = 08 26 40 (BCIS)
	iSn	A	29 38.5	D = 7.1
	eSg	A	30 30	
4.	eP	A	09 05 22	Near East Coast of Honshu, Japan
				39.45 N 142.90 E
				H = 08 53 08.0 h = 34 km MB = 4.6
				D = 80.64 Az = 331.0 (NEIC)
4.	ePKP	A	12 14 47.5	Fiji Region 19.3 S 178.2 W
				H = 11 55 07.4 h = 0 km MB = 4.8
				D = 147.80 Az = 348 (ISC)
				PKPV A 1.5s 15.1nm

Day	Phase		h m s	Remarks
4.	ePKP	A	13 21 42.5	<u>Samoa Islands Region</u> 16.27 S 172.68 W
				H = 13 02 06.4 h = 30 km MB = 4.7
				D = 145.53 Az = 355.2 (NEIC)
				PKPV A 1.5s 22.6nm
4.	ePKP	A	18 52 07	<u>Samoa Islands Region</u> 16.09 S 172.64 W
				H = 18 32 31.7 h = normal MB = 4.7
				D = 145.36 Az = 355.2 (NEIC)
4.	-iP diff	A	22 57 00	<u>Banda Sea</u> 7.43 S 125.58 E
	eiPKIKP	AB	23 00 56.5	H = 22 43 06.7 h = 377 km MB = 6.6
	+iPP	B	01 44	D = 110.82 Az = 321.5 (NEIC)
	+ipPP	B	03 08	P diff V A 1.8s 94.5nm
	eiSKS	B	06 58	PKIKPV A 1.8 149.0nm
	eSP	A	10 28	PPV B 7 1.52/um M = 6.3
	eSP	B	10 32	pPPV B 7 2.1/um
	iPKKP1	A	11 57	SPV A 2.5 338.0nm
	iPKKP2	A	12 08	PKKP1V A 1.4 97.6nm
	esPS	C	13 10	PKKP2V A 1.9 348.0nm
	eSS	C	16 20	LmH B 20 3.6/um
	e(ScPPKP)	A	19 43	LmV B 16 3.1/um
	eP'P'	A	20 21.5	
	LmH	B	44.7	
	LmV	B	51.7	
5.	eP	A	00 37 48	<u>Carlsberg Ridge</u> 5.19 N 61.92 E
				H = 00 27 31.4 h = normal MB = 5.2
				D = 61.69 Az = 326.2 (NEIC)
				PV A 1.9s 34.1nm M = 5.2
5.	eX	A	04 26 16.9	<u>Samoa Region</u> 15.91 S 172.8 W
				H = 04 06 26.0 h = 0 km MB = 4.3
				D = 145.17 Az = 355 (ISC)
				XV A 0.9s 9.7nm

Day	Phase		h m s	Remarks
5.	+iP	A	05 48 57	**Hokkaido, Japan Region** 41.96 N 142.33 E
				H = 05 37 02.9 h = 69 km MB = 5.2
				D = 78.23 Az = 330.5 (NEIC)
				PV A 1.0s 25.6nm M = 5.1
5.	eP	A	07 56 33	**Carlsberg** 5.08 N 62.00 E
				H = 07 46 10.9 h = normal MB = 4.8
				D = 61.82 Az = 326.2 (NEIC)
5.	eP	A	13 28 35	**Northern Sinkiang Prov., China**
				41.98 N 84.58 E
				H = 13 19 47.3 h = normal MB = 4.8
				D = 49.24 Az = 306.5 (NEIC)
5.	iP	A	18 58 57	**Carlsberg Ridge** 5.23 N 61.82 E
				H = 18 48 39.4 h = normal MB = 4.9
				D = 61.60 Az = 326.2 (NEIC)
				PV A 0.6s 9.6nm M = 5.1
5.	eP	A	22 46 39	**Tadzhik-Sinkiang Border Region**
	e	A	47 27	38.38 N 73.50 E
				H = 22 38 36.9 h = 118 km MB = 5.0
				D = 44.55 Az = 306.8 (NEIC)
				PV A 1.1s 12.1nm M = 4.6
6.	+eP	A	00 49 51.5	**Kurile Islands** 48.14 N 154.94 E
	e(sP)	A	50 21	H = 00 38 06.0 h = 68 km MB = 4.8
	LmH	C	01 19.5	D = 76.67 Az = 337.0 (NEIC)
	LmV	C	29	PV A 1.0s 43.3nm M = 5.4
				(sP)V A 1.4 32.6nm
				LmH C 28 0.79/um
				In the intervall 06d 10h 19m - 07d 06h 05m the seismograph SKM-III had been out of operation
6.	ePKP	A	14 54 04	**Solomon Islands** 5.86 S 154.35 E
				H = 14 36 01.9 h = 414 km MB = 5.0
				D = 125.61 Az = 331.7 (NEIC)

Day	Phase		h m s	Remarks
6.	eP	A	15 39 01.5	<u>Ecuador</u> 1.45 S 77.91 W H = 15 26 15.8 h = 178 km MB = 5.0 D = 90.81 Az = 40 (ISC)
6.	eP	A	22 52 22	<u>Alaska Peninsula</u> 54.79 N 162.85 W H = 22 40 42.4 h = normal MB = 4.7 D = 74.83 Az = 3.6 (NEIC)
7.	ePKIKP ePKP2 LmV LmH	A AB B B	00 23 07 23 46 01 43.4 43.8	<u>Macquarie Islands Region</u> 53.32 S 159.04 E H = 00 03 08.5 h = normal MB = 5.9 (NEIC) D = 159.85 LmH B 19.5s 2.23/um M = 5.9 LmV B 21 1.79/um 5.9
7.	eP	A	00 41 13.5	<u>Carlsberg Ridge</u> 5.19 N 61.91 E H = 00 30 55.8 h = normal MB = 5.1 D = 61.68 Az = 326.2 (NEIC)
7.	eP epP	A A	03 27 18 27 42	<u>Southern Alaska</u> 60.13 N 152.75 W H = 03 16 22.6 h = 98 km MB = 5.1 D = 68.86 Az = 10.6 (NEIC) h = 96 km
7.	e	A	13 16 19	traces
7.	ePKHKP ePKP2	A A	23 38 04.5 38 10.5	<u>Tonga Islands</u> 21.36 S 174.13 W H = 23 18 15.5 h = normal MB = 5.1 D = 150.43 Az = 352.6 (NEIC)
8.	eP1 eP2 eS LmV LmH	A A B C C	05 05 05 05 09 13 18 28.5 28.6	<u>Central Mid-Atlantic Ridge</u> 8.05 N 38.79 E H = 04 55 04.9 h = normal MB = 5.4 D = 59.39 Az = 34.8 (NEIC) P1V A 2.0s 51.3nm M = 5.3 P2V A 1.8 71.0nm 5.5 SH B 18 1.2/um 5.4 LmH C 18.5 2.1/um 5.3 LmV C 18 4.5/um 5.7

Day	Phase		h m s	Remarks
8.	eP	A	06 36 25	Off Coast of Oregon 42.65 N 126.32 W
	LmH	B	07 12.8	H = 06 24 13.7 h = 11 km MB = 5.6
	LmV	B	14.3	D = 80.11 Az = 25.7 (NEIC)
				PV A 2.4s 138.0nm M = 5.5
				LmH B 20 1.1 µm 5.2
				LmV B 14 1.1 µm 5.4
8.	+eiP	A	09 44 36.5	Szechwan Province, China
	LmH	C	10 15.0	29.62 N 101.77 E
				H'= 09 33 42.1 h = 53 km MB = 5.3
				D = 67.81 Az = 316.6 (NEIC)
				PV A 1.8s 67.6nm M = 5.3
8.	ePKIKP	A	10 19 53.5	New Britain Region 5.01 S 150.98 E
				H = 10 01 14.3 h = 157 km MB = 5.0
				D = 123.25 Az = 330.3 (NEIC)
9.	iPn	A	00 27 55	Austria 47.2 N 11.4 E
	e	A	27 59	H = 00 27 12 (VIE)
	iPg	A	28 10	D = 3.45 Az = 2 (ISC)
	iSn	A	28 46.5	
	iSg	A	29 02	
9.	+iP	AB	04 19 37	Northern Sinkiang Prov., China
	eS	C	26 38	42.18 N 84.65 E
	LmH	C	37.5	H = 04 10 50.7 h = normal MB = 5.9
	LmV	C	41.7	D = 49.16 Az = 306.4 (NEIC)
				PV A 1.1s 234.0nm M = 6.1
				LmH C 20 4.8 µm 5.5
				LmV C 15.5 6.2 µm 5.8
9.	eP	A	08 34 21.5	Kurile Islands 48.03 N 155.16 E
				H = 08 22 32.7 h = 45 km MB = 4.7
				D = 76.83 Az = 337.1 (NEIC)
				PV A 1.1s 24.2nm M = 5.1
9.	eP	A	10 52 44	Northern Sinkiang Prov., China
				41.99 N 84.64 E

Day	Phase		h m s	Remarks
cont. 9.	eP	A	10 52 44	H = 10 43 56.3 h = normal MB = 4.8
				D = 49.27 Az = 306.5 (NEIC)
				PV A 1.1s 16.1nm M = 5.0
9.	ePKP	A	15 51 40.5	<u>Loyalty Islands</u> 20.22 S 168.30 E
				H = 15 31 59.3 h = 51 km
				D = 144.49 Az = 334 (ISC)
				PKPV A 1.2s 12.2nm
9.	ePKIKP	A	20 59 21	<u>Tonga Islands</u> 18.79 S 173.78 W
	ePKHKP	A	59 24	H = 20 39 40.0 h = normal MB = 5.1
	ePKP2	A	59 27	D = 147.93 Az = 353.5 (NEIC)
				PKHKPV A 1.3s 39.3nm
10.	eP	A	00 59 05	<u>Off East Coast of Honshu, Japan</u>
	epP	A	59 15	33.78 N 142.18 E
				H = 00 46 29.1 h = 38 km MB = 4.9
				D = 85.30 Az = 331.0 (NEIC)
				h = 33 km
				PV A 1.5s 20.1nm M = 5.1
				pPV A 1.3 21.8nm
10.	+eP1	AB	02 14 10	<u>Southern Iran</u> 28.43 N 52.83 E
	+iP2	A	14 13	H = 02 06 53.2 h = normal MB = 6.1
	iP3	A	14 15.5	D = 38.09 Az = 317.2 (NEIC)
	ei	B	15 25	P1V A (2.0)s 248.0nm M = 5.7
	ePP	B	15 40	P1V B 9 9.6 µm 6.6
	e	B	17 00	P2V A 1.2 183.0nm 5.8
	iS	B	20 00	P3V A 1.8 690.0nm 6.2
	e	B	21 05	PPV B 15 11.3 µm 6.4
	LmH	B	33.7	SH B 13.5 26.9 µm 6.9
	LmV	B	33.8	LmH B 16 111.0 µm 6.8
				LmV B 16 138.3 µm 7.0
10.	eP	A	02 32 46	<u>Southern Iran</u> 28.0 N 53.3 E
				H = 02 25 18 h = 0 km MB = 4.3
				D = 38.63 Az = 317 (ISC)

Day	Phase		h m s	Remarks
10.	eP	A	02 41 49	Southern Iran 28.42 N 52.91 E H = 02 34 31.5 h = normal MB = 4.9 D = 38.15 Az = 317.1 (NEIC)
10.	eP	A	04 02 03	Southern Iran 28.52 N 52.83 E H = 03 54 55 h = 121 km MB = 4.7 D = 38.02 Az = 317 (ISC)
10.	eP	A	08 41 10.5	Southern Iran 28.25 N 53.08 E H = 08 33 51.8 h = normal MB = 4.6 D = 38.37 Az = 317.2 (NEIC)
10.	ePKP	A	16 07 12.5	Samoa Islands Region 16.36 S 172.87 W
	eX	A	07 18	H = 15 47 35.7 h = normal MB = 4.9
				D = 145.61 Az = 354.9 (NEIC)
				PKPV A 1.6s 33.0nm
				XV A 1.7 48.5nm
11.	+eiP1	AB	02 31 20.5	Central Mid-Atlantic Ridge
	+i	B	31 24.5	0.97 N 28.29 W
	iPP	B	33 38	H = 02 21 15.7 h = normal MB = 6.0
	ePPP	B	34 45	D = 59.92 Az = 28.2 (NEIC)
	eiS	B	39 36	PV A 1.9s 773.0nm M = 6.5
	e	C	45 42	PV B 4 4.7 /um 7.0
	iSSS	C	46 26	SH B 13 37.2 /um 7.1
	LmH	B	54.2	LmH B 17 32.3 /um 6.5
	LmV	B	56.7	LmV B 17 24.6 /um 6.4
11.	eP	A	11 15 20.5	Greece 39.27 N 21.43 E
	LmH	C	19.0	H = 11 12 14.0 h = 55 km MB = 4.3
				D = 13.32 Az = 331.9 (NEIC)
11.	e(P)	A	12 08 54	Cuba Region 19.03 N 80.73 W
	LmH	C	32.0	H = 11 56 56.3 h = 33 km MB = 4.8 (NEIC)
				D = 76.98

Day	Phase		h m s	Remarks
11.	eP	A	14 26 10	Hokkaido, Japan Region 42.31 N 142.87 E H = 14 14 15.6 h = 60 km MB = 5.3 D = 78.13 Az = 330.7 (NEIC) PV A 1.0s 23.6nm M = 5.1
12.	e(P)	A	05 59 02	Southern Iran 28.28 N 53.12 E H = 05 51 38.9 h = normal MB = 4.3 D = 38.38 Az = 317.2 (NEIC)
12.	+eiPKHKP	A	10 19 34.8	Tonga Islands 19.32 S 174.99 W
	e	A	20 13.5	H = 10 00 10.1 h = 203 km MB = 5.2
	epPKP	A	20 26.5	D = 148.30 Az = 352.0 (NEIC)
				PKHKPV A 1.1s 43nm
12.	ePKHKP	A	15 07 58.5	Fiji Islands Region 19.74 S 176.00 W
	ePKP2	A	08 04	H = 14 48 13.3 h = normal MB = 4.9
				D = 148.58 Az = 350.7 (NEIC)
12.	e(P)	A	18 45 03	Southern Iran 28.32 N 53.09 E H = 18 37 40.8 h = normal MB = 5.1 D = 38.32 Az = 317.2 (NEIC) (P)V A 1.4s 13.9nm M = 4.6
12.	eP	AB	23 15 08	Southern Iran 28.38 N 53.04 E
	eS	C	21 00	H = 23 07 49.9 h = normal MB = 5.0
	LmH	B	32.5	D = 38.25 Az = 317.1 (NEIC)
	LmV	B	32.6	PV A 1.4s 23.2nm M = 4.8
				PV B 5 0.4/um 5.5
				LmH B 18 0.6/um 4.5
				LmV B 16 0.5/um 4.5
13.	eP	A	01 26 05.5	Kurile Islands 48.31 N 155.1 E H = 01 14 11.2 h = 0 km MB = 4.2 D = 76.56 Az = 337 (ISC)
13.	eP	A	18 44 29	Southern Iran 28.26 N 53.00 E H = 18 37 10.3 h = normal MB = 4.5 D = 38.31 Az = 317.2 (NEIC) PV A traces

Day	Phase		h m s	Remarks
14.	LmH	B	03 31.9	Spain 38.3 N 0.8 W
	LmV	B	32.6	H = 03 22 13 (BCIS)
				D = 17.2
				LmH B 12s 0.3/um M = 3.8
14.	eP	A	05 55 09	South of Honshu, Japan
	e	A	55 16	33.65 N 140.85 E
				H = 05 42 41.6 h = 75 km MB = 4.7
				D = 84.87 Az = 330.3 (NEIC)
				PV A 1.5s 20.1nm M = 4.9
				In the intervall 14d 10h 01m - 18h 11m the short-period seismographs (type A) had been out of operation.
15.	eP	A	05 54 58	Eastern Sea of Japan
				43.57 N 138.98 E
				H = 05 43 37.8 h = 245 km MB = 4.5
				D = 75.59 Az = 328.5 (NEIC)
15.	+ePKHKP	A	07 47 52.5	Loyalty Islands Region 21.52 S 170.41 E
	ePKP2	A	47 54.5	H = 07 28 28.9 h = 159 km MB = 5.3
	e	A	48 42	D = 146.51 Az = 335.3 (NEIC)
	e	A	48 48	PKHKPV A 1.3s 39.2nm
15.	LmH	B	15 49.7	Aegean Sea 40.4 N 25.5 E
				H = 15 41 26 (BCIS)
				D = 14.15
				LmH B 12s 0.4/um M = 3.8
16.	eP	A	00 06 09	Romania 45.53 N 26.44 E
				H = 00 03 31.7 h = 136 km MB = 4.6
				D = 11.15 Az = 302.6 (NEIC)
				PV A 1.8s 64.1nm M = 5.0
16.	eP	A	00 27 36	Kurile Islands 46.55 N 152.51 E
				H = 00 15 45.6 h = 53 km MB = 4.5
				D = 77.47 Az = 335.7 (NEIC)

Day	Phase		h m s	Remarks
16.	eP	A	01 40 55	<u>Southern Sumatra</u> 1.84 S 99.83 E
				H = 01 27 55.7 h = 33 km MB = 5.3
				D = 90.28 Az = 320.5 (NEIC)
				PV A 1.5s 30.2nm
16.	eP	A	02 58 09	<u>Kurile Islands</u> 43.23 N 146.58 E
	e	A	58 13	H = 02 46 11.8 h = 54 km MB = 4.9
	epP	A	58 24	D = 78.62 Az = 333 (ISC)
				PV: traces
				pPV A (1.1)s 20.2nm
16.	eP1n	AB	10 11 07.5	<u>Austria</u> 47.73 N 16.12 E
	iP2n	A	11 09.8	H = 10 10 03.6 h = 18 km MB = 4.9
	iP3n	AB	11 13	D = 4.15 Az = 316.3 (NEIC)
	iPg	AB	11 21	P1nV A 1.0s 74.8nm
	iSn	B	11 54	P2nV A 1.2 220.0nm
	LmH	B	12.3	P3nV A 1.1 435.0nm
	LmV	B	12.3	LmH B 1.5 100.0 μm M = 6.1
				LmV B 1.0 43.3 μm
16.	e(Sn)	A	10 22 34	Probably <u>Austria</u>
	eSg	A	22 58.5	
16.	ePKHKP	A	10 39 54.5	<u>Tonga Islands</u> 20.27 S 173.51 W
	e	A	40 04.5	H = 10 20 05.8 h = normal MB = 4.7
				D = 149.42 Az = 353.6 (NEIC)
				PKHKPV A 1.5s 30.2nm
16.	ePn	A	10 41 08.5	<u>Austria</u> 47.7 N 16.01 E
	ePg	A	41 20	H = 10 40 01 (VIE)
	eiSg	A	42 12	D c. 4.12 Az = 317 (ISC)
16.	ePn	A	11 02 32	<u>Austria</u> 47.75 N 16.17 E
	ePg	A	02 48	H = 11 01 00 (BCIS)
	eSg	A	03 39	D = 4.16 Az = 316 (ISC)

Day	Phase		h m s	Remarks
16.	eP1n	A	11 05 49.5	Austria 47.73 N 16.02 E
	eP2n	A	05 51.5	H = 11 04 44.3 h = 19 km MB = 4.4
	eP3n	A	05 54	D = 4.11 Az = 316.9 (NEIC)
	iPg	A	06 08	P1nV A 0.5s 13.4nm
	eiSn	A	06 37	P2nV A 0.4 29.0nm
	eiSg	AB	06 55	P3nV A 1.0 118.5nm
	LmH	B	07.0	LmH B 1.0 22.7 µm M = 5.7
	LmV	B	07.0	LmV B 1.0 12.0 µm
16.	eP	A	13 59 08.5	Greenland Sea 79.46 N 4.09 E
				H = 13 53 09.2 h = normal MB = 4.7
				D = 29.06 Az = 170.1 (NEIC)
				PV A 1.0s 19.7nm M = 4.8
16.	e(Sn)	A	13 59 40	Austria 47.8 N 16.7 E
	eSg	A	00 02	H = 13 57 55 h = 0 km (ISC)
				D = 4.4
16.	ePn	A	14 02 47	Austria 47.7 N 16.1 E
	ePg	A	03 02	H = 14 01 39
	eSg	A	03 50	D = 4.17 Az = 316 (ISC)
17.	ePn1	A	01 42 00.5	Adriatic Sea 42.89 N 17.14 E
	ePn2	A	42 03.5	H = 01 39 55.6 h = normal MB = 4.6
	eSn	A	43 40.5	D = 8.63 Az = 335.9 (NEIC)
	eSg	A	44 30	Pn1V A 0.8s 19.4nm M = 5.3
	LmH	C	45.5	Pn2V A 0.7 28.8nm 5.5
	LmV	C	45.5	
17.	eP	A	04 16 05.5	Kurile Islands 49.32 N 152.96 E
				H = 04 04 25.8 h = normal MB = 5.1
				D = 75.06 Az = 335.7 (NEIC)
17.	eP	A	10 46 28	Burma-India Border Region
	epP	A	46 54.5	25.58 N 95.36 E
	esP	A	47 06	H = 10 35 45.7 h = 108 km MB = 5.0
				D = 66.81 Az = 316.5 (NEIC)
				h = 109 km

Day	Phase		h m s	Remarks
cont. 17.				PV A traces
				pPV A 1.2s 48.8nm
				sPV A 1.2 18.3nm
17.	+iP	AB	11 02 10.5	Taiwan Region 24.19 N 122.52 E
	eS	B	12 28	H = 10 49 42.7 h = 35 km MB = 5.8
	eSS	C	18 05	D = 83.87 Az = 323.3 (NEIC)
	LmH	B	44.2	PV A 2.0s 166.8nm M = 5.8
	LmV	B	45.1	PV B 10 1.8 /um 6.2
				SH B 17 2.2 /um 6.0
				LmH B 17.5 18.0 /um 6.5
				LmV B 16.5 19.8 /um 6.6
17.	eP	A	11 14 07.5	Taiwan Region 24.31 N 122.45 E
	e	A	14 26	H = 11 01 40.2 h = normal MB = 5.1
				D = 83.74 Az = 323.2 (NEIC)
				PV A 1.2s 14.2nm M = 5.0
17.	eP	A	14 15 40	Bonin Islands Region 27.89 N 142.77 E
	ePP	A	19 07	H = 14 02 38.7 h = normal MB = 5.3
	e	A	19 15.5	D = 90.70 Az = 331.4 (NEIC)
	LmH	B	56.2	PV A 1.4s 18.6nm M = 5.2
	LmV	B	15 03.7	LmH B 16 0.6 /um 5.1
				LmV B 14 0.4 /um 5.1
17.	ePKP	A	14 51 01	Solomon Islands 6.85 S 154.99 E
	e	A	51 07	H = 14 32 06.3 h = 47 km MB = 5.2
				D = 126.78 Az = 331.7 (NEIC)
17.	epP	A	15 20 20.5	Iran 31.95 N 59.34 E
				H = 15 12 43.5 h = 44 km MB = 4.5
				D = 39.63 Az = 312.4 (NEIC)
17.	LmH	B	15 28.0	LmH B 18s 0.7 /um
	LmV	B	28.0	LmV B 18 1.0 /um
18.	LmH	C	00 18.5	LmV C 16s 0.7 /um
	LmV	C	18.6	

Day	Phase		h m s	Remarks
18.	ePKIKP	A	02 08 24	Loyalty Islands Region 21.28 S 170.33 E
	iPKHKP	AB	08 25.5	H = 01 48 57.4 h = 118 km MB = 5.2
	ePKP2	A	08 27.5	D = 146.25 Az = 335.4 (NEIC)
	eX	A	08 44	PKHKPV A 1.1s 60.5nm
	eY	A	08 58.5	XV A 1.4 30.2nm
				YV A 1.4 27.9nm
18.	eP	A	05 16 34	Near East Coast of Kamchatka
				55.65 N 162.41 E
				H = 05 05 16.0 h = normal MB = 4.8
				D = 71.32 Az = 340.9 (NEIC)
18.	eP	A	05 56 43.5	North Atlantic Ocean 36.41 N 11.15 W
	eS	C	06 00 45	H = 05 51 52.8 h = normal MB = 4.7
	LmH	B	04.2	D = 21.67 Az = 41.8 (NEIC)
	LmV	B	06.8	LmH B 19s 1.8/um M = 4.5
				LmV B 14 0.67/um 4.3
18.	eP	A	15 17 26.5	Lake Tanganyika Region 2.98 S 28.71 E
				H = 15 07 49.1 h = 5 km MB = 5.4
				D = 55.39 Az = 346.9 (NEIC)
18.	eSS	C	19 00 20	Southwest of Africa 52.47 S 17.94 E
	LmH	C	28.6	H = 18 27 27.3 h = 33 km
	LmV	C	28.6	MB = 5.2 MS = 4.9 (NEIC)
				D = 102.88
				LmH C 22s 0.6/um M = 5.1
				LmV C 19 1.0/um 5.4
18.	eP	A	23 28 41.5	Caspian Sea 39.97 N 51.31 E
				H = 23 22 36.4 h = normal MB = 4.8
				D = 29.55 Az = 304.5 (NEIC)
				PV A 0.8s 9.7nm M = 4.6
19.	eP	A	14 52 04	Kurile Islands Region 44.47 N 151.57 E
				H = 14 40 00.5 h = normal MB = 5.0
				D = 79.08 Az = 335.3 (NEIC)
				PV A 1.2s 12.2nm M = 4.8

Day	Phase		h m s	Remarks
19.	eP	A	19 44 11	South of Honshu, Japan 33.29 N 140.79 E
				H = 19 31 40.1 h = 61 km MB = 4.7
				D = 85.16 Az = 330.3 (NEIC)
20.	eP	A	00 44 44.5	Northern Sinkiang Prov., China
				42.01 N 84.58 E
				H = 00 35 56.7 h = normal MB = 5.1
				D = 49.22 Az = 306.5 (NEIC)
				PV A 0.7s 15.3nm M = 5.1
20.	ePKHKP	AB	17 21 28.5	Tonga Islands Region 22.87 S 175.15 W
	ePKP2	A	21 36	H = 17 01 37.0 h = 37 km MB = 5.3
	eSS	B	44 38	D = 151.78 Az = 350.9 (NEIC)
	LmH	B	18 36.5	PKHKPV A 1.6s 38.4nm
	LmV	B	37.5	PKP2 A 2.0 60.0nm
				LmH B 19 0.9/um M = 5.5
				LmV B 18 1.0/um 5.6
21.	-eP	AB	01 39 44.5	Fox Islands, Aleutian Is.
	e	A	39 55	54.01 N 166.85 W
	epP	AB	40 09.5	H = 01 28 09.5 h = 103 km MB = 5.8
	ePP	B	42 32	D = 75.71 Az = 1.0 (NEIC)
	eS	C	49 16	h = 102 km
	eSP	C	49 55	PV A 1.6s 220.0nm M = 5.7
	e	C	02 00 10	PV B 7 1.3/um 5.8
				PPV B 8 0.8/um 6.1
21.	e	A	01 50 53	South of Java 10.58 S 113.72 E
				H = 01 32 05.8 h = 44 km MB = 5.3
				D = 105.78 Az = 319.7 (NEIC)
21.	eP	A	13 33 08.5	Taiwan Region 24.12 N 122.48 E
	epP	A	33 18	H = 13 20 41.2 h = 37 km MB = 5.1
				D = 83.90 Az = 323.3 (NEIC)
				h = 36 km
				PV A 0.9s 7.8nm M = 4.8

Day	Phase		h m s	Remarks
21.	e	A	13 36 40	Norwegian Sea 62.84 N 2.43 E
	e	A	38 48	H = 13 33 25.9 h = normal MB = 4.9
	e	A	41 12.5	D = 13.20 Az = 153.6 (NEIC)
				traces
21.	-iPKP	A	13 44 45.0	Tonga Islands 15.30 S 173.66 W
	e	A	44 58.5	H = 13 25 21.3 h = 125 km MB = 5.6
	ePP	A	48 05	D = 144.49 Az = 354.2 (NEIC)
	e(pPP)	B	48 25	PKPV A 1.6s 110.0nm
	e(SKKS)	C	54 50	PPV A 1.8s 47.3nm M = 5.4
	eSS	C	14 07 15	LmH B 21 1.33/um
	LmH	B	50	LmV B 18 1.02/um
	LmV	B	55	
21.	eP	A	21 28 33.5	Southern Sinkiang Prov., China
	LmH	C	49.0	34.99 N 81.03 E
	LmV	C	52.6	H = 21 19 29.5 h = normal MB = 4.8
				D = 51.40 Az = 310.3 (NEIC)
				PV A 1.4s 16.3nm M = 4.8
				LmH C 22 0.9/um 4.7
				LmV C 15 0.8/um 5.0
22.	eiPn	A	10 19 08	Austria 46.49 N 13.82 E
	iPg	A	19 24.3	H = 10 18 02.3 h = 38 km
	iSn	A	19 50.5	D = 4.41 Az = 341.4 (NEIC)
	iSg	A	20 09	
22.	e(Pn)	A	10 39 50	Northern Italy 46.8 N 12.8 E
	e(Sg)	A	40 51.5	H = 10 38 52 h = 0 km
				D = 3.91 Az = 349
23.	eX	A	03 43 09.5	XV A 0.8s 11.5nm
23.	eP	A	05 16 50	Greece 38.11 N 20.74 E
				H = 05 13 33.0 h = 39 km MB = 4.7
				D = 14.12 Az = 335.6 (NEIC)
				PV A 0.8s 13.5nm

Day	Phase		h m s	Remarks
23.	eP	A	16 17 08	North Atlantic Ridge 49.4 N 28.63 W
				H = 16 11 33.0 h = 0 km MB = 4.3
				D = 25.68 Az = 72 (ISC)
				traces
23.	eP	A	20 53 28.5	Southern Sumatra 5.74 S 104.22 E
				H = 20 40 10.0 h = 103 km MB = 5.5
				D = 96.05 Az = 320.2 (NEIC)
23.	ePKP	A	23 56 05	New Hebrides Islands 17.47 S 167.74 E
				H = 23 36 33.3 h = 17 km MB = 4.6
				D = 141.79 Az = 335.4 (NEIC)
24.	ePKP	A	01 39 55	Easter Island Region 31.36 S 111.06 W
	eX	A	40 02.5	H = 01 20 48.2 h = normal MB = 5.8
	eY	A	40 07	D = 133.90 Az = 48.0 (NEIC)
	ePKS	AB	43 38	XV A 1.4s 20.9nm
	eSS	C	02 00 10	YV A 2.0 94.0nm
	LmH	B	42.6	LmH B 18.5 2.2/um M = 5.9
	LmV	B	43.2	LmV B 18 2.6/um 6.0
24.	ePKP	A	01 58 37.5	North of New Zealand 16.84 S 173.84 W
	eX	A	58 42	H = 01 39 07 h = 131 km MB = 4.9
	epPKP	A	59 09	D = 145.99 Az = 354 (ISC)
				PV A 0.8s 19.2nm
				XV A 1.4 23.2nm
				pPKPV A 1.2 20.4nm
24.	ePKIKP	A	02 23 01.5	Fiji Islands Region 21.33 S 179.32 W
	iPKHKP	A	23 07	H = 02 04 25.0 h = 615 km MB = 5.5
	eiPKP2	A	23 14.5	D = 149.51 Az = 346.2 (NEIC)
	e	A	24 51	PKHKPV A 1.2s 36.6nm
	ePP	A	25 24	PKP2V A 1.2 106.0nm
	e	A	26 42	PPV A 2.5 107.8nm M = 5.4
24.	ePKP2	A	03 10 25	South of Kermadec Islands
				32.91 S 178.42 W
				H = 02 49 45.8 h = 33 km MB = 4.7 (NEIC)
				D = 160.80
				PKP2V A 1.6s 46.7nm

Day	Phase		h m s	Remarks
24.	+eiP1	AB	10 09 49	Taiwan 23.64 N 121.55 E
	+iP2	A	09 53	H = 09 57 21.7 h = normal MB = 6.0
	ePP	B	13 00	D = 83.78 Az = 323.0 (NEIC)
	ePPP	B	14 45	P1V A 1.7s 418.0nm M = 6.3
	eS	B	20 10	P1V B 12 29.0 μm 7.3
	ePS	B	21 05	P2V A 4 6000.0nm 7.1
	eSS	B	25 40	SH B 14 27.1 μm 7.2
	e	B	30 00	LmH B 15.5 237.0 μm 7.7
	LmH	B	52.2	LmV B 16 350.0 μm 7.9
	LmV	B	52.2	
24.	eP	A	11 44 38.5	Taiwan 23.53 N 121.55 E
	iX	A	44 44	H = 11 32 10.1 h = normal MB = 5.0
				D = 83.86 Az = 323.0 (NEIC)
				PV A 1.3s 30.6nm M = 5.3
				XV A 1.4 46.5nm
24.	eP	A	12 12 59	Taiwan 23.64 N 121.56 E
				H = 12 00 29.4 h = 19 km MB = 4.9
				D = 83.78 Az = 323.0 (NEIC)
				PV A 1.4s 27.9nm M = 5.3
24.	eP	A	14 48 27	Southern Iran 28.47 N 53.00 E
				H = 14 41 09.3 h = normal MB = 4.6
				D = 38.16 Az = 317.1 (NEIC)
24.	eP	A	18 10 59	Taiwan 23.72 N 121.65 E
	e	A	11 02.5	H = 17 58 31.5 h = normal MB = 5.3
	ePP	A	14 19	D = 83.77 Az = 323.0 (NEIC)
	eS	B	21 21	PV A 1.7s 36.4nm M = 5.3
	LmH	B	53.2	LmH B 15.5 2.4 μm 5.7
	LmV	B	53.2	LmV B 15 2.9 μm 5.8
24.	eP	A	18 44 50	Hokkaido, Japan Region 41.27 N 141.24 E
	e	A	44 54	H = 18 32 50.3 h = 26 km MB = 5.0
				D = 78.42 Az = 330.0 (NEIC)
				P : traces

Day	Phase		h m s	Remarks
25.	eP	A	00 46 48.5	Eastern Caucasus 43.24 N 45.25 E H = 00 41 36.7 h = 25 km MB = 3.8 D = 23.97 Az = 300 (ISC)
25.	eP	A	03 26 29.5	Kamchatka 54.43 N 158.8 E H = 03 15 04.6 D = 71.75 Az = 339 (ISC)
25.	eP	A	13 28 32.5	Southern Iran 28.36 N 53.18 E
	e	A	28 52	H = 13 21 14.8 h = 45 km MB = 5.0 D = 38.35 Az = 317.1 (NEIC) PV A 1.2s 16.3nm M = 4.8
25.	e	A	18 49 09	
	e	A	49 16.5	
25.	+iP	AB	19 43 10.5	Mindoro, Philippine Islands
	ei	A	43 19	13.37 N 120.31 E
	ePP	AB	46 51	H = 19 30 09.3 h = 50 km MB = 6.2
	ePPP	B	48 38	D = 91.23 Az = 322.9 (NEIC)
	eSKS	B	53 38	PV A 2.5s 3150.0nm M = 7.3
	iS	B	54 05	PV B 6 15.1/um 7.6
	ePS	B	55 09	PPV A 4.5 10130.0nm 7.6
	ePPS	B	56 00	PPV B 12 16.7/um 7.3
	eSS	B	20 00 17	SH B 15 35.7/um 7.5
	LmH	B	30.7	LmH B 17 253.0/um 7.7
	LmV	B	32.6	LmV B 17 303.0/um 7.8
25.	eP	A	20 50 47	Mindoro 13.55 N 120.41 E H = 20 37 48.7 h = 79 km MB = 5.0 D = 91.14 Az = 323 (ISC) PV A 1.2s 14.2nm M = 5.1
25.	eP	A	20 59 40	Mindoro, Philippine Islands 13.52 N 120.52 E H = 20 46 35.9 h = normal MB = 5.4 D = 91.23 Az = 323.0 (NEIC) PV A 1.5s 35.2nm M = 5.5

Day	Phase		h m s	Remarks
25.	eP	A	21 38 23	Mindoro 13.29 N 120.36 E H = 21 25 30 h = 129 km MB = 4.7 D = 91.32 Az = 323 (ISC) PV A 1.5s 15.1nm M = 4.9
26.	ePKP eSKP	A A	01 51 48 54 31.5	Fiji Islands Region 18.22 S 173.04 E H = 01 33 18.5 h = 593 km MB = 5.1 D = 144.42 Az = 339.6 (NEIC) PKPV A 0.8s 38.5nm SKPV A 1.6 60.5nm
26.	e	A	02 02 37	traces
26.	eP LmH LmV	A B B	04 43 19.5 05 26.2 27.0	Mindoro, Philippine Islands 13.51 N 120.52 E H = 04 30 19.5 h = 72 km MB = 5.1 D = 91.23 Az = 323.0 (NEIC) PV A 1.5s 20.1nm M = 5.3 LmH B 20 0.6 µm 5.0 LmV B 16 0.5 µm 5.0
26.	eP Lm1H LmV Lm2H	A B B B	06 31 49 07 14.5 15.5 17.6	Mindoro, Philippine Islands 13.55 N 120.55 E H = 06 18 44.9 h = normal MB = 5.2 D = 91.23 Az = 323.0 (NEIC) PV A 1.2s 20.3nm M = 5.4 Lm1H B 17.5 2.4 µm 5.7 LmV B 17.5 2.1 µm 5.7 Lm2H B 16.5 2.5 µm 5.8
26.	eP e eX LmH LmV	A A A B B	06 33 57 34 04 34 09 39.2 40.5	Turkey 39.50 N 26.35 E H = 06 30 23.9 h = 26 km MB = 5.0 D = 15.22 Az = 321.9 (NEIC) XV A 2.1s 139.0nm LmH B 13.5 3.0 µm M = 4.6 LmV B 11 3.2 µm 4.9

Day	Phase		h m s	Remarks
26.	eP	A	06 56 43	Mindoro, Philippine Islands
				13.49 N 120.60 E
				H = 06 43 42.9 h = 63 km MB = 5.0
				D = 91.31 Az = 323.0 (NEIC)
				PV A 1.8s 33.8nm M = 5.4
				In the intervall 26d 09h 47m - 10h 00m the short-period seismographs (type A) had been out of operation.
26.	ePKIKP	A	12 40 38.5	South of Fiji Islands 23.23 S 179.15 E
	iPKHKP	A	40 44.5	H = 12 21 55.4 h = 567 km MB = 5.0
	iPKP2	A	40 55	D = 150.96 Az = 343.5 (NEIC)
	epPKP2	A	43 04.5	PKHKPV A 1.1s 86.7nm
				PKP2V A 1.3 61.1nm
26.	eP	A	16 03 20	Turkey 39.51 N 26.34 E
	eX	A	03 26	H = 15 59 45.0 h = 28 km MB = 4.8
	LmH	B	08.5	D = 15.21 Az = 321.9 (NEIC)
	LmV	B	10.0	XV A 1.7s 87.8nm
				LmH B 13 3.1 µm M = 4.6
				LmV B 11.5 2.7 µm 4.8
26.	eP	A	16 09 04.5	Mindoro, Philippine Islands
	e	A	09 17	13.22 N 120.34 E
				H = 15 55 59.3 h = 35 km MB = 5.1
				D = 91.36 Az = 322.9 (NEIC)
				PV A (1.3)s 17.4nm M = 5.3
26.	eP	A	16 14 15.5	Mindoro, Philippine Islands
				13.23 N 120.30 E
				H = 16 01 13.3 h = 56 km MB = 5.1
				D = 91.33 Az = 322.9 (NEIC)
				PV A 1.4s 18.6nm M = 5.3
26.	ePn	A	16 43 04	Austria 47.7 N 16.0 E
	e	A	43 18	H = 16 42 00 h = 7 km
	eSg	A	44 06.5	D = 4.18 Az = 318 (ISC)

Day	Phase		h m s	Remarks
26.	eP	A	17 48 10.5	Mindoro, Philippine Islands
	ePP	A	51 48.5	13.43 N 120.58 E
	LmH	B	18 32.0	H = 17 35 08.4 h = 53 km MB = 5.2
	LmV	B	33.2	D = 91.33 Az = 323.0 (NEIC)
				PV A 1.6s 27.5nm M = 5.4
				LmH B 16 0.5/um
				LmV B 15.5 0.6/um
26.	eP	A	21 17 38	Greece 38.28 N 21.97 E
				H = 21 14 06.3 h = normal MB = 4.4
				D = 14.38 Az = 332.6 (NEIC)
				PV A 1.4s 18.6nm M = 4.5
26.	e	A	21 51 36	LmH B 20s 0.5/um
	LmH	B	22 34.0	LmV B 17 0.4/um
	LmV	B	34.4	
26.	eP	A	24 00 51.5	Taiwan 23.78 N 121.49 E
				H = 23 48 27.0 h = normal MB = 4.6
				D = 83.64 Az = 323.0 (NEIC)
27.	eP	A	01 42 37	Mindoro, Philippine Islands
	eX	A	42 44	13.51 N 120.69 E
	eY	A	42 50	H = 01 29 38.2 h = 72 km MB = 5.4
	ePP	A	46 17.5	D = 91.33 Az = 323.0 (NEIC)
	eS	C	53 00	PV A 1.6s 55.0nm M = 5.7
	eSP	C	54 32	XV A 1.6 41.2nm
	LmV	B	02 26.2	YV A 1.8 57.4nm
	LmH	B	28.4	LmH B 16.5 3.3/um
				LmV B 16.5 2.4/um
27.	ePn	A	02 49 04.5	Austria 47.64 N 16.47 E
	eSg	A	50 04	H ≙ 02 47 55 h = 14 km
				D = 4.39 Az = 315 (ISC)
27.	eP	A	06 03 58	Southern Sumatra 0.58 S 99.64 E
				H = 05 51 06.4 h = 54 km MB = 5.3
				D = 89.19 Az = 320.5 (NEIC)

Day	Phase		h m s	Remarks
27.	eP	A	06 57 31	Mindoro, Philippine Islands
	LmH	B	07 40.2	13.44 N 120.44 E
	LmV	B	44.4	H = 06 44 29.7 h = 56 km MB = 5.2
				D = 91.25 Az = 322.9 (NEIC)
				PV A 1.2s 14.2nm M = 5.2
				LmH B 16 1.5/um
				LmV B 16 1.2/um
27.	eP	A	14 50 19.5	Mindoro 13.19 N 120.43 E
	e	A	50 25	H = 14 37 16.7 h = 45 km MB = 5.4
				D = 91.44 Az = 323 (ISC)
27.	eP	A	15 48 40.5	Mindoro 13.33 N 120.34 E
	e	A	48 50	H = 15 35 39.8 h = 65 km MB = 5.1
	LmH	B	16 35.1	D = 91.27 Az = 323 (ISC)
	LmV	B	35.1	PV A 1.5s 17.5nm
				LmH B 16 0.2/um
				LmV B 16 0.4/um
27.	eP	A	19 33 22.5	Taiwan Region 24.07 N 122.51 E
	e	A	33 32	H = 19 20 51.6 h = 15 km MB = 5.6
	LmH	B	15.2	D = 83.96 Az = 323.3 (NEIC)
	LmV	B	16.1	PV A 1.2s 32.5nm M = 5.4
				LmH B 15 1.4/um 5.5
				LmV B 16 1.2/um 5.4
27.	ePKIKP	A	20 00 57	South of Fiji Islands 23.20 S 179.07 E
	iPKHKP	A	01 03	H = 19 42 14.5 h = 574 km MB = 5.2
	eiPKP2	A	01 13.5	D = 150.91 Az = 343.5 (NEIC)
				PKHKPV A 1.1s 72.6nm
				PKP2V A 1.4 65.2nm
27.	eP	A	22 58 35	Mindoro, Philippine Islands
				13.26 N 120.50 E
				H = 22 45 32.5 h = 49 km MB = 4.7
				D = 91.42 Az = 322.9 (NEIC)
				PV A 1.6s 19.2nm M = 5.3

Day	Phase		h m s	Remarks
27.	LmV	C	23 06.5	LmV C 16s 0.3 μm
28.	eP	A	01 02 36	Tibet 31.26 N 84.90 E
	LmH	C	25.3	H = 00 52 56.8 h = normal MB = 5.1
	LmV	C	28.9	D = 56.29 Az = 312.9 (NEIC)
				PV A 1.3s 30.6nm M = 5.2
				LmH C 24 0.6 μm 4.6
				LmV C 17 0.6 μm 4.8
28.	iPg	A	07 59 40.5	Czechoslovakia 50°45.7' N 14°25.5' E
	iSg	A	08 00 05.5	explosion
	LmH	A	00 12	H = 07 59 00 yield 20 t (PRU)
				D = 1.79 Az = 267 (ISC)
28.	eP	A	11 41 45	Burma 17.04 N 94.77 E
	eX	A	41 52	H = 11 30 18.1 h = normal MB = 5.4
	ePP	A	44 29	D = 72.71 Az = 318.5 (NEIC)
	LmH	B	12 19.0	PV A 1.5s 30.2nm M = 5.1
	LmV	B	19.0	XV A 1.7 66.7nm
				PPV A 2.0 25.6nm 5.1
				LmH B 20 0.6 μm 4.8
				LmV B 20 0.6 μm 4.9
28.	e	A	12 47 14	
28.	-eP diff	B	23 47 00	Solomon Islands 5.08 S 154.23 E
	epP diff	B	48 32	H = 23 32 10.2 h = 409 km MB = 6.0
	ePKIKP	AB	50 23.5	D = 124.87 Az = 331.9 (NEIC)
	+i	A	52 09.5	PKIKPV A 1.6s 220.0nm
	+ePP	B	52 14	PKIKPV B 12 2.06 μm
	-iPP	B	52 20	PPV B 12 4.6 μm M = 6.8
	e	B	53 00	PKKPV A 1.5 75.4nm
	epPP	B	53 43	LmH B 17.5 11.0 μm
	ePPP	B	55 16	LmV B 16 6.5 μm
	e(SKS)	B	56 51	
	i(SKKS)	B	58 36	
	iPKKP	A	24 00 08.5	
	iSKSP	B	01 40	

Day	Phase		h m s	Remarks
cont. 28.	eSPP	B	24 03 03	
	eSS	B	08 32	
	esSS	B	11 06	
	esPSS	B	11 23	
	LmH	B	37.1	
	LmV	B	43.8	
29.	eP	A	05 07 42.5	Mindoro, Philippine Islands
	epP	A	07 52	13.5 N 120.7 E
				H = 04 54 39 MB = 5.3 MS = 4.5
				D = 91.3 (ANUSSR)
				h = 34 km
				PV A 1.4s 16.3nm M = 5.2
				pPV A 1.4 16.3nm
29.	eP	A	08 29 17	Off East Coast of Honshu, Japan
				33.66 N 141.07 E
				H = 08 16 45.2 h = 52 km MB = 4.8
				D = 84.95 Az = 330.4 (NEIC)
29.	e	A	15 05 34	
29.	eP	A	16 11 42	Southern Iran 28.29 N 52.97 E
				H = 16 04 21.1 h = normal MB = 4.9
				D = 38.28 Az = 317.2 (NEIC)
29.	+iP	A	18 33 52.5	Crete 34.81 N 24.63 E
	ei	A	33 56	H = 18 29 38.3 h = 47 km MB = 5.1
	eS	C	37 10	D = 18.44 Az = 333.0 (NEIC)
	eScP	A	41 53	PV A 1.0s 86.6nm M = 4.9
	LmV	B	42.4	LmH B 18 1.1 μm 4.2
	LmH	B	42.5	LmV B 13 1.2 μm 4.5
29.	ePKIKP	A	23 25 37	Solomon Islands 6.52 S 155.01 E
				H = 23 06 39.0 h = 54 km MB = 5.5
				D = 126.50 Az = 331.8 (NEIC)

Day	Phase		h m s	Remarks
30.	eP	A	00 26 30.5	Mindoro, Philippine Islands
	e	A	26 44.5	13.29 N 120.41 E
	ePP	A	30 18	H = 00 13 26.3 h = normal MB = 5.2
				D = 91.35 Az = 322.9 (NEIC)
				PV A 1.6s 33.0nm M = 5.4
				PPV: traces
30.	eP	A	10 37 49.5	Crete 35.07 N 23.76 E
				H = 10 33 45.0 h = 81 km MB = 4.3
				D = 17.89 Az = 334.1 (NEIC)
				PV A 0.8s 11.0nm M = 4.1
30.	eP	A	15 28 35	Mindoro, Philippine Islands
	e	A	28 54	13.54 N 120.54 E
	ePP	A	32 14	H = 15 15 34.3 h = 56 km MB = 5.5
	eS	B	39 20	D = 91.23 Az = 323.0 (NEIC)
	eSP	B	40 28	PV A 1.3s 65.5nm M = 5.9
	ePS	B	40 36	PPV A 1.6 22.0nm 5.3
	eSS	C	45 35	LmH B 20 3.9/um
	LmH	B	16 11.4	LmV B 20 2.7/um
	LmV	B	11.8	
30.	eP	A	16 20 08	Mindoro, Philippine Islands
				13.61 N 120.50 E
				H = 16 07 08.6 h = 66 km MB = 5.3
				D = 91.15 Az = 322.9 (NEIC)
				PV A 1.2s 24.2nm M = 5.4

Day	Phase		h m s	Remarks
1.	eP	A	03 53 46.5	New Hebrides Islands 15.37 S 167.66 E
				H = 03 34 33.0 h = 132 km MB = 4.5 (NEIC)
				D = 139.85
1.	ePKP2	A	10 55 03.5	South of Kermadec Islands
	LmH	C	12 19.3	32.72 S 178.47 W
	LmV	C	28.9	H = 10 34 24.3 h = 33 km MB = 4.9 (NEIC)
				D = 160.61
				PKP2V A 1.6s 33.0nm
1.	eP	A	17 36 12.5	Bonin Islands Region 27.97 N 142.77 E
	epP	A	36 23.5	H = 17 23 11.5 h = normal MB = 5.1 (NEIC)
	ePP	A	39(57)	h = 36 km D = 90.62
	LmH	C	18 16.0	
	LmV	C	21.0	
1.	eP	A	19 59 08	Southern Sumatra 4.36 S 103.11 E
				H = 19 45 57.5 h = 90 km MB = 5.4 (NEIC)
				D = 94.29
				PV A 1.1s 8.1nm M = 5.0
1.	eP	A	20 13 49	Taiwan 23.71 N 121.74 E
	e	A	14 02	H = 20 01 15.8 h = 19 km MB = 4.7 (NEIC)
	LmH	C	51.3	D = 83.83
	LmV	C	55.1	PV A traces
2.	eP	A	01 03 24	Greenland Sea 73.45 N 7.19 E
				H = 00 58 21.3 h = normal MB = 5.0 (NEIC)
				D = 22.97
				PV A 1.6s 27.5nm M = 4.5
2.	eP	A	03 23 27.5	Mindanao, Philippine Islands
	ePP	A	27 40	5.57 N 124.60 E
				H = 03 10 24.1 h = 387 km MB = 5.2 (NEIC)
				D = 99.97
				PV A traces

Day	Phase		h m s	Remarks
2.	ePKHKP	A	06 24 39.5	North of New Zealand 21.06 S 175.63 W
				H = 06 04 46.1 h = 0 km
				D = 149.92 Az = 351 (ISC)
				PKHKPV A 1.2s 12.2nm
2.	-eiP	AB	07 10 04.5	East Central Pacific Ocean
	ePP	AB	14 04.5	5.21 N 100.34 W
	eSKS	B	20 48	H = 06 56 23.4 h = normal MB = 5.8 (NEIC)
	eS	B	21 36	D = 99.64
	ePS	B	23 06	PV A 1.6s 55.0nm M = 5.8
	eSS	C	28 32	PPV A 2.0 42.7nm 5.6
	LmV	B	52.3	LmH B 18 1.9/um 5.6
	LmH	B	52.6	LmV B 18 1.9/um 5.7
2.	eP	A	17 05 34	Southern Iran 28.50 N 52.64 E
				H = 16 58 19.7 h = 59 km MB = 4.5
				D = 37.93 Az = 317 (ISC)
2.	eP	A	19 27 18	Southern Nevada 37.15 N 116.20 W
				H = 19 15 01.8 h = 20 km MB = 5.0 (NEIC)
				D = 81.27
				PV A 1.4s 11.6nm M = 4.7
3.	eP	A	05 01 00	Andreanof Islands, Aleutian Is.
	LmH	C	31.0	51.45 N 179.21 W
	LmV	C	31.0	H = 04 49 06.8 h = 56 km MB = 5.3 (NEIC)
				D = 77.87
				PV A 1.1s 16.1nm M = 4.9
				LmV C 28 0.4/um
3.	eP	A	08 58(40)	North Atlantic Ridge 44.39 N 28.29 W
	LmH	B	09 08.6	H = 08 52 53.0 h = 33 km MB = 4.1 (NEIC)
	LmV	B	09.0	D = 27.41
				PV A traces
				LmH B 18s 0.5/um M = 4.1
				LmV B 18 0.6/um 4.4

Day	Phase		h m s	Remarks
3.	ePKHKP	A	18 31 52	Tonga Islands 19.99 S 173.76 W
	LmH	B	20 15.0	H = 18 12 05.1 h = 33 km MB = 4.6 (NEIC)
	LmV	B	15.0	D = 149.11
				LmH B 18s 0.5/um M = 4.1
				LmV B 18 0.6/um 4.4
4.	eP	AB	04 03 01.5	Off East Coast of Honshu, Japan
	epP	AB	03 15.5	33.47 N 141.03 E
	esP	B	03 22	H = 03 50 28.9 h = 45 km MB = 5.5 (NEIC)
	ePP	A	06 18	D = 85.10 h = 45 km
	eS	B	13 30	PV A 1.7s 72.7nm M = 5.5
	LmH	C	44.2	
	LmV	C	44.2	
4.	e	A	04 29 39	South of Java 10.73 S 113.70 E
	e(PP)	A	30 06	H = 04 11 35.0 h = normal MB = 5.8 (NEIC)
				D = 105.88
4.	e	A	06 39 54	Chiapas, Mexico 15.99 N 93.08 W
				H = 06 26 40.6 h = 115 km MB = 5.1 (NEIC)
				D = 86.81
4.	eP diff	C	08 05 00	New Hebrides Islands 15.92 S 167.51 E
	+eiPKP	B	07 34	H = 07 48 17.3 h = 45 km MB = 6.1 (NEIC)
	-ePKP	A	07 36	D = 140.29
	+iPKIKP1	A	07 42.5	PKPV A 1.0s 25.6nm
	ei(PKIKP2)	A	07 45	PKIKP1V A 1.6 318.0nm
	ei(pPKIKP)	A	07 56.5	PKIKP2V A 1.4 437.0nm
	eiPP	AB	10 36	pPKIKPV A 2.0 547.0nm
	ei	B	10 44	ScPPKPV A 3.0 237.0nm
	eScPPKP	A	19 21	PcSPKPV A 3.0 342.0nm
	ePcSPKP	A	19 41	LmH B 19 15.4/um M = 6.7
	eSKSP	C	20 40	LmV B 18 11.2/um 6.6
	ePS	C	20 55	
	eSPP	C	22 40	
	eSKKS2	C	24 20	
	eSS	C	28 52	
	LmH	B	09 16.8	
	LmV	B	16.8	

Day	Phase		h m s	Remarks
4.	ePKHKP	A	09 29 26	Tonga Islands 21.56 S 173.90 W
	e	A	29 38	H = 09 09 34.8 h = 22 km MB = 5.0 (NEIC)
	e	A	29 43	D = 150.65
				PKHKPV A 2.5s 123.0nm
4.	eX	A	17 15 59.5	XV A 0.8s 9.6nm
4.	+eiP	AB	21 44 04	Crete 35.12 N 23.61 E
	+i	A	44 06.8	H = 21 40 00.9 h = 46 km MB = 5.9 (NEIC)
	eiS	BC	47 11	D = 17.79
	LmH	B	52.4	PV B 8s 18.0 μm M = 6.3
	LmV	B	52.4	SH B 10 18.2 μm 6.8
				LmH B 12 135.0 μm 6.4
				LmV B 15 138.0 μm 6.5
5.	+iP	AB	10 28 38.5	Taiwan 22.98 N 121.54 E
	e	A	28 43.5	H = 10 16 11.1 h = 55 km MB = 5.2 (NEIC)
	eS	C	39 00	D = 84.30
	LmH	B	11 11.2	PV A 1.6s 71.5nm M = 5.4
	LmV	B	11.3	LmH B 15 2.5 μm
				LmV B 15 3.7 μm
5.	LmH	C	20 37.7	South Sandwich Islands Region
	LmV	C	37.7	55.28 S 29.17 W
				H = 19 38 33.4 h = 33 km MB = 5.3 (NEIC)
				D = 110.89
				LmH C 21s 0.5 μm M = 5.1
				LmV C 22 0.6 μm 5.4
5.	ePKIKP	AB	23 35 25	New Britain Region 4.16 S 152.68 E
	epPKIKP	A	35 32.5	H = 23 16 27.9 h = 32 km MB = 5.6 (NEIC)
	eiPP	B	37 07	D = 123.33 h = 27 km
	+ePP	A	37 12	PKIKPV A 1.2s 24.4nm
	ePKKP	A	45 15	pPKIKPV A 2.0 85.5nm
	ePS	B	46 57	PPV B 8 2.8 μm M = 6.9
	ePPS	C	48 40	PPV A 2.0 239.0nm 6.4
	eSS	B	53 46	LmH B 19.5 13.0 μm 6.6
	LmH	B	24 31.5	LmV B 19.5 12.4 μm 6.6
	LmV	B	31.5	

Day	Phase		h m s	Remarks
6.	e	A	04 49 39	Balleny Islands Region 62.18 S 164.80 E
	LmH	B	06 07.0	H = 04 28 33.9 h = 33 km (NEIC)
	LmV	B	07.4	D = 161.37
				LmH B 18.5s 0.7 µm M = 5.4
				LmV B 20 0.7 µm 5.5
6.	ePKHKP	A	15 06 40.5	Easter Island Cordillera 56.1 S 124.2 W
				H = 14 46 43.2
				D = 153.45 Az = 83 (ISC)
7.	ePn	A	14 45 50.5	Yugoslavia 43.29 N 17.65 E
	i	A	45 52.0	H = 14 43 49.9 h = 57 km MB = 4.4 (NEIC)
	eiSn	A	47 27	D = 8.43
	e	A	47 29	PnV A 0.7s 30.6nm
	eSb	A	48 07	
	eSg	A	48 31	
	LmH	B	49.3	
	LmV	B	49.4	
7.	ePKIKP	AC	16 58 32	New Ireland Region 4.79 S 153.78 E
	ePP	B	17 00 14	H = 16 39 42 h = 103 km MB = 5.5 (NEIC)
	ePP	AC	00 18	D = 124.41
	epPP	A	00 41	PKIKPV A 2.5s 38.5nm
	ePS	C	10 08	PPV B 10 0.35 µm M = 5.3
	eSS	C	17 05	PPV A 2.0 30.0nm 4.9
	LmV	C	41.7	LmH C 38 1.2 µm
	LmH	C	42.1	LmV C 40 2.8 µm
7.	ePKIKP	AB	22 26 28	South Pacific Cordillera
	ePKP2	B	27 02	53.71 S 134.21 W
	ePKP2	A	27 08.5	H = 22 06 30.1 h = normal MB = 5.4 (NEIC)
	eSKP	C	30 00	D = 158.95
	ePP	B	30 47	PKIKPV A 2.0s 42.7nm
	eSKKS	B	37 28	PKP2V B 8 1.1 µm
	e	C	41 36	PKP2V A 1.8 74.4nm
	ePPS	C	44 08	PPV B 7 1.1 µm M = 6.1
	eSS	C	50 44	LmH B 20 5.3 µm 6.3
	eSSP	C	51 42	LmV B 20 5.3 µm 6.4

Day	Phase		h m s	Remarks
cont. 7.	eSSS	C	22 56 44	
	LmV	B	23 32.6	
	LmH	B	32.7	
8.	eP	A	00 12 24	Greenland Sea 73.52 N 7.57 E
	e	A	12 29.5	H = 00 07 20.9 h = normal MB = 4.5 (NEIC)
				D = 23.03
				PV A 1.3s 10.9nm M = 4.2
8.	e(PKP)	A	06 37 45	Near N. Coast of West New Guinea
	ePP	A	38 19	1.10 S 138.35 E
	ePS	B	48 00	H = 06 18 58 h = 33 km MB = 5.7 (NEIC)
	LmH	B	07 30.5	D = 113.29
	LmV	B	30.5	LmH B 18s 1.3/um M = 5.6
				LmV B 18 1.5/um 5.6
8.	e	A	09 02 21	Greece - Bulgaria Border Region
	e	A	03 51	41.48 N 23.6 E
				H = 08 58 16.3 h = 51 km (ISC)
				D = 12.2
8.	eP1	A	09 23 50	Greece - Bulgaria Border Region
	eP2	A	23 54	41.64 N 23.53 E
	eS	A	26 25	H = 09 20 54.5 h = 10 km MB = 5.0 (NEIC)
	LmH	B	29.1	D = 12.21
	LmV	B	29.1	P2V A 2.2s 81.8nm M = 5.2
				LmH B 10.5 3.1/um 4.6
				LmV B 10.5 3.1/um
8.	eP	A	13 37 13	Taiwan 23.72 N 121.34 E
	LmH	C	14 19.4	H = 13 24 41.6 h = 5 km (NEIC)
	LmV	C	19.4	D = 83.60
				LmV C 16s 0.45/um M = 5.0
8.	ePKP	A	16 34 58.5	Fiji Islands Region 19.44 S 176.33 W
	e	AB	35 01.5	H = 16 15 16.7 h = 71 km MB = 5.4 (NEIC)
	e	A	35 08	D = 148.23
	e	A	35 24	PKPV A 1.5s 37.7nm

Day	Phase		h m s	Remarks
cont. 8.	e	A	16 35 29	LmH B 23s 2.5 µm
	e	C	36 52	LmV B 23 1.7 µm
	ePPS	C	51 10	
	eSS	C	57 40	
	LmH	B	17 37.0	
	LmV	B	37.0	
8.	eP	A	19 42 37	Near East Coast of Kamchatka
				52.98 N 159.87 E
				H = 19 31 06.2 h = 30 km (NEIC)
				D = 73.33
				PV A traces
9.	+iPKIKP	AB	12 38 59.0	Fiji Islands Region 17.80 S 178.86 W
	eiPKHKP	A	39 01	H = 12 20 22.6 h = 568 km MB = 5.8 (NEIC)
				D = 146.17
				PKIKPV A 1.3s 113.5nm
				PKIKPV B 5.0 2.2 µm
				PKHKPV A 1.5 537.0nm
9.	-eiPKP	AB	13 29 04	Tonga Islands 15.55 S 175.02 W
	epPKP	AB	30 09	H = 13 09 59.0 h = 267 km MB = 5.4 (NEIC)
				D = 144.59 h = 259 km
				PKPV A 1.2s 146.0nm
9.	eP	AB	17 44 03	Turkey 39.58 N 26.29 E
	e	A	44 13.5	H = 17 40 25.1 h = 37 km MB = 4.7 (NEIC)
	LmH	B	50.1	D = 15.13
	LmV	B	50.6	PV A 1.8s 47.8nm M = 4.4
				LmH B 9.5 2.2 µm 4.6
				LmV B 11 3.5 µm 4.9
10.	ePKP	A	06 07 19	New Britain Region 4.06 S 150.20 E
	ePP	A	08 57	H = 05 48 22.0 h = 33 km MB = 5.5 (NEIC)
	eSS	C	25 25	D = 122.04
	LmV	B	07 02.1	PKPV A 1.6s 16.5nm
	LmH	B	02.2	LmH B 20 1.0 µm M = 5.4
				LmV B 20 1.2 µm 5.5

Day	Phase		h m s	Remarks
10.	eP	A	17 43 38	Ecuador 2.81 S 78.67 W
				H = 17 30 35.0 h = 89 km MB = 5.4 (NEIC)
				D = 92.34
10.	ePP	A	22 21 30	Northern Celebes 0.24 S 122.32 E
	LmH	C	23 03.8	H = 22 03 10.6 h = 59 km MB = 5.7 (NEIC)
	LmV	C	07.3	D = 103.20
				PPV A 1.6s 19.3nm M = 5.5
				LmH C 24 0.45/um
				LmV C 20 0.3/um
11.	+iP	AB	00 56 53.0	Hokkaido, Japan Region
	+ipP	A	57 12	42.76 N 144.54 E
	eS	B	01 06 40	H = 00 44 58.1 h = 68 km MB = 5.5 (NEIC)
	esSP	B	08 04	D = 78.33 h = 74 km
	LmH	B	28.2	PV A 2.5s 655.0nm M = 6.1
	LmV	B	35.6	LmH B 23 3.6/um
				LmV B 18 2.0/um
11.	eP	A	03 10 34.5	Southern Greece 37.77 N 21.97 E
				H = 03 07 11.7 h = 127 km (NEIC)
				D = 14.84
11.	ePKP	A	04 08(20)	New Hebrides Islands 16.02 S 167.73 E
	epPKP	A	08 32	H = 03 48 50.7 h = 61 km MB = 5.2 (NEIC)
	LmH	C	05 02.3	D = 140.47
	LmV	C	05.5	LmH C 29s 0.35/um
				LmV C 24 0.6/um
11.	eP	A	04 40 53	Southern Greece 37.03 N 24.59 E
	e	A	40 56	H = 04 36 59.3 h = 12 km
	LmH	C	46.4	D = 16.48 Az = 330 (ISC)
11.	LmH	C	20 01.0	LmH C 16s 0.4/um
11.	ePKIKP	AB	21 37 38.5	Kermadec Islands Region
	ePKHKP	A	37 49	27.61 S 179.15 W
	ePKP2	A	38 07.5	H = 21 18 29.6 h = 387 km MB = 5.8 (NEIC)
	epPKIKP	A	39 16	D = 155.58 h = 407 km

Day	Phase		h m s	Remarks
cont. 11.	eSKKS	C	21 47 52	PKIKPV A 1.9s 87.1nm
	eSKSP	C	51 16	PKHKPV A 1.3 43.6nm
	eSS	B	22 01 00	PKP2V A 1.3 61.1nm
	eSSS	B	06 58	
12.	eP	AB	01 33 49.5	<u>Taiwan</u> 24.62 N 121.64 E
	epP	A	34 05	H = 01 21 28.7 h = 62 km MB = 5.3 (NEIC)
	eS	B	44 00	D = 83.04 h = 58 km
	eSS	C	49 24	PV A 1.3s 22.0nm M = 4.9
	LmH	B	02 15.5	PV B 8 0.45 µm 5.5
	LmV	B	15.9	pPV A 1.4 37.2nm
				pPV B 6 0.55 µm
				LmH B 19 2.6 µm
				LmV B 18 2.9 µm
12.	e(Sg)	A	12 51 54	<u>Poland</u> 50.35 N 18.82 E
				H = 12 49 23.0 M = 3.1 (WAR)
				D = 4.65
12.	iPg	A	13 00 26.4	<u>Czechoslovakia</u> 50°10.5' N 13°17.5' E
	iSg	A	00 43.8	Explosion yield 17 t (PRU)
				D = 1.15
12.	LmH	B	18 10.1	<u>Bolivia</u> 17.28 S 66.62 W
	LmV	B	10.8	H = 17 16 35.9 h = 102 km MB = 5.0 (NEIC)
				D = 95.95
				LmH B 20s 0.9 µm
				LmV B 20 0.8 µm
13.	+iP	A	06 43 16	<u>North of New Zealand</u> 16.67 S 174.42 W
				H = 06 23 48.9 h = 116 km MB = 4.4
				D = 145.77 Az = 353 (ISC)
				PV A 1.6s 30.2nm
13.	LmH	C	07 08.0	<u>New Britain Region</u> 5.66 S 151.18 E
	LmV	C	08.0	H = 05 57 07.8 h = 69 km MB = 5.1 (NEIC)
				D = 123.90

Day	Phase		h m s	Remarks
13.	ePKIKP	A	09 43 28	South of Kermadec Islands
	ePKP2	A	44 08.5	32.03 S 178.62 W
	LmH	C	10 51.4	H = 09 23 34.7 h = 41 km MB = 5.6 (NEIC)
	LmV	C	51.4	D = 159.91
				PKIKPV A 2.0s 51.3nm
				PKP2V A 1.5 25.1nm
				LmV C 25 0.6 um M = 5.3
13.	eP	A	13 42 20	Off East Coast of Kamchatka
				52.35 N 160.44 E
				H = 13 30 45.2 h = 33 km MB = 4.5
				D = 74.04 Az = 340 (ISC)
13.	eP	A	14 04 45	North Atlantic Ocean 52.96 N 35.34 W
	LmH	B	19.3	H = 13 58 50.6 h = 33 km MB = 4.4 (NEIC)
	LmV	B	19.4	D = 28.73
				PV A traces
				LmH B 16s 2.0 um M = 4.8
				LmV B 15 1.8 um 4.9
13.	eP	A	15 12 18	North Atlantic Ridge 45.08 N 28.06 W
	LmH	B	22.4	H = 15 06 38.2 h = 33 km MB = 4.8 (NEIC)
	LmV	B	22.9	D = 26.95
				PV A 1.6s 16.5nm M = 4.4
				LmH B 15.5 1.2 um 4.6
				LmV B 17 1.4 um 4.7
13.	eP	A	16 42 55	North Atlantic Ridge 45.09 N 28.14 W
				H = 16 37 13.3 h = 33 km MB = 4.8 (NEIC)
				D = 27.00
				PV A 2.0s 51.3nm M = 4.8
13.	eP1	A	16 46 03	North Atlantic Ridge 45.02 N 28.22 W
	eP2	AB	46 07	H = 16 40.22.0 h = 33 km MB = 5.0 (NEIC)
	e	A	46 13	D = 27.08
	eS	BZ	50 30	P1V A 1.4s 27.9nm M = 4.7
	eS	BE	50 38	P2V A 3.0 263.0nm 5.3
	LmH	B	56.1	SH B 13 1.1 um 5.3

Day	Phase		h m s	Remarks
cont. 13.	LmV	B	16 56.6	LmH B 16.5s 9.4 /um M = 5.4
				LmV B 16.5 9.6 /um 5.6
13.	LmH	C	22 49.0	Near Coast of Central Chile
	LmV	C	50.0	32.69 S 71.78 W
				H = 21 44 56.4 h = 40 km MB = 5.4 (NEIC)
				D = 110.66
				LmH C 20s 0.15 /um M = 4.5
				LmV C 18 0.3 /um 4.8
14.	+iPKHKP	A	01 07 23.3	North of New Zealand 21.8 S 175.2 W
				H = 00 47 31 h = 0 km MB = 4.9
				D = 150.73 Az = 351 (ISC)
				PKHKPV A 0.8s 13.5nm
14.	+eP	AB	12 13 27.5	Off East Coast of Honshu, Japan
	e	A	16 18	40.29 N 143.37
	e	A	16 26	H = 12 01 16.1 h = 12 km MB = 5.3 (NEIC)
	ePP	B	16 31	D = 80.08
	e	A	16 50	PV A 2.0s 115.5nm M = 5.6
	eS	B	23 30	LmH B 19 4.9 /um 5.9
	eSS	B	28 47	LmV B 13.5 2.9 /um 5.8
	LmH	B	47.7	
	LmV	B	13 01.5	
14.	eP	AB	19 46 20	Andreanof Islands, Aleutian Is.
	eS	C	56.4	51.23 N 179.19 W
	eSS	C	20 01 30	H = 19 34 24.2 h = 46 km MB = 5.1 (NEIC)
	LmV	B	28.5	D = 80.08
	LmH	B	28.7	PV A 1.0s 21.6nm M = 5.1
				LmH B 16.5 0.5 /um 5.0
				LmV B 18 0.6 /um 5.0
15.	eiPKIKP	AB	09 51 40.5	South of Fiji Islands 23.63 S 179.90 E
	-iPKHKP	A	51 47.5	H = 09 32 54.6 h = 553 km MB = 5.6 (NEIC)
	eiPKP2	A	51 58.5	D = 151.53 h = 584 km
	epPKIKP	A	53 52.5	PKIKPV A 1.3s 39.2nm
	epPKHKP	A	53 57.5	PKHKPV A 1.2 118.0nm

Day	Phase		h m s	Remarks
cont. 15.	epPKP2	A	09 54 08	PKP2V A 1.4s 102.0nm
	eSS	B	10 14 12	pPKHKPV A 1.5 131.0nm
15.	LmH	B	11 15.0	Chile Argentina Border Region
	LmV	B	16.3	29.66 S 69.48 W
				H = 10 09 40.0 h = 39 km MB = 5.6 (NEIC)
				D = 107.09
				LmH B 18s 0.7 /um M = 5.3
				LmV B 18 0.4 /um 5.0
15.	ePKHKP	A	11 41 31.5	South Pacific Cordillera
	e	A	41 37	55.30 S 128.45 W
	ePKP2	A	41 46	H = 11 21 29.3 h = 33 km MB = 5.0 (NEIC)
				D = 155.76
				PKHKPV A 1.4s 14.0nm
15.	LmH	B	12 46.8	Probably: Central Mid.-Atlantic Ridge
	LmV	B	46.8	3.96 N 31.60 W
				H = 12 12 43.3 h = 33 km MB = 4.7 (NEIC)
				D = 58.95
				LmH B 18s 0.4 /um M = 4.6
				LmV B 18.5 0.4 /um 4.6
15.	eP	A	13 16 13	Kurile Islands 45.40 N 151.89 E
				H = 13 04 14.9 h = 33 km MB = 4.4 (NEIC)
				D = 78.33
				PV A 1.1s 10.8nm M = 4.8
16.	eP	A	01 46 11.5	Crete 34.37 N 25.75 E
	LmH	B	52.9	H = 01 41 42.0 h = 33 km MB = 3.7 (NEIC)
	LmV	B	54.2	D = 19.26
				LmH B 13s 0.4 /um M = 4.3
				LmV B 17.5 0.4 /um 4.3
16.	eP	A	11 07 08	Southern Iran 28.35 N 52.62 E
	LmH	B	26.3	H = 10 59 52.6 h = 37 km MB = 5.0 (NEIC)
	LmV	B	29.9	D = 38.02

Day	Phase		h m s	Remarks
16.	eP	A	16 19 08.5	Hokkaido, Japan Region
				42.47 N 142.99 E
				H = 16 07 20.4 h = 119 km MB = 4.8 (NEIC)
				D = 78.03
17.	eP	AB	05 40 56.5	Philippine Islands Region
	iX	A	41 06.7	13.36 N 119.88 E
	ePP	AB	44 32	H = 05 27 51.6 h = 37 km MB = 5.7 (NEIC)
	eSKS	C	51 26	D = 90.98
	eS	C	51 52	PV A 1.1s 16.1nm M = 5.3
	eSS	C	58 00	XV A 2.0 132.0nm
	eSSS	C	06 01 25	LmH B 14 6.8/um 6.2
	LmH	B	27.8	LmV B 14 5.9/um 6.2
	LmV	B	27.8	
17.	iPn	A	08 14 33.7	Federal Republic of Germany
	iPg	A	14 43.7	48.36 N 8.93 E
	iSn	A	15 08.7	H = 08 13 50.6 h = 33 km (NEIC)
	iSg	A	15 20.2	H = 08 13 49.5
				$48^{o}18.3'$ N $9^{o}02.4'$ E (BCIS)
				D = 2.88
17.	eP	A	09 45 59	Pakistan 32.91 N 69.91 E
				H = 09 37 39.8 h = 37 km
				D = 45.70 Az = 311 (ISC)
				PV A 1.2s 10.2nm M = 4.6
17.	eP	A	09 48 00	Pakistan 33.13 N 70.14 E
				H = 09 39 35.6 h = 0 km MB = 4.8
				D = 45.70 Az = 311 (ISC)
				PV A 1.1s 20.2nm M = 5.0
17.	eP	A	10 14 32	West Pakistan 33.50 N 71.54 E
	eS	B	21 16	H = 10 06 05.8 h = 33 km MB = 5.3 (NEIC)
	eSS	B	24 48	D = 46.36
	LmH	B	38.5	PV A 1.2s 18.3nm M = 4.9
	LmV	B	39.1	LmH B 17.5 1.5/um 5.0
				LmV B 14 1.8/um 5.2

Day	Phase		h m s	Remarks
17.	eP	A	14 22 24	<u>Kurile Islands</u> 49.73 N 155.33 E H = 14 10 49.8 h = 92 km MB = 5.0 D = 75.30 Az = 337 PV A 1.4s 18.6nm M = 4.7
18.	eP	A	02 50 32.5	<u>Southern Iran</u> 27.96 N 55.78 E H = 02 42 58.0 h = 49 km MB = 4.7 D = 40.21 Az = 317 (ISC) PV A 1.8s 13.5nm M = 4.5
18.	+iP	AB	02 54 31.5	<u>Near East Coast of Honshu, Japan</u>
	ipP	A	54 43.5	38.62 N 142.56 E
	eS	B	03 05.0	H = 02 42 17.7 h = 39 km MB = 5.2 (NEIC)
	LmH	B	32.2	D = 81.23 h = 39 km
	LmV	B	34.0	PV A 1.4s 37.2nm M = 5.2
				LmH B 17.5 1.9 μm 5.5
				LmV B 15 1.1 μm 5.4
18.	iPn	A	08 11 46	<u>Federal Republic of Germany</u>
	iPg	AB	11 55.5	48.31 N 8.97 E
	iSn	A	12 20	H = 08 11 02.0 h = normal (NEIC)
	iSg	AB	12 31.5	H = 08 11 01.5
	LmH	B	12.8	$48^{o}17.4'$ N $9^{o}04.0$ E (BCIS)
	LmV	B	12.8	D = 2.91
				LmH B 6.5s 3.2 μm M = 3.8
				LmV B 6 1.3 μm
19.	e(P)	A	01 17 29	<u>Tunisia</u> 35.17 N 9.11 E
	e	A	17 36	H = 01 13 43.5 h = 50 km MB = 4.4 (NEIC)
	LmH	B	23.8	D = 15.58
	LmV	B	24.0	(PV) A 1.8s 23.6nm M = (4.1)
				LmH B 15.5 2.2 μm 4.4
				LmV B 15.5 1.5 μm 4.4
19.	eP	A	07 17 50	<u>Greenland Sea</u> 73.98 N 9.60 E H = 07 12 45.1 h = 33 km MB = 4.8 (NEIC) D = 23.44

Day	Phase		h m s	Remarks
19.	LmH	B	10 49.0	Volcano Islands Region
	LmV	B	55.0	22.25 N 144.18 E
				H = 09 55 24.2 h = 70 km MB = 5.2 (NEIC)
				D = 96.24
				LmV B 20s 0.2/um
19.	iPn	A	11 00 19.5	Yugoslavia $43^{o}10'$ N $17^{o}42'$ E
	e	A	02 20.5	H = 10 58 12.6 (Sarajevo, BCIS)
	eSg	A	02 49	D = 8.6
	LmH	B	03.1	PnV A 0.7s 17.2nm M = 5.4
	LmV	B	03.7	LmH B 10 0.6/um 3.6
19.	e	A	11 43 44	
20.	ePKIKP	BC	02 54 07	New Hebrides Islands 15.63 S 167.09 E
	e	A	54 15	H = 02 34 45.6 h = 33 km MB = 5.2 (NEIC)
	ePP	AB	57 07	D = 139.87
	eSKP	B	57 58	LmH B 20s 1.0/um M = 5.5
	ePS	C	03 07.4	LmV B 20 1.0/um 5.6
	eSS	B	15 30	
	LmH	B	58.0	
	LmV	B	58.5	
20.	eP	A	03 07 48.5	Off East Coast of Honshu, Japan
	e	A	08 00	33.51 N 142.26 E
				H = 02 55 12.1 h = 32 km MB = 4.5 (NEIC)
				D = 85.57
				PV A 1.6s 22.0nm M = 5.2
20.	eP	A	21 28 32	Off East Coast of Honshu, Japan
				33.56 N 142.33 E
				H = 21 15 52.8 h = 16 km MB = 4.5 (NEIC)
				D = 85.55
				PV A 1.0s 9.9nm M = 5.0
20.	eP	AB	21 37 53.5	South of Alaska 53.53 N 160.05 W
	eS	B	47 38	H = 21 26 07.2 h = normal MB = 5.2 (NEIC)
	LmH	B	22 18.8	D = 75.96

Day	Phase		h m s	Remarks
cont. 20.	LmV	B	22 21.4	PV A 1.7s 60.6nm M = 5.3
				LmH B 16 0.2/um 4.5
				LmV B 16 0.2/um 4.6
21.	eP	A	00 20 35	Greenland Sea 73.45 N 7.0 E
				H = 00 15 26.7 h = 0 km MB = 4.4
				D = 22.97 Az = 172 (ISC)
				PV A 1.0s 11.8nm M = 4.4
21.	LmH	C	04 45.5	Near Coast of Peru 16.03 S 73.87 W
	LmV	C	45.5	H = 03 50 30.9 h = 85 km MB = 5.4 (NEIC)
				D = 99.41
				LmH C 20s 0.2/um
				LmV C 22 0.4/um
21.	ePKIKP	AB	06 21 41	North of New Zealand 27.14 S 174.95 E
	ePKHKP	A	21 47	H = 06 01 54.4 h = 33 km MB = 5.5 (NEIC)
	ePKP2	A	22 02	D = 153.30
	ePP	C	25 35	PKP2V A 1.6s 43.0nm
	eSS	C	45 10	LmH B 16 0.2/um M = 4.8
	LmH	B	07 52.3	LmV B 16 0.2/um 5.1
	LmV	B	55.0	
21.	eP	A	08 02 08	Greenland Sea 73.60 N 7.58 E
	e	A	02 14.5	H = 07 57 05.2 h = normal MB = 4.5 (NEIC)
	LmH	B	11.7	D = 23.10
	LmV	B	11.7	PV A 1.2s 28.4nm M = 4.6
				LmH B 14 0.2/um 3.7
				LmV B 14 0.2/um 3.9
21.	eP	A	11 51 11.5	Northwest of Kurile Islands
				52.45 N 153.04 E
				H = 11 40 30.6 h = 437 km MB = 4.7 (NEIC)
				D = 72.23
22.	eP1	AB	06 16 56.5	Luzon, Philippine Islands
	eiP2	A	16 58.5	16.60 N 122.29 E
	ePP	C	20 32	H = 06 04 00.1 h = 34 km MB = 5.7 (NEIC)

Day	Phase		h m s	Remarks
cont. 22.	iS	B	06 27 48	D = 89.81
	i	C	28 26	P1V A 1.4s 37.2nm M = 5.5
	eiSS	C	33 57	P1V B 12 8.3μm 6.9
	eiSSSS	C	40 44	P2V A 1.4 107.0μm 6.0
	LmV	B	07 03.4	SH B 14 18.0μm 6.1
	LmH	B	03.5	LmH B 17 85.5μm 7.3
				LmV B 17 117.0μm 7.4
22.	eP	A	06 24 10.5	<u>Luzon, Philippine Islands</u>
	e	A	24 27.5	16.72 N 122.43 E
				H = 06 11 13.4 h = 33 km MB = 5.5
				D = 89.79 Az = 323 (ISC)
				PV A 1.6s 60.4nm M = 5.7
22.	LmH	C	12 16.0	<u>Northern Celebes</u> 1.41 N 120.30 E
	LmV	C	(16.0)	H = 11 20 54 h = 33 km MB = 5.2 (NEIC)
				D = 100.66
				LmH C 30s 0.6μm M = 4.9
				LmV C disturbance
22.	-iPKIKP	AB	21 05 11.5	<u>Tonga Islands</u> 17.69 S 175.19 W
	-iPKHKP	AB	05 14.5	H = 20 45 57.3 h = 227 km MB = 6.2 (NEIC)
	+iPKP2	A	05 16.4	D = 146.67
	+eipPKP	B	06 08	PKIKPV A 2.6s 1520.0nm
	+ipPKP	A	06 10.3	PKHKPV A off scale
	+i	A	06 13.4	PKHKPV B 8 96.0μm
	eiPP	B	08 33	pPKP B 11 58.9μm
	LmH	B	22 14.5	pPKP A off scale
	LmV	B	15.5	PPV B 7 17.9μm M = 7.4
				LmH B 19 8.95μm
				LmV B 18 9.9μm
23.	eP	A	00 32 22	<u>Luzon, Philippine Islands</u>
				16.78 N 122.38 E
				H = 00 19 28.6 h = 45 km MB = 5.0 (NEIC)
				D = 89.72

Day	Phase		h m s	Remarks
23.	e(P)	A	03 17 37	Greece-Bulgaria Border Region
	e	A	20 08	41.71 N 23.64 E
				H = 03 14 28.2 h = 5 km MB = 4.3 (NEIC)
				D = 12.20
23.	eP	A	07 43 49	Southern Sumatra 2.20 S 99.57 E
	epP	A	43 57.5	H = 07 30 49.4 h = normal MB = 4.8 (NEIC)
				D = 90.39 h = 29 km
				PV A (1.4)s 11.6nm M = 5.0
				pPV A 1.6 16.5nm
23.	-eP	A	10 11 21	Azores Islands Region 37.58 N 31.93 W
	LmH	C	24.0	H = 10 04 44.1 h = normal MB = 5.0 (NEIC)
	LmV	C	24.2	D = 33.40
				PV A 1.6s 38.5nm M = 5.1
				LmV C 16 0.5 µm 4.5
23.	ePn	A	11 07 29	Southern Italy 39.0 N 17.1 E
	ePg	A	08 34	H = 11 04 24 (BCIS)
	eSg	A	11 15	D = 12.35
	LmH	B	13.7	LmH B 13s 0.4 µm M = 3.6
	LmV	B	13.7	LmV B 13 0.5 µm
23.	e(pP)	A	12 04 41.5	South of Honshu, Japan 33.53 N 140.74 E
				H = 11 51 50.7 h = 64 km MB = 4.4 (NEIC)
				D = 85.1
23.	eP	A	17 34 20	Off East Coast of Honshu, Japan
				34.0 N 142.1 E
				H = 17 21 48 h = 53 km MB = 4.4
				D = 85.12 Az = 331 (ISC)
23.	eP	A	18 16 00	Off East Coast of Honshu, Japan
	LmH	B	19 02.0	33.49 N 142.32 E
	LmV	B	02.9	H = 18 03 21.7 h = 19 km MB = 5.1 (NEIC)
				D = 85.61
				PV A 1.4s 37.2nm M = 5.4
				LmH B 12 0.2 µm 4.7
				LmV B 12 0.2 µm 4.8

Day	Phase		h m s	Remarks
23.	e	A	18 27 26.5	Afganistan - USSR Border Region 38.49 N 70.17 E H = 18 17 14.1 h = 33 km MB = 4.9 (NEIC) D = 42.39
24.	eP	A	10 22 03	Arabian Sea 13.06 N 57.59 E H = 10 12 49.2 h = normal MB = 4.7 D = 52.82 Az = 324.9 (NEIC)
24.	ePKHKP	A	11 45 43	Loyalty Islands Region 22.31 S 171.21 E
	e	A	45 45	H = 11 26 01.1 h = 38 km MB = 5.6
	e	B	45 48	D = 147.53 Az = 335.6 (NEIC)
	eX	A	45 52.5	XV A 1.6s 55.0nm
	eY	B	46 44	YV B 9 1.9 μm
	LmH	B	12 54.8	LmH B 18.5 1.9 μm M = 5.8
	LmV	B	54.8	LmV B 20 2.1 μm 5.9
25.	eP	A	00 53 10	South Indian Ocean 27.37 S 66.77 E H = 00 40 04.3 h = normal MB = 5.0 D = 91.67 Az = 328.5 (NEIC)
25.	eP	A	02 30 27	Off East Coast of Honshu 34.11 N 142.26 E H = 02 17 56 h = 47 km MB = 4.2 D = 85.05 Az = 331 (ISC)
26.	eX	A	02 33 30	XV A 1.0s 9.8nm
26.	+eP	A	08 51 22	Mindoro, Philippine Islands
	epP	A	51 32.5	13.29 N 120.41 E H = 08 38 18.8 h = 38 km MB = 5.2 D = 91.35 Az = 322.9 (NEIC) h = 36 km PV A 1.4s 37.2nm M = 5.6 pPV A 1.6 33.0nm
26.	LmH	C	11 37.5	South Atlantic Ridge 47.36 S 12.49 W
	LmV	C	43.4	H = 10 48 52.2 h = 33 km MB = 4.9 (NEIC) D = 99.81 LmH C 40s 1.2 μm M = 5.1

Day	Phase		h m s	Remarks
26.	e	A	12 21 28	
26.	LmV	B	18 36.8	South Atlantic Ridge 47.45 S 12.33 W
	LmH	B	38.2	H = 17 41 54.3 h = 33 km MB = 4.8 (NEIC)
				D = 99.87
				LmH B 18.5s 2.2 µm M = 5.7
				LmV B 20 1.7 µm 5.6
27.	ePKP	A	03 53 31	Loyalty Islands Region 22.45 S 171.3 E
				H = 03 33 45 h = 40 km
				D = 147.68 Az = 336 (ISC)
				traces
27.	-iP	AB	04 17 24.8	Kamchatka 54.99 N 156.27 E
	-iPcP	A	17 47	H = 04 06 50.5 h = 409 km MB = 5.7
	eipP	AB	18 55	D = 70.67 Az = 337.0 (NEIC)
	ePP	B	20 07	h = 428 km
	iS	B	26 05	PV A 1.6s 728.0nm M = 6.1
	iSP	B	26 40	PV B 3.5 1.87 µm 6.1
	esS	B	28 48	PcPV A 1.0 122.0nm
	esSS	B	33 08	pPV A 1.5 286.0nm
28.	ePKIKP	A	01 13 30.5	South of Fiji Islands 22.04 S 179.71 W
	eiPKHKP	A	13 36	H = 00 54 52.0 h = 610 km MB = 5.0
	ePKP2	A	13 44	D = 150.10 Az = 345.5 (NEIC)
				PKHKPV A 1.0s 33.5nm
28.	eP diff	B	02 09 42	South of Sumbawa Island
	ePP	AB	14 06	11.02 S 117.06 E
	eiPP	AB	14 10	H = 01 55 21.9 h = normal MB = 6.3
	iPPP	B	16 28	D = 108.24 Az = 319.7 (NEIC)
	eiPPPP	B	18 16	PPV B 10s 2.9 µm M = 7.0
	iSKS	B	20 24	LmH B 24.5 10.1 µm 6.3
	iS	B	21 46	LmV B 18.5 5.8 µm 6.2
	iPS	B	23 32	
	ePPS	B	24 30	
	ePKKP	A	25 15	
	eSS	B	29 00	

Day	Phase		h m s	Remarks
cont. 28.	iSSS	B	02 33 24	
	eSSSS	B	36 46	
	LmH	B	56.8	
	LmV	B	03 05.8	
28.	eP	A	03 18 41	Turkey 38.95 N 29.99 E H = 03 14 36.3 h = normal MB = 4.7 D = 17.46 Az = 318.0 (NEIC) PV A 1.8s 40.6nm M = 4.3
28.	ePn	A	06 10 36.5	Yugoslavia 46.5 N 14.9 E
	eSg	A	11 55	H = 06 09 25 (Ljubljana, BCIS) D = 4.68
28.	eP	A	10 52 36.5	Kurile Islands 43.37 N 147.82 E H = 10 40 35.7 h = 36 km MB = 4.7 D = 78.90 Az = 333.3 (NEIC) PV A 1.2s 12.2/um M = 4.8
28.	+ePKIKP	A	18 31 07.5	Fiji Islands Region 18.02 S 179.40 W
	-eiPKHKP	A	31 09.9	H = 18 12 36.2 h = 621 km MB = 5.1
	eiPKP2	A	31 12.4	D = 146.28 Az = 347.3 (NEIC)
	epPKP2	A	33 35	PKIKPV A 1.6s 27.4nm PKHKPV A 1.1 56.5nm pPKP2V A 1.9 37.9nm
28.	e(Sg)	A	21 15 27	Yugoslavia 46.0 N 15.0 E H = 21 12 57 h = 0 km (ISC) D = 5.2
28.	LmH	C	24 01.0	Revilla Gigedo Islands Region
	LmV	C	01.8	19.35 N 108.32 W H = 23 11 36.7 h = 33 km MB = 4.9 (NEIC) D = 92.64 LmH C 30s 0.5/um M = 4.8 LmV C 28 1.4/um 5.3

Day	Phase		h m s	Remarks
29.	eP	A	08 01 24.5	<u>Kurile Islands</u> 43.47 N 147.68 E
				H = 07 49 22.1 h = 30 km MB = 4.4
				D = 78.77 Az = 333.2 (NEIC)
29.	eP	A	19 15 42	<u>Hindu Kush Region</u> 36.38 N 70.76 E
	ePP	A	17 29	H = 19 07 56.0 h = 236 km MB = 5.2
				D = 44.04 Az = 308.2 (NEIC)
29.	ePn	A	20 00 28	<u>Central Italy</u> 43.66 N 13.86 E
	e	A	00 37	H = 19 58 43.8 h = normal MB = 4.3
	eSn	A	01 47	D = 7.16 Az = 348.5 (NEIC)
	eSg	A	02 35	PnV A 1.0s 13.8nm M = 4.9
30.	ePKP	A	04 16 01	<u>Tonga Islands</u> 15.79 S 173.09 W
				H = 03 56 24.5 h = normal MB = 4.4
				D = 145.02 Az = 354.8 (NEIC)
				PKPV A 1.8s 37.2nm
30.	eP	A	06 46 07.5	<u>Tadzhik SSR</u> 38.34 N 69.49 E
				H = 06 38 16.8 h = normal MB = 5.1
				D = 42.05 Az = 306.4 (NEIC)
30.	eP	A	19 40 42	<u>North Atlantic Ocean</u> 56.97 N 33.70 W
	LmH	B	51.6	H = 19 35 00.6 h = normal MB = 4.0
	LmV	B	51.6	D = 27.07 Az = 84.0 (NEIC)
				PV A 1.6s 22.0nm M = 4.5
30.	eP	A	20 00 11	<u>North Atlantic Ocean</u> 56.85 N 33.77 W
	e	A	00 15.5	H = 19 54 28.9 h = normal MB = 4.2
	e	A	00 45	D = 27.13 Az = 83.7 (NEIC)
	LmH	B	11.0	PV A 1.5s 25.1nm M = 4.6
	LmV	B	11.0	LmH B 16 0.55 µm 4.2
				LmV B 16 0.7 µm 4.4
30.	LmH	B	20 16.3	LmH B 14s 0.8 µm
	LmV	B	16.4	LmV B 13 0.8 µm

Day	Phase		h m s	Remarks
30.	iPn	A	23 58 03.2	Central Italy 43.79 N 13.39 E
	iSn	A	59 20	H = 23 56 21.5 h = normal MB = 4.1
	eSg	A	24 00(15)	D = 6.96 Az = 350.7 (NEIC)
31.	eSg	A	09 00 59	France 49.1 N 6.7 E
				H = 08 59 05 (BCIS)
				D = 3.54
31.	e(Sg)	A	11 43 47	Switzerland 47.62 N 7.6 E
				H = 11 41 43 h = 13 km
				D = 4.01 Az = 40
31.	e(Pg)	A	17 37(36)	Poland 50.31 N 18.87 E
	e(Sg)	A	38 29	H = 17 36 02.7 M = 3.5 (WAR)
				D = 4.65 Az = 277 (ISC)

Day	Phase		h m s	Remarks
1.	eP	A	03 20 27	Fox Islands, Aleutian Is.
	epP	A	20 36.5	51.05 N 169.57 W
	e	A	20 44	H = 03 08 25.1 h = normal MB = 5.0
				D = 78.68 Az = 359.2 (NEIC)
				h = 35 km
				PV A 1.3s 28.4nm M = 5.1
				pPV A 1.2 36.6nm
1.	ePKP	A	04 23 41	North of New Zealand 15.31 S 173.08 W
				H = 04 04 02.9 h = 0 km MB = 4.9
				D = 144.55 Az = 355 (ISC)
				PKPV A 1.6s 13.7nm
1.	ePKIKP	A	06 43 27	Tonga Islands 21.15 S 174.36 W
	+ePKHKP	A	43 33	H = 06 23 44.2 h = normal MB = 5.0
				D = 150.19 Az = 352.3 (NEIC)
				PKIKPV A traces
				PKHKPV A 1.4s 32.6nm
1.	LmH	B	07 30.5	Probably Luzon, Philippine Islands
	LmV	B	30.5	16.58 N 122.44 E
				H = 06 31 15.5 h = 45 km MB = 5.0 (NEIC)
				D = 89.91
				LmH B 16s 0.2/um M = 4.7
				LmV B 16 0.2/um 4.7
1.	eP	A	09 45 13.5	Kurile Islands 47.45 N 152.40 E
				H = 09 33 31.0 h = 85 km MB = 4.1
				D = 76.61 Az = 335.6 (NEIC)
2.	eiPg	A	14 45 29	Adelebsen, Federal Rep. of Germany
	eSg	A	45 50	51°36.56'N 09°44.77'E
				Explosion, yield 8.4 t (Hannover)
				D = 1.5
2.	e	A	15 15 30.5	South of Honshu, Japan
	e	A	15 40.5	33.46 N 140.76 E
				H = 15 02 50.1 h = 59 km MB = 4.8
				D = 85.00 Az = 330.3 (NEIC)

Day	Phase		h m s	Remarks
2.	eP	A	21 58 40.5	Andreanof Islands, Aleutian Is.
				51.77 N 174.61 W
				H = 21 46 47.0 h = 50 km MB = 4.9
				D = 77.82 Az = 355.9 (NEIC)
				PV A 1.3s 10.9nm M = 4.7
2.	+eP	A	23 03 17	Northern Sumatra 3.58 N 96.73 E
	e	A	03 26	H = 22 50 47.4 h = normal MB = 5.5
	e	A	03 42.5	D = 84.15 Az = 320.4 (NEIC)
				PV A 1.0s 15.7nm M = 5.1
3.	LmH	B	03 07.0	Southwestern Ryukyu Islands
	LmV	B	13.8	23.54 N 125.53 E
				H = 02 16 51.1 h = 33 km MB = 5.2 (NEIC)
				D = 86.02
				LmH B 17s 0.6/um M = 5.0
				LmV B 18 0.4/um 4.9
3.	e	A	11 03 11.5	North Atlantic Ridge 20.12 N 45.71 W
				H = 10 53 40.3 h = normal MB = 4.9
				D = 54.07 Az = 41.4 (NEIC)
3.	eP	A	18 49 09.5	North Atlantic Ridge 27.13 N 44.56 W
	epP	A	49 14	H = 18 40 28.8 h = normal MB = 5.2
	esP	A	49 25	D = 48.27 Az = 45.1 (NEIC)
	ePcP	A	50 36	PV A 1.2s 22.4nm M = 5.1
	LmH	B	19 06.2	pPV A 1.5 25.2nm
	LmV	B	06.2	LmH B 19 0.8/um 4.7
				LmV B 18 0.9/um 4.9
3.	LmH	B	19 56.5	Probably Santa Cruz Islands
	LmV	B	56.5	12.6 S 136.7 E
				H = 18 31 19.9 and H = 18 46 31.0 (NEIC)
				D = 136.7
				LmH B 18s 0.7/um M = 5.4
				LmV B 18 0.6/um 5.4

Day	Phase		h m s	Remarks
4.	eP	A	05 17 20.5	North Atlantic Ocean 57.71 N 33.53 W
	LmV	B	28.5	H = 05 11 40.5 h = normal MB = 4.5
	LmH	B	29.2	D = 26.92 Az = 85.5 (NEIC)
				PV A traces
				LmH B 14s 0.1 µm M = 3.7
				LmV B 14 0.3 µm 4.2
4.	eP	A	10 36 04	Carlsberg Ridge 0.60 S 67.40 E
				H = 10 24 58.2 h = normal MB = 4.9
				D = 69.51 Az = 325.8 (NEIC)
				PV A 3.0s 105.0nm M = 5.4
4.	e(P)	A	18 50 35	Near Coast of Oaxaca, Mexico
				15.73 N 94.45 W
				H = 18 37 50.1 h = 81 km MB = 4.6
				D = 87.82 Az = 37.8 (NEIC)
4.	ePKP	A	22 32 11.5	Tonga Islands 15.25 S 173.29 W
	e	A	32 20	H = 22 12 37.8 h = normal MB = 5.5
	e	A	32 25	D = 144.47 Az = 354.6 (NEIC)
	e	A	33 26.5	PKPV A 1.6s 24.7nm
5.	eP	A	10 48 24	Southern Greece 37.82 N 21.36 E
	e	A	48 32	H = 10 44 59.2 h = 69 km MB = 4.2
	LmH	B	55.0	D = 14.58 Az = 334.7 (NEIC)
	LmV	B	55.0	LmH B 13.5s 0.7 µm
				LmV B 12 0.8 µm
5.	eP	A	19 07 13	North of Franz Josef Land
				86.45 N 38.87 E
				H = 19 00 12.1 h = normal MB = 4.4
				D = 36.40 Az = 209.4 (NEIC)
				PV A traces
6.	eP	A	02 31 35	Andreanof Islands, Aleutian Is.
	LmH	C	03 10.0	51.58 N 178.27 W
				H = 02 19 41.6 h = 51 km MB = 5.3
				D = 77.81 Az = 353.6 (NEIC)

Day	Phase		h m s	Remarks
cont. 6.	LmV	C	03 10.2	PV A 1.4s 27.9nm M = 5.1
				LmH C 20 0.2/um
				LmV C 21 0.4/um
6.	e	A	04 51 51	<u>Southern Sumatra</u> 2.10 S 100.78 E
				H = 04 38 37 h = 84 km MB = 4.8
				D = 91.08 Az = 320 (ISC)
6.	+eiP	AB	05 33 35	<u>North Atlantic Ridge</u> 32.94 N 39.86 W
	e	AB	33 42	H = 05 25 50.2 h = normal MB = 5.5
	e	B	34 36	D = 41.34 Az = 48.9 (NEIC)
	eS	B	39 49	PV A 2.6s 242.0nm M = 5.5
	eSS	B	43 08	PV B 4.5 0.55/um 6.1
	LmH	C	47.7	SH B 12.5 1.8/um 5.8
	LmV	C	47.8	LmH C 20 3.8/um 5.3
				LmV C 18 9.9/um 5.8
6.	eP	A	10 54 50.5	<u>Off East Coast of Kamchatka</u>
				55.91 N 163.78 E
				H = 10 43 32.6 h = normal MB = 4.7
				D = 71.32 Az = 341.7 (NEIC)
				PV A 1.3s 8.7nm M = 4.6
6.	eP	A	18 02 11	<u>Southern Iran</u> 26.86 N 53.37 E
	LmH	B	23.7	H = 17 54 43.3 h = 54 km MB = 4.9
	LmV	B	23.7	D = 39.57 Az = 318.3 (NEIC)
				PV A 1.8s 16.9nm M = 4.6
				LmH B 15 0.3/um
6.	e	A	18 13 42	<u>France</u> 47 1/4 N 2 1/2 W
	e	A	14 11.5	H = 18 08 23 (BCIS)
				D = 9.9
6.	ePKIKP	A	20 15 34.5	<u>Tonga Islands</u> 16.27 S 173.27 W
	epPKIKP	A	15 45.5	H = 19 55 57.7 h = normal MB = 4.8
				D = 145.48 Az = 354.5 (NEIC)
				h = 39 km
				PKIKPV A 1.2s 18.3nm
				pPKIKPV A 0.9 15.6nm

Day	Phase		h m s	Remarks
7.	+iP	A	01 35 46.5	Eastern Kazakh SSR 49.76 N 78.18 E
	ePn	A	37 18.5	H = 01 27 57.1 h = 0 km MB = 5.5
				D = 41.30 Az = 297.7 (NEIC)
				Underground explosion MB = 5.7 (UPP)
				PV A 1.0s 108.0nm M = 5.5
7.	LmV	B	06 24.1	Luzon, Philippine Islands
	LmH	B	24.2	12.48 N 123.70 E
				H = 05 24 19.6 h = 13 km MB = 5.1 (NEIC)
				D = 93.9
7.	eSg	A	10 28 22	Yugoslavia 44.0 N 16.6 E
				H = 10 24 19 h = 0 km (ISC)
				D = 7.5
7.	eSn	A	10 53 22	Yugoslavia 44.22 N 16.59 E
	eSb	A	53 52	H = 10 50 12.9 h = 57 km (ISC)
				D = 7.3
7.	ePKP2	A	12 00 02.5	Kermadec Islands Region 31.94 S 178.19 W
				H = 11 39 26.0 h = 33 km MB = 5.3 (NEIC)
				D = 159.95
				PKP2V A 1.1s 26.2nm
7.	eP	A	12 15 07.5	Halmahera 2.51 N 128.18 E
	ePP	A	19 30	H = 12 01 17.8 h = 152 km MB = 5.7
	LmH	B	13 49.0	D = 104.53 Az = 324.0 (NEIC)
	LmV	B	54.6	PV A 2.0s 34.2nm M = 6.0
				PPV A 2.0 34.2nm 5.5
				LmH B 17 0.8/um
				LmV B 14 0.3/um
				Possibly the surface waves do not belong to this earthquake but to
				Kermadec Islands Region 31.9 S 178.2 W
				H = 11 39 26.0 h = normal MB = 5.3 (NEIC)
7.	-iP	A	16 51 43.8	Kamchatka 54.96 N 157.79 E
				H = 16 40 54.2 h = 281 km MB = 4.7

Day	Phase		h m s	Remarks
cont. 7.	epP	A	16 52 55.5	D = 71.04 Az = 338.0 (NEIC) h = 323 PV A 1.1s 34.3nm M = 5.0
7.	LmH	B	24 03.0	Near Coast of Chiapas, Mexico
	LmV	B	03.0	14.49 N 93.08 W H = 23 09 04.4 h = 54 km MB = 4.6 (NEIC) D = 87.98 LmH B 18s 0.3/um LmV B 18 0.25/um
8.	LmH	B	04 05.6	Off Coast of Peru 10.70 S 79.04 W
	LmV	B	06.0	H = 03 02 37.5 h = 29 km MB = 5.0 (NEIC) D = 98.60 LmH B 16s 0.3/um M = 4.9 LmV B 18 0.4/um 5.0
8.	eP	A	06 25 22	Fox Islands, Aleutian Is. 51.11 N 169.53 W H = 06 13 20.5 h = normal MB = 4.6 D = 78.62 Az = 359.3 (NEIC) PV A 1.0s 15.8nm M = 5.0
8.	-iP	AB	09 26 42.5	Taiwan Region 21.06 N 120.18 E
	ePP	A	30 00	H = 09 14 08.3 h = normal MB = 5.4
	eX	A	30 14.5	D = 85.06 Az = 322.7 (NEIC)
	eSKS	B	37 00	PV A 2.3s 256.0nm M = 6.0
	eSSS	C	47 12	PV B 6 0.76/um 6.1
	LmH	B	10 08.4	PPV A 2.0 42.7nm 5.5
	LmV	B	10.2	XV A 2.3 97.6nm LmH B 14 4.3/um 6.0 LmV B 13 4.5/um 6.1
8.	eP	A	09 45 31	Western Iran 34.09 N 46.23 E H = 09 39 21.4 h = 18 km MB = 4.9 D = 30.11 Az = 313.9 (NEIC) H = 09 39 32 34.4 N 46.6 E (BCIS)

Day	Phase		h m s	Remarks
8.	eP	A	10 30 18	Taiwan Region 21.01 N 120.17 E
	e	A	30 29.5	H = 10 17 43.5 h = normal MB = 4.9
				D = 85.09 Az = 322.7 (NEIC)
				PV A 1.5s 25.2nm M = 5.2
8.	ePg	A	11 49 23	Austria 47.5 N 10.8 E
	iSn	A	49 47.5	H = 11 48 22 (BCIS)
	iSg	A	50 04	D = 3.18
8.	eiPKHKP	A	16 35 08	Fiji Islands Region 19.38 S 177.08 W
	e	A	35 29	H = 16 15 56.1 h = 290 km MB = 5.3
	e	A	36 32	D = 148.05 Az = 349.5 (NEIC)
	e	A	36 45	PKHKPV A 1.8s 47.3nm
8.	eP	A	16 56 58.5	Taiwan Region 21.02 N 120.34 E
				H = 16 44 23.5 h = normal MB = 4.7
				D = 85.18 Az = 322.8 (NEIC)
				PV A 1.4s 11.6nm M = 4.9
8.	eSg	A	17 20 12.5	Poland 50.28 N 19.22 E
				H = 17 17 32.1 M = 2.8 (WAR)
				D = 4.85
8.	eP	A	17 31 15	Eastern Caucasus 43.23 N 47.21 E
	LmH	B	41.7	H = 17 25 52.2 h = 51 km MB = 4.5
	LmV	B	42.6	D = 25.22 Az = 299.6 (NEIC)
8.	eP	AB	18 10 54.5	Jujuy Province, Argentina
	epPP	B	16 00	22.58 S 66.17 W
	esPP	B	16 20	H = 17 57 38.6 h = 259 km MB = 5.4
	iSKS	B	21 12	D = 99.78 Az = 39.2 (NEIC)
	eS	B	22 00	PV A 1.2s 20.3nm M = 5.4
	e	B	23 04	LmH B 18.5 0.4/um
	LmH	B	55.2	LmV B 16 0.4/um
	LmV	B	55.2	
8.	eP diff	AB	19 08 05	Near Coast of Central Chile
	ePKIKP	B	12 03	30.46 S 71.80 W

Day	Phase		h m s	Remarks
cont. 8.	iPP	B	19 12 34	H = 18 53 41.7 h = 39 km MB = 6.2
	ei	B	15 57	D = 109.02 Az = 42.0 (NEIC)
	eSKS	B	18 40	P diff V A 2.3s 54.8nm
	eS diff	B	20 21	PPV A 2.5 492.0nm M = 6.8
	ePS	B	22 00	PPV B 6 1.95 µm 7.0
	ePKKP	A	23 34	LmH B 19 26.1 µm 6.8
	eSS	B	27 59	LmV B 19 22.1 µm 6.8
	eSSS	B	32 00	
	LmV	B	58.7	
	LmH	B	58.8	
8.	ePP	A	19 41 00	Near Coast of Central Chile
				30.51 S 71.79 W
				H = 19 22 09.5 h = 57 km MB = 5.5
				D = 109.05 Az = 42.0 (NEIC)
				PPV A 2.0s 85.5nm M = 6.1
9.	+iPKHKP	AB	00 39 47.5	Tonga Islands 21.13 S 174.04 W
	ePKP2	A	39 58	H = 00 19 56.8 h = normal MB = 5.5
	e	C	40 08	D = 150.21 Az = 352.7 (NEIC)
	LmV	B	47.1	PKHKPV A 1.6s 52.2nm
	LmH	B	47.5	PKP2V A 2.0 162.5nm
				LmH B 21 0.8 µm M = 5.4
				LmV B 20 0.7 µm 5.5
9.	eP	AB	07 46 45	Crete 34.84 N 26.53 E
	eX	A	47 28	H = 07 42 19.9 h = normal MB = 4.9
	eS	B	50 19	D = 19.16 Az = 330.0 (NEIC)
	i	B	50 38	H = 07 42 28 34.9 N 26.5 E (BCIS)
	LmH	B	55.6	PV A 1.4s 23.2nm M = 4.2
	LmV	B	56.1	XV A 1.5 52.8nm
				LmH B 14 2.0 µm 4.6
				LmV B 10 1.9 µm 4.9
9.	iPg	A	11 51 03	Czechoslovakia 50°33.8'N 14°00.4'E
	iSg	A	51 24	Explosion yield 16.5 t (PRU)
				D = 1.51

Day	Phase		h m s	Remarks
10.	eP	AB	03 42 09.5	<u>Southern Yukon Territory, Canada</u>
	ePP	A	44 37	61.52 N 140.21 W
	eS	B	50 58	H = 03 31 24.1 h = normal MB = 5.3
	eSS	C	55 05	D = 65.93 Az = 19.2 (NEIC)
	LmV	B	04 13.3	PV A 2.1s 100.5nm M = 5.5
	LmH	B	13.5	LmH B 16.5 0.8 µm 5.0
				LmV B 18 0.6 µm 4.9
10.	eP	A	08 15 31.5	<u>Kurile Islands</u> 44.64 N 148.12 E
				H = 08 03 48.1 h = 142 km MB = 5.1
				D = 77.86 Az = 333.4 (NEIC)
				PV A 1.8s 40.6nm M = 4.9
10.	ePKHKP	A	10 17 59.5	<u>North of New Zealand</u> 20.65 S 174.54 W
				H = 09 58 06.2 h = 0 km MB = 4.9
				D = 149.67 Az = 352 (ISC)
10.	e(P)	A	11 37 46	<u>West Pakistan</u> 28.16 N 66.47 E
	LmH	B	12 02.5	H = 11 29 10.9 h = 17 km MB = 4.5
	LmV	B	04.2	D = 46.71 Az = 314.3 (NEIC)
10.	ePKHKP	A	12 41 52	<u>Tonga Islands</u> 21.24 S 174.27 W
	ePKP2	A	41 59	H = 12 22 06.6 h = 62 km MB = 5.2
	LmH	B	13 51.0	D = 150.29 Az = 352.4 (NEIC)
	LmV	B	56.0	
10.	eSg	A	21 37 43	<u>Austria</u> 47.9 N 16.95 E
				H = 21 35 21 h = 7 km M = 2.6 (VIE)
				D = 4.45
11.	+iPKHKP	A	04 51 05	<u>Tonga Islands</u> 20.01 S 175.27 W
				H = 04 31 19.3 h = normal MB = 5.1
				D = 148.95 Az = 351.5 (NEIC)
				PKHKPV A 1.8s 40.6nm
11.	iPKHKP	AB	14 49 40	<u>Tonga Islands</u> 20.95 S 175.43 W
	epPKHKP	A	50 02	H = 14 29 56.8 h = 75 km MB = 5.6
				D = 149.84 Az = 351.1 (NEIC) h = 77 km
				PKHKPV A 1.4s 69.7nm

Day	Phase		h m s	Remarks
11.	eP1	A	16 54 16.5	<u>Celebes Sea</u> 3.94 N 124.32 E
	+iP2	A	54 21.5	H = 16 41 01.0 h = 325 km MB = 5.8
	iP3	AB	54 25.5	D = 101.10 Az = 323.2 (NEIC)
	eiPP1	B	58 30	P1V A 1.7s 139.0nm M = 6.2
	+iPP2	A	58 36	P3V A 2.4 3590.0nm 7.4
	eiPP3	A	58 41.5	PP1V B 6.5 3.9µm 6.6
	LmH	B	17 38.6	PP2V A 2.4 3380.0nm 7.0
	LmV	B	47.4	PP3V A 2.5 10600.0nm 7.4
				PP3V B 10 7.3µm 6.7
				LmH B 20 156.0µm
				LmV B 16 129.0µm
11.	eX	A	17 36 47	XV A 1.9s 60.6nm
	e	A	36 54	
11.	eP	A	18 34 00	<u>Near Coast of Guatemala</u>
				13.46 N 90.58 W
				H = 18 21 16.7 h = 52 km MB = 5.1
				D = 87.28 Az = 38.5 (NEIC)
				PV A 1.2s 10.2nm M = 4.9
11.	ePKIKP	A	18 55 25	<u>East New Guinea Region</u>
				5.50 S 146.92 E
				H = 18 36 49.3 h = 154 km MB = 5.2
				D = 121.60 Az = 328.3 (NEIC)
				PKIKPV A 0.9s 11.7nm
11.	eP	A	21 24 49.5	<u>Near East Coast of Honshu, Japan</u>
				40.11 N 141.46 E
				H = 21 12 52.0 h = 98 km MB = 4.6
				D = 79.51 Az = 330.2 (NEIC)
12.	ePKIKP1	A	01 04 30.5	<u>Santa Cruz Islands</u> 12.38 S 166.97 E
	e(PKIKP2)	A	04 40.5	H = 00 45 45.8 h = 245 km MB = 5.5
	eiX	A	04 43	D = 136.86 Az = 337.2 (NEIC)
	e(pPKIKP2)	A	05 40	PKIKP1V A 0.7s 9.6nm
	ePP	A	07 29	PKIKP2V A 1.0 21.6nm
	eSKP	A	07 55	XV A 1.1 36.3nm
				PPV A 1.7 21.2nm M = 4.8
				SKPV A 2.5 77.0nm

Day	Phase		h m s	Remarks
12.	ePKP2	A	09 50 23.5	South of Fiji Islands 21.56 S 175.64 E
				H = 09 30 32.6 h = normal MB = 4.9
				D = 148.56 Az = 340.4 (NEIC)
12.	eP	AB	10 09 02	South of Panama 5.22 N 78.18 W
	e	A	09 11	H = 09 56 22.6 h = 27 km MB = 5.6 (NEIC)
	eS	B	19 30	D = 85.88
	LmH	B	50.2	PV A 2.2s 131.0nm M = 5.8
	LmV	B	54.1	PV B 5 5.45/um 6.1
				LmH B 17 0.3/um 4.8
				LmV B 18 0.4/um 4.9
12.	eP	A	11 32 43	Luzon, Philippine Islands
	eSKS	B	43 15	16.60 N 122.41 E
	eS	B	43 30	H = 11 19 44.4 h = normal MB = 5.2
	ePS	B	44 34	D = 89.87 Az = 323.5 (NEIC)
	LmV	B	12 19.0	PV A 1.4s 9.3nm M = 4.9
	LmH	B	19.1	LmH B 17 0.7/um 5.2
				LmV B 17.5 1.1/um 5.4
12.	eP	B	13 40 15	Iran-Iraq Border Region
	e	A	40 16	33.11 N 46.32 E
	eS	B	45 16	H = 13 34 00.7 h = normal MB = 5.4
	LmV	B	55.5	D = 30.85 Az = 315.0 (NEIC)
	LmH	B	55.8	PV B 6s 0.55/um M = 5.6
				SH B 11 1.15/um 5.3
				LmH B 13 3.3/um 5.2
				LmV B 15 2.4/um 5.1
12.	iP	AB	19 59 23.8	Fox Islands, Aleutian Is.
	ePP	B	20 02 14	53.35 N 166.79 W
	ePPP	B	04 20	H = 19 47 37.2 h = 44 km MB = 5.8
	ePPPP	B	05 28	D = 76.37 Az = 1.0 (NEIC)
	eS	B	09 00	PV A 1.6s 220.0nm M = 5.9
	eSKS	B	09 25	PV B 8 1.6/um 6.1
	ePS	B	09 45	LmH B 18 5.0/um 5.9
	e(PPS)	C	10 25	LmV B 19 4.4/um 5.8
	eSS	C	14 15	

Day	Phase		h m s	Remarks
cont. 12.	LmH	B	20 37.3	
	LmV	B	38.2	
13.	eP	AB	01 01 52.5	Iran-Iraq Border Region
	eS	B	06 56	33.10 N 46.26 E
	LmH	B	17.5	H = 00 55 37.4 h = 27 km MB = 5.1
	LmV	B	18.8	D = 30.82 Az = 315.1 (NEIC)
				LmH B 13.5s 0.9/um M = 4.6
				LmV B 12.5 0.8/um 4.7
13.	eP	A	10 55 21.5	Eastern Russia 54.92 N 126.38 E
	e	A	55 29	H = 10 45 05.3 h = normal MB = 4.8
	e	A	55 45	D = 61.60 Az = 318.9 (NEIC)
	ePcP	A	56 02	PV A 1.8s 60.8nm M = 5.4
	ePP	A	57 33	LmH B 16 2.4/um 5.5
	LmH	B	11 20.1	LmV B 14 1.3/um 5.3
	LmV	B	24.2	
13.	ePKIKP	AB	17 18 28	Kermadec Islands 29.69 S 177.34 W
	eiPKP2	AB	19 02	H = 16 58 36.1 h = 46 km MB = 5.5
	e	A	19 40	D = 158.01 Az = 344.6 (NEIC)
	ePP	AB	22 40	PKIKPV A 1.8s 37.2nm
	eSKSP	C	33 00	PKP2V A 1.5 100.5nm
	LmV	B	41.3	PPV B 9 0.6/um M = 5.7
	LmH	B	42.0	LmH B 18 0.7/um 5.4
				LmV B 17.5 0.8/um 5.6
13.	LmH	B	20 38.5	Near Coast of Chiapas, Mexico
	LmV	B	38.6	14.45 N 92.83 W
				H = 19 44 35.3 h = 127 km MB = 4.4 (NEIC)
				D = 87.86
				LmH B 18s 0.3/um
				LmV B 18 0.3/um
14.	eP	A	00 56 00	Caspian Sea 40.08 N 51.92 E
	e	A	56 12	H = 00 49 54.4 h = 47 km MB = 4.7
				D = 29.88 Az = 304.2 (NEIC)
				PV A 1.6s 19.2nm M = 4.6

Day	Phase		h m s	Remarks
14.	eiP	A	01 03 23.5	<u>Southern Alaska</u> 60.50 N 153.41 W
	epP	A	03 56	H = 00 52 35.7 h = 152 km MB = 5.2
	esP	A	04 15.5	D = 68.56 Az = 10.2 (NEIC)
				h = 144 km
				PV A 0.9s 38.9nm M = 5.2
14.	eP	A	04 40 42	<u>Iran-Iraq Border Region</u>
	eS	B	45 52	33.03 N 46.10 E
	LmH	B	55.5	H = 04 34 28.1 h = normal MB = 5.3
	LmV	B	57.8	D = 30.78 Az = 315.2 (NEIC)
				LmH B 18s 0.6 µm M = 4.3
				LmV B 13 0.7 µm 4.6
14.	ePKHKP	A	07 10 32	<u>West of Macquarie Island</u>
	ePKP2	A	10 36	53.67 S 140.77 E
				H = 06 50 45.2 h = normal
				D = 149.20 Az = 285.4 (NEIC)
				PKHKPV A traces
				PKP2V A 1.8s 20.2nm
14.	iPg	A	14 10 32.9	<u>Czechoslovakia</u> 49°34.6'N 12°47'E
	eSg	A	10 49.5	Explosion H = 14 10.0 (PRU)
				D = 1.32
14.	eP	A	17 16 50	<u>South of Honshu, Japan</u>
				33.45 N 140.65 E
				H = 17 04 22.4 h = 72 km MB = 4.5
				D = 84.96 Az = 330.2 (NEIC)
				PV A 1.8s 13.5nm M = 4.7
14.	eP	A	18 39 43	<u>Near East Coast of Honshu, Japan</u>
	epP	A	39 54	36.78 N 141.31 E
				H = 18 27 25.0 h = 48 km MB = 4.6
				D = 82.35 Az = 330.4 (NEIC)
				h = 45 km
				PV A traces
				pPV A 1.8s 27.0nm

Day	Phase		h m s	Remarks
14.	ePn	AB	18 57 37	Central Italy 43.71 N 13.44 E
	Pn max	A	57 42	H = 18 55 53.3 h = 14 km MB = 4.9
	ePg	A	58 14.5	D = 7.05 Az = 350.5 (NEIC)
	iSn	A	58 55	Pn max A 1.3s 109.0nm M = 5.5
	eSg	A	59 45	LmH B 10 7.2 µm 4.6
	LmV	B	19 00.0	LmV B 12 3.4 µm
	LmH	B	00.4	
14.	iPn	A	21 02 44.4	Central Italy 43.67 N 13.47 E
	ePg	A	03 24	H = 21 00 59.6 h = 6 km MB = 4.7
	iSn	A	04 01	D = 7.09 Az = 350.4 (NEIC)
	eSg	A	04 54.5	PnV A 1.2s 32.6nm M = 5.4
	LmH	B	05.5	LmH B 14 1.0 µm 3.6
	LmV	B	07.9	LmV B 6 0.4 µm
14.	ePKP	A	21 30 01	North of New Zealand 17.63 S 174.89 W
				H = 21 10 17.4 h = 0 km MB = 5.0
				D = 146.65 Az = 352 (ISC)
15.	eP	AB	00 36 48.5	Greece 38.34 N 22.20 E
	eiX	A	36 55	H = 00 33 23.6 h = 26 km MB = 4.9
	eS	B	39 44	D = 14.41 Az = 332.0 (NEIC)
	LmH	B	43.0	PV A 0.8s 9.6nm M = 4.4
	LmV	B	43.1	XV A 1.4 55.8nm
				LmH B 14 4.1 µm 4.7
				LmV B 11 3.8 µm
15.	eiPKP	AB	01 33 21	Tonga Islands 17.69 S 174.68 W
	ePP	B	36 40	H = 01 13 41.7 h = normal MB = 5.2
	LmH	B	02 41.8	D = 146.74 Az = 352.7 (NEIC)
	LmV	B	45.3	PKPV A 1.9s 121.0nm
				PKPV B 7.5 1.0 µm
				PPH B traces
15.	ePKIKP	AC	04 10 18.5	Kermadec Islands 29.81 S 177.04 W
	e	A	11 02	H = 03 50 22.8 h = normal MB = 4.1
	ePP	B	14 28	D = 158.19 Az = 345.1 (NEIC)
	LmH	C	05 32.7	LmH C 19s 0.2 µm M = 5.0
	LmV	C	32.7	LmV C 18 0.7 µm 5.5

Day	Phase		h m s	Remarks
15.	ePg	A	05 33 04.5	<u>Central Italy</u> 43.9 N 12 1/4 E
				H = 05 30 37 (BCIS)
				D = 6.80
15.	e(Sn)	A	09 16 38	<u>Central Italy</u> 43.8 N 13.6 E
				H = 09 13 46 (BCIS)
				D = 7.00
15.	e(P)	A	09 28 57	<u>Greece</u> 38.22 N 22.68 E
	LmH	B	35.4	H = 09 25 26.0 h = 76 km MB = 4.0
	LmV	B	35.5	D = 14.70 Az = 331.2 (NEIC)
				LmH B 12.5s 0.3 μm M = 3.6
				LmV B 10 0.2 μm
15.	eP	A	14 22(53)	<u>Greece</u> 38.5 N 24.2 E
	e	A	23 13	H = 14 19 19 (BCIS)
				<u>Crete</u> 34.25 N 24.0 E
				H = 14 18 30.1 h = 0 km
				D = 18.72 Az = 335 (ISC)
15.	+iP	AB	15 33 12.5	<u>Hokkaido, Japan Region</u>
	epP	A	33 32	42.39 N 142.98 E
	e	A	34 10	H = 15 21 20.3 h = 76 km MB = 5.5
	LmH	B	16 06.7	D = 78.10 Az = 330.8 (NEIC)
	LmV	B	13.2	h = 78 km
				PV A 1.2s 42.7nm M = 5.3
				LmH B 18.5 0.4 μm
				LmV B 13 0.3 μm
15.	e(PcP)	A	17 37 38	<u>Carlsberg Ridge</u> 0.68 N 67.05 E
	eS	B	46 19	H = 17 26 15.36 h = 33 km MB = 4.7
	eSS	B	50 32	D = 68.26 Az = 325.6
	LmV	B	18 08.4	(PcP)V A 1.9s 15.2nm
	LmH	B	08.7	LmH B 18 0.2 μm M = 4.3
				LmV B 18 0.2 μm 4.4

Day	Phase		h m s	Remarks
16.	iPg	A	08 00 28	Czechoslovakia $50^{o}29'$N $13^{o}57.2'$E
	iSg	A	00 48	Explosion H = 08 00.0 yield 8.8 t (PRU)
				D = 1.48
16.	eP	A	08 46 31.5	South of Panama 5.16 N 78.12 W
				H = 08 33 52.1 h = normal MB = 4.7
				D = 85.88 Az = 39.7 (NEIC)
				PV A 2.0s 25.6nm M = 5.1
16.	eP	A	19 05 54	Hindu Kush Region 35.98 N 69.24 E
				H = 18 57 51.6 h = 40 km MB = 4.5
				D = 43.31 Az = 308.4 (NEIC)
17.	eX	A	01 00 09.5	XV A 1.7s 21.2nm
	e	A	00 22	
17.	+ePKP	AB	01 08 29.5	New Hebrides Islands 20.42 S 169.25 E
	+ipPKP	AB	08 42	H = 00 48 59.1 h = 64 km MB = 4.7
	LmH	B	02 16.0	D = 145.05 Az = 335.0 (NEIC)
	LmV	B	16.0	h = 44 km
				PKPV A 1.8s 94.5nm
				PKPV B 5 0.45/um
				pPKPV A 1.4 112.0nm
				pPKPV B 1.6 0.3/um
				LmH, LmV traces
17.	ePKHKP	A	06 17 15	Tonga Islands 21.42 S 174.38 W
	ePKP2	A	17 20	H = 05 57 23.1 h = 41 km MB = 4.8
				D = 150.46 Az = 352.3 (NEIC)
				PKHKPV A 1.6s 11.0nm
				PKP2V A 1.5 20.5nm
17.	ePn	A	09 01 55.5	Northern Italy 46.3 N 12.6 E
	ePg	A	02 14	H = 09 00 48 (BCIS)
	iSn	A	02 46.5	D = 4.40
	iSg	A	03 13.5	

Day	Phase		h m s	Remarks
17.	iPn	A	09 03 34	Austria 48.5 N 14.6 E
	iPg	A	03 41	H = 09 02 48 (BCIS)
	iSg	A	04 16.5	D = 2.42
	LmH	B	04.3	PnV A 0.5s 577.0nm
	LmV	B	04.6	LmH B 4 4.29 µm M = 4.0
				LmV B 5 1.76 µm
17.	ePKP2	A	22 54 37.5	Kermadec Islands Region
				31.28 S 178.96 W
				H = 22 34 05 h = 32 km
				D = 159.11 Az = 341 (ISC)
18.	ePKIKP	A	01 19 47	South Pacific Ocean 37.21 S 95.16 W
	ePPS	C	33 20	H = 01 00 43.9 h = normal MB = 5.6
	eSS	C	39 00	D = 127.66 Az = 50.4 (NEIC)
	LmV	B	02 15.9	PKPV A 2.4s 55.3nm
	LmH	B	16.0	LmH B 17.5 0.5 µm M = 5.3
				LmV B 19.5 0.7 µm 5.4
18.	LmH	C	05 02.5	LmH C 22s 0.5 µm
18.	eP	A	12 19 58	Central Alaska 62.39 N 153.14 W
	LmH	C	45.0	H = 12 09 06.2 h = 17 km MB = 4.8
	LmV	C	45.5	D = 66.67 Az = 10.5 (NEIC)
				PV A 1.8s 20.3nm M = 5.0
				LmH, LmV traces
18.	e	A	17 40 38	
18.	+iPn	A	23 33 35	Austria 48.6 N 14.6 E
	iPx	A	33 36.5	H = 23 32 49 (BCIS)
	ePb	A	33 41.5	D = 2.81
	iPg	A	33 44	PnV A 0.5s 21.2nm
	e(Sb)	A	34 11.5	PxV A 0.6 46.0nm
	iSg	A	34 19	
19.	eP	A	01 14 41.5	Rat Islands, Aleutian Is.
	LmH	B	47.0	52.19 N 175.03 E

Day	Phase		h m s	Remarks
cont. 19.	LmV	B	01 58.0	H = 01 02 53.9 h = 53 km MB = 5.3
				D = 76.59 Az = 349.2 (NEIC)
				PV A 1.6s 27.5nm M = 5.0
				LmH B 21.5 1.0/um
				LmV B 16 0.7/um
19.	eP	A	01 45 21	Mindoro, Philippine Islands
	epP	A	45 32	13.34 N 120.34 E
	esP	A	45 39	H = 01 32 19.0 h = 49 km MB = 5.3 MS=4.7
	e	A	49 15	D = 91.27 Az = 322.9 (NEIC)
	LmV	B	02 32.8	h = 46 km
	LmH	B	32.8	PV A 2.4s 55.2nm M = 5.6
				pPV A 2.3 61.0nm
				sPV A 1.8 47.3nm
				LmH B 18.5 1.1/um 5.4
				LmV B 15 1.0/um 5.3
19.	eP	A	01 55 19	Komandorsky Islands Region
				54.44 N 168.55 E
				H = 01 43 48.0 h = normal MB = 5.0
				D = 73.52 Az = 344.9 (NEIC)
19.	ePg	A	04 12 10	France 44.29 N 6.44 E
	eSb	A	13 19	H = 04 09 50.1 h = 85 km
	eSg	A	13 40	D = 7.26 Az = 27.0 (NEIC)
19.	LmH	B	04 41.0	LmH B 13s 0.2/um
	LmV	B	41.0	LmV B 14 0.3/um
19.	eP	AB	06 06 29	North Atlantic Ocean 57.31 N 33.37 W
	Pm	A	06 34.5	H = 06 00 50.4 h = normal MB = 5.0
	eX	B	07 00	D = 26.86 Az = 84.9 (NEIC)
	eS	B	11 00	PmV A 1.4s 72.1nm M = 5.1
	LmH	B	17.2	PV B 10 0.6/um 5.1
	LmV	B	17.4	XH B 13 0.9/um
				XV B 13 0.7/um
				LmH B 14.5 3.6/um 5.1
				LmV B 15 3.6/um 5.2

Day	Phase		h m s	Remarks
19.	ePn	A	07 21 22.5	Austria 48.6 N 14.6 E
	i(Pb)	A	21 29.5	H = 07 20 36 (BCIS)
	ePg	A	21 31	D = 2.81
	i(Sb)	A	21 59	
	iSg	A	22 06	
19.	e	A	12 09 25	New Hebrides Islands 13.42 S 166.55 E
				H = 11 49 48.7 h = 35 km MB = 5.1
				D = 137.64 Az = 336.4 (NEIC)
19.	eP	A	12 20 05	North Atlantic Ocean 57.34 N 33.35 W
	Pm	A	20 10	H = 12 14 26.2 h = normal MB = 5.1
	LmH	B	30.8	D = 26.85 Az = 85.0 (NEIC)
	LmV	B	31.0	PV A 1.2s 20.3nm M = 4.6
				PmV A 1.6 43.9nm 4.8
				LmH B 16 1.0/um 4.5
				LmV B 15 1.0/um 4.6
19.	LmH	B	13 16.0	LmH B 18s 0.5/um
	LmV	B	17.2	LmV B 19 0.6/um
19.	+eP	AB	15 24 57	Aleutian Islands Region
	eS	B	35 00	49.68 N 173.63 W
	eSS	C	40 32	H = 15 12 50.8 h = normal MB = 5.4
	LmH	B	16 04.7	D = 79.76 Az = 356.6 (NEIC)
	LmV	B	09.9	PV A 2.1s 88.2nm M = 5.4
				LmH B 20 0.5/um 4.9
				LmV B 16 0.4/um 4.8
19.	e	A	18 01 02	
19.	eP	A	18 20 00	Kurile Islands Region 43.83 N 151.53 E
				H = 18 07 53.4 h = normal MB = 4.5
				D = 79.66 Az = 335.4 (NEIC)
20.	ePKP2	AB	02 02 20	West of Macquarie Islands
	e(PP)	B	06 20	60.60 S 154.11 E
	eSS	B	25 56	H = 01 41 53.0 h = 33 km MB = 5.4 (NEIC)

Day	Phase		h m s	Remarks
cont. 20.	LmV	B	03 24.6	D = 156.92
	LmH	B	24.8	PKP2V A 2.0s 25.6nm
				LmH B 20 1.0/um M = 5.5
				LmV B 20 1.0/um 5.6
20.	eP	A	02 44 48	**Crete** 34.86 N 23.43 E
				H = 02 40 34 h = 0 km
				D = 17.97 Az = 335 (ISC)
20.	eP	A	04 26 54	**Southern Alaska** 59.56 N 153.56 W
	epP	A	27 19.5	H = 04 15 53.9 h = 98 km MB = 5.1
	e	A	27 37	D = 69.50 Az = 10.0 (NEIC)
				h = 100 km
				PV A 1.2s 16.3nm M = 4.7
				pPV A 0.9 19.4nm
20.	+eP	A	05 34 09.5	**Afghanistan-USSR Border Region**
	e(sP)	A	34 43	36.44 N 71.56 E
				H = 05 26 07.1 h = 110 km MB = 5.1
				D = 44.50 Az = 308.2 (NEIC)
				h = (100) km
				PV A 1.1s 22.2nm M = 4.9
20.	ePKP	A	17 22 16	**Tonga Islands** 15.09 S 173.86 W
	epPKP	A	22 26	H = 17 02 40.3 h = 45 km MB = 4.6
				D = 144.26 Az = 354.0 (NEIC)
				pPKPV A 1.9s 22.7nm
20.	ePP	A	18 43 55	**Talaud Islands** 2.55 N 125.83 E
	LmH	B	19 32.6	H = 18 25 43.1 h = 84 km MB = 5.3 (NEIC)
	LmV	B	32.6	D = 103.11
				LmH B 16s 0.3/um
				LmV B 16 0.2/um
21.	eP	A	03 50 22	**North of Svalbard** 80.04 N 0.88 E
	e	A	50 31	H = 03 44 16.1 h = normal MB = 4.3
				D = 29.74 Az = 166.2 (NEIC)

Day	Phase		h m s	Remarks
21.	ePKHKP	A	04 59 08	Tonga Islands 17.96 S 174.68 W
	epPKP	A	59 43.5	H = 04 39 39.2 h = 132 km MB = 5.2
				D = 147.00 Az = 352.6 (NEIC)
				PKHKPV A 1.2s 24.4nm
				pPKPV A 1.6 19.2nm
21.	e	A	07 54 52.5	
21.	ePn	A	15 08 37	Central Italy 43.79 N 13.29 E
	eSn	A	09 50.5	H = 15 06 53.26 h = 4 km MB = 4.4
				D = 6.96 Az = 351.2 (NEIC)
21.	eP	AB	15 42 10	Kyushu, Japan 32.00 N 131.78 E
	epP	A	42 20.5	H = 15 29 50.7 h = 41 km MB = 5.4
	esP	A	42 27.5	D = 82.25 Az = 326.3 (NEIC)
	ePP	B	45 16	h = 43 km
	eS	B	52 36	PV A 1.9s 45.4nm M = 5.2
	ePS	C	53 18	LmH B 17 12.9/um 6.4
	eSS	C	58 00	LmV B 16.5 15.9/um 6.5
	LmH	B	16 22.8	
	LmV	B	22.9	
22.	eP	A	02 47 36.5	Kurile Islands 48.3 N 155.1 E
				H = 02 35 47 MPV = 5.0
				D = 76.6 (ANUSSR)
				PV A 1.1s 18.1nm M = 5.0
22.	eP	A	05 59 50	North Atlantic Ridge 46.98 N 27.56 W
	e	A	06 00 10	H = 05 54 19.2 h = normal MB = 4.1
	LmH	B	09.4	D = 25.85 Az = 67.3 (NEIC)
	LmV	B	09.4	PV A 1.9s 30.3nm M = 4.6
				LmH B 15 0.9/um 4.4
				LmV B 18 0.9/um 4.5
22.	eP	A	06 08 24.5	Southern Alaska 61.42 N 147.49 W
				H = 05 57 34.2 h = 48 km MB = 4.5
				D = 67.04 Az = 14.3 (NEIC)

Day	Phase		h m s	Remarks
22.	eP	A	06 21 12	<u>Kyushu, Japan</u> 31.81 N 131.76 E
	esP	A	21 28.5	H = 06 08 54.3 h = 49 km MB = 4.7
	LmH	B	07 01.9	D = 82.40 Az = 326.3 (NEIC)
	LmV	B	02.0	h = 45 km
				PV A 2.0s 42.7nm M = 5.1
				LmH B 17.5 1.2 μm 5.3
				LmV B 16 1.5 μm 5.5
22.	i	A	12 47 21	
	i	A	48 04.5	
22.	i	A	14 08 38	
	ei	A	09 16.5	
23.	eP	A	04 22 20.5	<u>Iceland Region</u> 66.07 N 17.9 W
				H = 04 17 30.8 h = 0 km MB = 3.5
				D = 21.55 Az = 121
23.	+eP	A	06 54 47.5	<u>Turkey</u> 36.69 N 32.42 E
	i	A	54 53.5	H = 06 50 11.0 h = normal MB = 4.3
	e	A	54 57.5	D = 20.43 Az = 319.6 (NEIC)
				PV A 1.1s 34.3nm M = 4.6
23.	eP	A	09 00 54	<u>South of Honshu, Japan</u>
	ePcP	A	01 01	33.28 N 140.66 E
				H = 08 48 22.8 h = 66 km MB = 4.8
				D = 85.12 Az = 330.2 (NEIC)
				PV traces
				PcPV A 1.6s 11.0nm
23.	eP	A	14 24 10	<u>Fox Islands, Aleutian Is.</u>
	e	A	24 25.5	53.87 N 165.50 W
	e	A	24 32	H = 14 12 23.5 h = 23 km MB = 5.0
	e	A	24 39	D = 75.83 Az = 1.9 (NEIC)
	LmH	B	15 01.4	PV A 1.1s 22.2nm M = 5.1
	LmV	B	07.1	LmH B 20 0.3 μm 4.6
				LmV B 18 0.3 μm 4.6

Day	Phase		h m s	Remarks
23.	ePKHKP	A	16 24 46	Fiji Islands Region 20.30 S 178.40 W H = 16 06 03.8 h = 595 km MB = 4.4 D = 148.70 Az = 347.7 (NEIC)
23.	eP ePP	A A	16 42 10 45 38.5	Southeast of Shikoku, Japan 30.26 N 136.79 E H = 16 30 22.2 h = 508 km MB = 4.7 D = 86.03 Az = 328.6 (NEIC)
23.	ePKP	A	17 40 57	Fiji Region 17.9 S 176.8 W H = 17 21 20 h = 0 km MB = 4.6 D = 146.65 Az = 350 (ISC) PKPV A 1.1s 10.1nm
24.				In the intervall 05h 03m - 06h 19m the short-period seismographs (type A) had been out of operation.
24.	ePn eSn eSg LmH LmV	A A A B B	07 19 49 21 17 22 03 22.2 23.0	Yugoslavia 43.71 N 16.86 E H = 07 17 55.5 h = normal MB = 5.3 D = 7.81 Az = 334.6 (NEIC) PnV A 1.1s 52.4nm M = 5.5 LmH B 5 2.3/um 4.5 LmV B 14 1.2/um
24.	ePKP	A	10 04 22	Tonga Islands 16.89 S 173.02 W H = 09 44 41.3 h = normal MB = 4.2 D = 146.12 Az = 354.7 (NEIC) PKPV A 1.7s 12.1nm
24.	ePKP	A	15 24 12	Tonga Islands 15.14 S 173.60 W H = 15 04 36.2 h = normal MB = 4.8 D = 144.34 Az = 354.3 (NEIC)
24.	-iP1 -iP2 ePP1 iPP2	AB AB BC AB	15 37 22 37 30 39 00 39 12	Hindu Kush Region 36.23 N 69.67 E H = 15 29 22.2 h = 47 km MB = 6.0 D = 43.44 Az = 308.2 (NEIC) P1V A 1.9s 288.0nm M = 5.7

Day	Phase		h m s	Remarks
cont. 24.	iPP3	B	15 39 22	P2V A 2.2s 456.0nm M = 5.9
	eS	B	43 36	P2V B 10 4.0/um 6.1
	eSS	C	46 40	PP1V B 4 1.55/um 6.2
	LmH	B	16 00.2	PP2V B 7 3.6/um 6.3
	LmV	B	00.2	PP3V B 12 7.3/um 6.4
				SH B 13 7.4/um 6.3
				LmH B 15 41.9/um 6.5
				LmV B 14 60.3/um 6.7
24.	eP	A	23 33 49	Halmahera 1.46 N 127.52 E
	epP	A	34 43	H = 23 20 03.4 h = 200 km MB = 5.6
	ePP	A	38 14.5	D = 104.98 Az = 323.6 (NEIC)
	epPP	A	39 07.5	h = 218 km
	eS	C	45 25	PV A 1.6s 16.5nm M = 5.8
	eSS	C	53 00	PPV A 1.6 52.2nm 5.5
	LmH	B	24 20.4	LmH B 16.5 0.6/um
	LmV	B	26.7	LmV B 12 0.3/um
25.	+iPn	A	05 01 04.5	Yugoslavia 44.04 N 15.76 E
	-iPx	A	01 10	H = 04 59 18.6 h = normal MB = 4.4
	eSn	A	02 26	D = 7.18 Az = 338.4 (NEIC)
	eSg	A	03 16	PnV A 1.0s 39.4nm M = 5.3
				PxV A 0.6 38.3nm
25.	eP	A	08 03 45	Hindu Kush Region 36.32 N 69.60 E
	e	A	05 41.5	H = 07 55 45.3 h = 46 km MB = 4.7
	LmH	B	26.7	D = 43.33 Az = 308.1 (NEIC)
	LmV	B	26.8	PV A 1.2s 16.3nm M = 4.6
				LmH B 14 0.6/um 4.6
				LmV B 14.5 0.8/um 4.9
25.	ePn	A	17 12 15.5	Northern Italy 44.50 N 10.51 E
	ePg	A	12 44	H = 17 10 43.1 h = 17 km MB = 4.3
	eSn	A	13 23	D = 6.20 Az = 6.5 (NEIC)
	eSg	A	14 06.5	PnV A 0.6s 7.7nm M = 4.7

Day	Phase		h m s	Remarks
26.	ePKP	A	00 44 19.5	Fiji Region 18.6 S 177.0 W
				H = 00 24 39.0 h = 0 km MB = 4.6
				D = 147.26 Az = 350 (ISC)
26.	eP	A	08 20 59.5	Taiwan Region 21.09 N 120.29 E
	LmH	B	09 00.5	H = 08 08 24.8 h = normal MB = 5.0
	LmV	B	00.6	D = 85.10 Az = 322.7 (NEIC)
				PV A 1.5s 30.2nm M = 5.3
				LmH B 14 0.7 μm 5.2
				LmV B 16 0.4 μm 4.9
26.	-eP	AB	10 17 12.5	Kamchatka 51.91 N 156.04 E
	-iP	A	17 12.8	H = 10 05 59.3 h = 189 km MB = 5.6
	eS	B	26 24	D = 73.47 Az = 337.3 (NEIC)
	ePS	B	27 26	PV A 1.7s 254.0nm M = 5.7
	esSS	C	32 30	PV B 2 0.55 μm 5.9
	esSSS	C	35 40	
	LmH	B	49.5	
	LmV	B	56.5	
26.	eP	A	12 36 14	Dodecanese Islands 35.58 N 27.04 E
				H = 12 31 58.4 h = 79 km MB = 4.0
				D = 18.75 Az = 328.2 (NEIC)
				PV A 1.0s 11.8nm M = 4.1
26.	eP	A	16 14 06	Central Mid-Atlantic Ridge
	eS	C	22 24	4.67 N 32.64 W
	LmV	B	37.3	H = 16 04 07.7 h = normal MB = 5.2
	LmH	B	37.8	D = 58.88 Az = 31.3 (NEIC)
				PV A 2.5s 108.0nm M = 5.5
				LmH B 18 0.4 μm 4.6
				LmV B 19 0.5 μm 4.7
26.	eP	A	16 41 07	Central Mid-Atlantic Ridge
	LmH	C	17 00.2	4.56 N 32.56 W
	LmV	C	04	H = 16 31 08.6 h = normal MB = 5.1
				D = 68.93 Az = 31.2 (NEIC)
				PV A 2.0s 42.7nm M = 5.2

Day	Phase		h m s	Remarks
27.	ePn	A	02 47 38	Central Italy 43.79 N 13.1 E
	e(Sn)	A	49 04	H = 02 45 51.3 h = 0 km
				D = 6.94 Az = 352 (ISC)
27.	eP	A	06 48 26	West Pakistan 29.69 N 70.27 E
	LmH	B	07 13.3	H = 06 39 44.4 h = 12 km MB = 5.5
	LmV	B	16.0	D = 48.07 Az = 313.0 (NEIC)
				PV A 1.4s 58.1nm M = 5.5
				LmH B 14.5 0.3/um 4.4
				LmV B 15 0.2/um 4.4
27.	eP	A	09 16 51	Burma 26.16 N 96.57 E
	e	A	17 05	H = 09 05 52.6 h = 23 km
				D = 67.14 Az = 316.5 (NEIC)
27.	eP	A	10 57 38	West Pakistan 29.75 N 70.27 E
				H = 10 48 55.6 h = 8 km MB = 5.4
				D = 48.03 Az = 313.0 (NEIC)
				PV A 1.3s 35.0nm M = 5.3
27.	iPn	A	12 58 09.2	Czechoslovakia 49°43'N 13°26.5'E
	i	A	58 10	Explosion H = 12 57.6 yield 12.5 t (PRU)
	eSg	A	58 30	D = 1.48
27.	eP	A	16 07 35	Hindu Kush Region 36.27 N 69.52 E
	e	A	07 41	H = 15 59 35.0 h = 53 km MB = 5.1
	e	A	09 33	D = 43.32 Az = 308.2 (NEIC)
	LmH	B	30.3	PV traces
	LmV	B	30.3	LmH B 15s 0.6/um
				LmV B 14 0.9/um
27.	LmH	C	20 33.7	Guatemala 13.98 N 91.75 W
	LmV	C	34.5	H = 19 39 36.1 h = 74 km MB = 5.1 (NEIC)
				D = 87.58
27.	e	A	24 07 28	South of Honshu, Japan
				33.32 N 140.94 E
				H = 23 54 49.1 h = 46 km MB = 4.1
				D = 85.20 Az = 330.4 (NEIC)

Day	Phase		h m s	Remarks
28.	ePn	A	01 46 17	Yugoslavia 43.04 N 20.48 E
	eSg	A	49 03.5	H = 01 43 56.5 h = normal MB = 4.9
	LmH	B	49.5	D = 9.73 Az = 324.5 (NEIC)
	LmV	B	50.5	PV A 1.4s 14.0nm M = 5.0
				LmH B 10 0.7/um 3.8
				LmV B 8 0.45/um
28.	eP	A	02 02 57	Near East Coast of Honshu, Japan
	LmH	C	37.0	37.45 N 141.33 E
				H = 01 50 43.9 h = 70 km MB = 5.4
				D = 81.77 Az = 330.3 (NEIC)
				PV A 1.4s 21.0nm M = 4.9
28.	ePKP	A	04 33 49	North of New Zealand 17.1 S 174.97 W
				H = 04 14 07.3 h = 0 km MB = 5.0
				D = 146.09 Az = 352 (ISC)
				PV A 1.0s 7.9nm
28.	eP	A	08 21 47.5	Egypt 27.8 N 33.9 E
				H = 08 15 48 h = 0 km MB = 4.2
				D = 28.39 Az = 329 (ISC)
28.	eP	A	09 40 19.5	South of Honshu, Japan 33.39 N 140.78 E
				H = 09 27 49.5 h = 69 km MB = 5.2
				D = 85.06 Az = 330.3 (NEIC)
				PV A 1.3s 13.1nm M = 4.8
28.	-iP1	AB	09 55 31.3	United Arab Republic 27.65 N 33.76 E
	eiP2	AB	55 35	H = 09 49 34.9 h = 15 km MB = 5.6
	eS	B	10 00 24	D = 28.47 Az = 329.8 (NEIC)
	ei	B	00 40	P1V A 1.6s 247.0nm M = 5.7
	eSS	B	01 38	P1V B 4.5 1.9/um 6.2
	LmH	B	12.9	P2V A 1.4 274.0nm 5.8
	LmV	B	15.5	P2V B 6 1.4/um 5.9
				LmH B 13 1.8/um 4.9
				LmV B 13 1.4/um 4.9
28.	LmH	B	16 41.7	LmH B 15s 0.3/um
	LmV	B	42.1	LmV B 15 0.3/um

Day	Phase		h m s	Remarks
28.	e(Sb)	A	17 26 36	Poland 50.33 N 18.93 E H = 17 24 03.0 M = 2.6 (WAR) D = 4.75 PV A 1.3s 6.5nm
28.	LmH LmV	B B	21 16.6 16.7	South of Africa 49.19 S 30.73 E H = 20 14 20.2 h = 33 km MB = 5.1 (NEIC) D = 100.80 LmH traces LmV B 18s 0.25 um M = 4.8
29.	eP e	A A	03 40 07 40 11	Afghanistan-USSR Border Region 38.89 N 71.45 E H = 03 32 11.2 h = 53 km MB = 4.9 D = 42.96 Az = 306.1 (NEIC)
29.	eP ePP	A A	03 57 31 04 01 23.5	Samar, Philippine Islands 11.38 N 125.70 E H = 03 44 10.0 h = 67 km MB = 5.1 (NEIC) D = 95.94 PV A 1.6s 13.7nm M = 5.2 PPV A 1.3 8.7nm 5.1
29.	e e	A A	04 45 58 46 35	
29.	iPn eSg	A A	15 05 30.1 05 51.5	Dorheim/Hessen, Fed. Republik of Germany 50°57.88'N 09°13.15'E Explosion yield 3.5 t (Hannover) H = 15 05 00.87 D = 1.50
29	eP	A	17 28 02	Kurile Islands 50.64 N 154.01 E H = 17 16 47.6 h = 212 km MB = 4.8 D = 74.13 Az = 336.2 (NEIC) PV A 1.4s 20.4nm M = 4.7
30.	ePKIKP	A	06 03 32	New Guinea 4.96 S 144.58 E

Day	Phase		h m s	Remarks
cont. 30.	e	A	06 03 37.5	H = 05 44 48.4 h = 83 km MB = 5.2
				D = 119.90 Az = 327.5 (NEIC)
30.	ePKP	A	09 09 12	North of New Zealand 16.25 S 174.71 W
				H = 08 49 32.5 h = 0 km M = 4.6
				D = 145.31 Az = 353 (ISC)
				PKPV A 0.7s 11.5nm
30.	ePKIKP	A	12 58 34.5	Fiji Islands Region 17.86 S 178.69 W
	iPKHKP	A	58 37	H = 12 40 00.3 h = 600 km MB = 5.3
	iPKP2	A	58 38.9	D = 146.27 Az = 348.2 (NEIC)
	e	A	58 44.5	PKIKPV A 1.2s 20.3nm
				PKHKPV A 1.4 74.4nm
				PKP2V A 1.2 61.0nm
30.	LmH	B	13 39.4	Near Coast of Chiapas, Mexico
	LmV	B	39.4	14.54 N 92.90 W
				H = 12 45 20.2 h = 33 km MB = 4.9 (NEIC)
				D = 87.83
				LmH B 19s 0.3/um M = 4.7
				LmV B 18 0.3/um 4.7
30.	ePKP	A	15 55 02	New Hebrides Islands 18.98 S 169.19 E
				H = 15 35 49.1 h = 180 km MB = 4.8
				D = 143.73 Az = 335.8 (NEIC)
				PKPV A 1.7s 24.2nm
30.	e	A	17 57 27	Southern Iran 27.20 N 56.82 E
				H = 17 49 33.4 h = normal MB = 4.6
				D = 41.40 Az = 316.9 (NEIC)
30.	eP	A	19 10 07.5	Taiwan 24.26 N 121.13 E
	eX	A	10 12	H = 18 57 43.4 h = normal MB = 4.9
	LmH	B	47.4	D = 83.04 Az = 322.8 (NEIC)
	LmV	B	51.8	PV traces
				XV A 1.4s 23.3nm
				LmH B 13.5 0.7/um M = 5.2
				LmV B 14.5 0.6/um 5.1

Day	Phase		h m s	Remarks
1.	e(Sn)	A	00 57 36	Poland 50.28 N 18.95 E
	e	A	57 41.5	H = 00 55 08.3 M = 2.5 (WAR)
				D = 4.7
1.	ePKHKP	A	06 04 37	Tonga Islands 21.90 S 175.05 W
	ePKP2	A	04 44.5	H = 05 44 48.2 h = 41 km MB = 5.0
	e	A	04 51.5	D = 150.83 Az = 351.3 (NEIC)
	LmH	B	07 17.9	PKHKPV A 1.6s 30.2nm
	LmV	B	21.0	LmH B 16 0.3 µm M = 5.1
				LmV B 16 0.4 µm 5.2
1.	ePn	A	11 09 30	Czechoslovakia 50°32.5'N 14°03.2'E
	iSg	A	09 56	H = 11 09.0
				Explosion yield 36 t (PRU)
				D = 1.54
1.	LmV	C	18 43.5	New Hebrides Islands 13.10 S 166.06 E
				H = 17 17 49.9 h = 19 km MB = 4.4 (NEIC)
				D = 137.16
2.	ePKHKP	A	01 07 26.5	Tonga Islands 20.84 S 174.19 W
	LmH	C	02 13.5	H = 00 47 37.7 h = normal MB = 4.6
	LmV	C	22.5	D = 149.91 Az = 352.6 (NEIC)
				PKHKPV A 2.0s 42.7nm
				LmH C 22 0.2 µm M = 4.8
				LmV C 16 0.3 µm 5.2
2.	LmH	B	01 31.4	Near Coast of Guerrero, Mexico
	LmV	B	31.7	16.36 N 98.52 W
				H = 00 34 42.8 h = 53 km MB = 4.9 (NEIC)
				D = 89.69
				LmH B 16s 0.3 µm
				LmV B 17.5 0.4 µm
2.	eP1	AB	13 03 04	Iran 30.10 N 50.85 E
	eiP2	A	03 06	H = 12 56 06.7 h = 31 km MB = 5.4 MS=5.3
	epP	A	03 16	D = 35.69 Az = 316.4 (NEIC)
	e	A	03 37	P1V A 1.8s 23.6nm M = 4.8

Day	Phase		h m s	Remarks
cont. 2.	e	A	13 03 52	P2V A 2.0s 111.0nm M = 5.4
	ePP	B	04 36	SH B 11 1.2 µm 5.5
	e	A	06 14	ScS B 9 1.6 µm
	eS	B	08 40	LmH B 17 2.8 µm 5.1
	eScS	B	13 20	LmV B 18 1.4 µm 4.9
	LmH	B	19.5	
	LmV	B	19.7	
3.	eP	A	01 32 06.5	<u>Mariana Islands</u> 18.73 N 145.07 E
	e(sP)	A	35(18)	H = 01 19 26.7 h = 618 km MB = 5.7
	e	A	35 30	D = 99.72 Az = 332.0 (NEIC)
	ePP	AB	36 20	PV A 1.6s 38.5nm M = 5.6
	ePPP	A	38 42	PPV A 1.5 80.4nm 5.7
	ePKKP	A	48(24)	LmH B 14 0.3 µm
	LmH	B	02 35.0	LmV B 16 0.3 µm
	LmV	B	35.0	
3.	eP	A	02 16 58.5	<u>Iran</u> 30.11 N 50.79 E
				H = 02 10 00.4 h = 38 km MB = 5.0
				D = 35.65 Az = 316.4 (NEIC)
				PV A 2.0s 25.6nm M = 4.8
3.	eP	A	05 33 07.5	<u>Socotra Region</u> 13.90 N 53.57 E
				H = 05 24 15.3 h = normal MB = 5.0
				D = 49.92 Az = 326.2 (NEIC)
				PV A 2.0s 34.2nm M = 5.0
3.	eP	A	05 45 05	<u>Mindanao, Philippine Islands</u>
	LmH	B	06 28.0	5.79 N 125.94 E
	LmV	B	36.8	H = 05 31 27.5 h = 96 km MB = 5.6
				D = 100.59 Az = 323.8 (NEIC)
				PV A traces
				LmH B 18s 0.2 µm
				LmV B 18 0.2 µm
3.	eP	A	21 45 15	<u>Iran</u> 30.05 N 50.99 E
	e	A	45 32	H = 21 38 22.2 h = 43 km MB = 5.1
	LmH	B	22 03.8	D = 35.81 Az = 316.4 (NEIC)

Day	Phase		h m s	Remarks
cont. 3.	LmV	B	22 03.8	LmH B 14.5s 0.3/um M = 4.2
				LmV B 16 0.4/um 4.3
3.	e(PKP)	A	22 16 38	Off Coast of Southern Chile
	LmH	B	23 07.6	45.06 S 79.63 W
	LmV	B	07.7	H = 21 57 35.9 h = 33 km MB = 4.8 (NEIC)
				e(PKP) traces
				D = 123.63
				LmH B 20.5s 0.8/um M = 5.4
				LmV B 21.5 1.1/um 5.5
4.	eiP	AB	01 16 47	Near East Coast of Honshu, Japan
	epP	A	17 00	38.51 N 141.97 E
	eS	B	26 52	H = 01 04 35.1 h = 52 km MB = 5.3
	LmH	B	55.8	D = 81.10 Az = 330.6 (NEIC)
	LmV	B	55.9	h = 46 km
				PV A 1.4s 79.0nm M = 5.5
				PV B 6 0.5/um 5.7
				pPV A 1.9 91.0nm
				LmH B 16.5 1.7/um
				LmV B 16.5 1.8/um
4.	e(P)	A	01 53 29.5	South of Honshu, Japan
	e	A	53 39	33.35 N 140.86 E
				H = 01 40 50.7 h = 42 km MB = 4.5
				D = 85.13 Az = 330.3 (NEIC)
4.	+eP	AB	10 29 33	Negros, Philippine Islands
	e	A	32 43	9.41 N 122.52 E
	ePP	AB	33 27	H = 10 16 12.3 h = 58 km MB = 5.5
	eSKS	B	40 04	D = 95.68 Az = 323.3 (NEIC)
	eS	B	40 50	PV A 2.2s 142.0nm M = 6.1
	ePS	B	42 14	PV B 7 0.6/um 6.2
	eSS	C	47 20	PPV A 1.8 94.6nm 6.0
	LmH	B	11 20.4	LmH B 17 4.4/um
	LmV	B	20.6	LmV B 17 5.9/um

Day	Phase		h m s	Remarks
4.	LmV	B	16 50.9	Solomon Islands 9.77 S 159.38 E
	LmH	B	54.2	H = 15 32 25.4 h = 63 km MB = 5.0 (NEIC)
				D = 131.36
				LmH B 18s 0.3/um
				LmV B 20 0.4/um
4.	ePKP2	A	24 04 51.5	Kermadec Islands Region
				31.23 S 179.35 W
				H = 23 44 44.3 h = 229 MB = 3.9 (NEIC)
				D = 158.95
				PKP2V A 1.3s 10.9nm
5.	eP	A	01 18 13	Northern Sinkiang Prov., China
	e	A	18 43.5	44.64 N 81.12 E
	e	A	19 19	H = 01 09 52.9 h = normal MB = 4.6
	LmH	B	34.7	D = 45.64 Az = 303.5 (NEIC)
	LmV	B	38.1	LmH B 16s 0.9/um M = 4.8
				LmV B 16 0.4/um 4.5
5.	ePKP	A	02 05 29.5	New Hebrides Islands 18.67 S 169.33 E
				H = 01 46 26.1 h = 243 km MB = 4.9
				D = 143.50 Az = 336.1 (NEIC)
5.	eP	A	04 13 27.5	North Atlantic Ridge 35.68 N 36.63 W
				H = 04 06 14.8 h = normal MB = 4.7
				D = 37.54 Az = 51.2 (NEIC)
				PV A traces
5.	eP	A	04 18 44	Northern Sinkiang Prov., China
				43.62 N 87.94 E
				H = 04 09 49.0 h = normal MB = 4.3
				D = 50.25 Az = 306.4 (NEIC)
5.	ePKP	A	06 43 34	Fiji Region 15.85 S 177.2 W
				H = 06 23 59.3 h = 0 km MB = 4.7
				D = 144.58 Az = 350 (ISC)
				PKPV A 1.4s 9.3nm

Day	Phase		h m s	Remarks
5.	eP	AB	10 28 13.5	Vancouver Island Region
	ipP	AB	28 22	49.55 N 127.21 W
	isP	A	28 26	H = 10 16 38.4 h = 27 km MB = 5.8
	ePP	B	31 06	D = 74.16 Az = 25.8 (NEIC)
	ePPP	B	32 55	PV A 1.2s 81.3nm M = 5.6
	ePPPP	B	33 48	pPV A 1.2 77.4nm
	eS	B	37 48	sPV B 6.5 1.5/um
	ePS	C	38 28	YV A (2.2) 358.0nm
	eSS	C	42 44	PPV B 8.5 0.65/um 5.9
	LmH	B	11 03.0	SH B 14 2.2/um 6.2
	LmV	B	07.4	LmH B 16 4.9/um 5.9
				LmV B 13 4.2/um 6.0
5.	e(Sn)	A	17 27 57.5	Poland 50.28 N 18.88 E
				H = 17 25 30.6 M = 3.0 (WAR)
				D = 4.65
5.	eP	A	18 08 34	Southern Greece 36.92 N 21.52 E
	LmH	B	15.5	H = 18 04 54.1 h = 17 km MB = 4.7 (NEIC)
	LmV	B	15.5	D = 15.46
				PV A traces
				LmH B 11.5s 0.4/um M = 3.8
				LmV B 12 0.4/um 4.0
5.	e(P)	A	19 12 19	Ionian Sea 37.56 N 20.62 E
	e	A	12 25	H = 19 08 44.4 h = normal MB = 3.8
				D = 14.58 Az = 336.7 (NEIC)
5.	LmV	B	21 25.0	Near Coast of Peru 15.68 S 75.19 W
				H = 20 21 55.8 h = 33 km MB = 5.1 (NEIC)
				D = 99.96
				LmV B 16s 0.3/um M = 4.9
6.	eP	A	01 10 47	Eastern Kazakh SSR 49.72 N 77.98 E
				H = 01 02 57.74 h = 0 km MB = 4.4
				D = 41.21 Az = 297.7 (NEIC)
				Artificial (BCIS)
				PV A 0.8s 7.7nm M = 4.5

Day	Phase		h m s	Remarks
6.	ePn	A	13 00 11.5	<u>Czechoslovakia</u> 50°37' N 14°11' E
	eSg	A	00 34	H = 12 59.6
				Explosion yield 5.8 t (PRU)
				D = 1.63
7.	eP	A	12 15 36	<u>Burma</u> 20.47 N 98.09 E
	eX	A	16 03	H = 12 04 11.6 h = 27 km MB = 5.0
	eSS	B	29.5	D = 72.26 Az = 318.1 (NEIC)
	eSSS	B	33.2	XV A 1.3s 21.8nm
	LmH	B	45.2	LmH B 20 1.6 µm M = 5.3
	LmV	B	52.4	LmV B 15 0.9 µm 5.2
7.	eP	A	12 49 39.5	<u>Philippine Islands Region</u>
	LmH	B	13 37.9	11.86 N 126.22 E
	LmV	B	40.2	H = 12 36 14.1 h = normal MB = 5.4
				D = 95.86 Az = 324.4 (NEIC)
				PV A 1.5s 17.6nm M = 5.3
				LmH B 17 0.6 µm 5.1
				LmV B 16 0.5 µm 5.1
7.	eP	A	13 39 18	<u>Alaska Peninsula</u> 57.49 N 155.00 W
				H = 13 28 01.1 h = 61 km MB = 3.8
				D = 71.66 Az = 8.9 (NEIC)
8.	eP	A	01 00 32.5	<u>Mindanao, Philippine Islands</u>
	LmH	B	49.1	7.30 N 126.71 E
	LmV	B	49.3	H = 00 47 02.7 h = 107 km MB = 4.8
				D = 99.82 Az = 324.2 (NEIC)
				PV A traces
				LmH B 18s 0.3 µm
				LmV B 15 0.9 µm
8.	eSn	A	03 02 40.5	<u>Northern Italy</u> 44.3 N 9.8 E
	eSg	A	03 36.5	H = 03 00 07 (BCIS)
				D = 6.39
8.	e	A	05 49 23	<u>Greece-Bulgaria Border Region</u>
	e	A	50 53	41.60 N 23.60 E
				H = 05 46 13.6 h = 28 km MB = 4.7
				D = 12.28 Az = 321.6 (NEIC)

Day	Phase		h m s	Remarks
8.	+iP	AB	06 57 23.5	Afghanistan-USSR Border Region
	e(PcP)	A	59 05	36.34 N 71.37 E
	e(PP)	A	59 11.5	H = 06 49 22.8 h = 121 km MB = 5.6
	e	A	59 19	D = 44.44 Az = 308.2 (NEIC)
	eS	B	07 03 51	PV A 1.5s 224.0nm M = 5.7
	eScS	B	07 16	PV B 2 0.45/um 5.8
				PH A 1.5 124.0nm 5.7
				SH B 7 0.5/um 5.3
8.	eP	A	08 41 26	Kurile Islands Region 46.15 N 154.62 E
				H = 08 29 27.3 h = normal MB = 4.9
				D = 78.42 Az = 337.0 (NEIC)
				PV A 1.0s 15.8nm M = 5.0
8.	e	A	11 29 14	
8.	eP	A	12 23 43.5	Oaxaca, Mexico 16.30 N 96.89 W
	e(pP)	A	23 53	H = 12 10 54.7 h = 62 km MB = 5.6
	LmH	B	13 05.5	D = 88.80 Az = 37.1 (NEIC)
	LmV	B	08.2	PV A 1.0s 11.8nm M = 5.1
				pPV A 1.3 30.6nm
				LmH B 18.5 0.7/um
				LmV B 17 1.0/um
9.	eP	A	07 03 34.5	Southwestern Russia 49.78 N 35.4 E
				H = 06 59 57.9 h = 0 km MB = 4.8
				D = 15.27 Az = 282 (ISC)
				Probably underground explosion MB = 5.1 (UPP)
				PV A 1.3s 13.1nm M = 5.0
9.	ePKIKP	A	11 27(36)	Fiji Islands Region 20.22 S 176.34 W
	ePKHKP	A	27 40.5	H = 11 08 23.7 h = 278 km MB = 4.6
	ePKP2	A	27 44	D = 148.99 Az = 350.2 (NEIC)
				PKHKPV A 1.9s 19.0nm
9.	-eiP	A	14 05 18.5	Sea of Okhotsk 53.13 N 154.43 E
	epP	A	06 47	H = 13 54 35.3 h = 390 km MB = 5.2

Day	Phase		h m s	Remarks
cont. 9.	ePP	A	14 07 56	D = 71.96 Az = 336.1 (NEIC)
	e(P'P')	A	34 27	h = 408
	e	A	34 33	PV A 1.6s 71.5nm M = 5.0
				PPV A 1.2 14.2nm 4.8
10.	ePKHKP	A	06 49 54.5	<u>South of Fiji Islands</u> 22.75 S 176.19 W
				H = 06 30 03.0 h = normal MB = 5.1
				D = 151.50 Az = 349.6 (NEIC)
				PKHKPV A 1.5s 25.1nm
10.	e(Sn)	A	07 39 03	<u>France</u> 47.3 N 6.5 E
				H = 07 36 23 h = 0 km (ISC)
				D = 4.75
10.	eP	A	08 31 37	<u>South of Honshu, Japan</u> 30.79 N 137.50 E
				H = 08 19 45.9 h = 461 km MB = 4.6
				D = 85.89 Az = 328.9 (NEIC)
				PV A traces
10.	eP	A	12 38 00	<u>Off East Coast of Kamchatka</u>
				53.57 N 161.65 E
				H = 12 26 31.0 h = normal MB = 4.1
				D = 73.14 Az = 340.6 (NEIC)
10.	e(P)	A	19 12 35	<u>Northern Sinkiang Prov., China</u>
	e	A	12 44.5	43.44 N 88.63 E
	LmH	B	35.5	H = 19 03 33.0 h = normal MB = 4.7
	LmV	B	35.6	D = 50.76 Az = 306.8 (NEIC)
				LmH B 16s 0.3 μm M = 4.3
				LmV B 14 0.3 μm 4.6
11.	e	A	03 33 07.5	
11.	+iP	AB	06 02 32.5	<u>Hindu Kush Region</u> 36.38 N 70.73 E
	epP	A	03 18	H = 05 54 44.2 h = 206 km MB = 5.4
	ePP/PcP	A	04 19	D = 44.01 Az = 308.2 (NEIC)
				h = 216 km
				PV A 1.6s 55.0nm M = 4.8

Day	Phase		h m s	Remarks
11.	eP	A	07 10 05	Kurile Islands 48.35 N 154.45 E H = 06 58 21.4 h = 62 km MB = 5.2 D = 76.35 Az = 336.7 (NEIC) PV A 1.2s 20.3nm M = 5.1
11.	ePKIKP	A	16 05 09.5	New Britain Region 6.03 S 153.00 E
	ePP	A	07 00	H = 15 46 12.0 h = 50 km MB = 5.5 D = 125.11 Az = 330.9 (NEIC) PKIKPV A 1.2s 16.3nm PPV A 1.6 27.4nm M = 5.3
11.	e(pP)	A	20 57 23.5	Burma-India Border Region 24.29 N 94.70 E H = 20 46 11.6 h = 48 km MB = 4.6 D = 67.33 Az = 316.8 (NEIC) (pP)V A 1.2s 16.3nm
11.	e	A	22 55 15	Iran-Iraq Border Region
	e	A	55 44	36.07 N 45.71 E
	ePP	A	55 52	H = 22 49 02.3 h = normal MB = 4.7
	LmH	B	23 09.0	D = 28.46 Az = 311.5 (NEIC)
	LmV	B	10.0	Surface waves traces
12.	eP	A	00 26 10	Kurile Islands 49.28 N 155.43 E
	LmH	C	59.0	H = 00 14 26.9 h = normal MB = 5.2
	LmV	C	01 03.4	D = 75.74 Az = 337.2 (NEIC) PV A 1.0s 19.7nm M = 5.1 LmH C 28 0.2/um 4.3 LmV C 20 0.3/um 4.6
12.	ePKP	A	11 13 30.5	North of New Zealand 16.5 S 173.2 W H = 10 53 57 h = 69 km D = 145.72 Az = 355 (ISC)
12.	eP	A	19 45 21.5	Southern Greece 37.29 N 21.92 E
	e	A	45 35	H = 19 41 48.2 h = 92 km MB = 4.1
	e	A	45 44	D = 15.25 Az = 334.3 (NEIC)
	LmH	B	50.7	LmH B 16s 0.9/um M = 4.0
	LmV	B	52.8	LmV B 14 0.3/um 3.8

Day	Phase		h m s	Remarks
12.	ePKP	A	20 24 05	Fiji Islands Region 18.81 S 179.06 W
				H = 20 05 23.7 h = 509 km MB = 4.6
				D = 145.17 Az = 348.1 (NEIC)
				PKPV A 1.8s 27.0nm
13.	eP	AB	02 12 30	Atlantic-Indian Rise 28.28 S 63.12 E
	e	A	12 39	H = 01 59 29.4 h = normal MB = 5.3
	e	A	12 45	D = 90.79 Az = 330.1 (NEIC)
	e	A	12 49.5	PV A 1.9s 79.5nm M = 5.5
	eSKS	B	23 00	LmH B 21 0.3/um 4.8
	ePS	B	24 39	LmV B 20 0.3/um 4.7
	eSS	B	29 23	
	LmH	B	51.0	
	LmV	B	56.6	
13.	iPg	A	12 00 43	Czechoslovakia 50.41 N 13.84 E
	iSg	A	01 03	Explosion H = 12 00.5 11.0 t (KHC)
				D = 1.44 Az = 280 (ISC)
				Explosion D c. 1.6
13.	eP	A	15 22 44.5	Celebes Sea 3.76 N 124.61 E
				H = 15 09 28.7 h = 346 km MB = 5.0
				D = 101.42 Az = 323.3 (NEIC)
13.	LmH	B	17 25.6	LmH B 16s 0.2/um
13.	ePn	A	22 22 58	Central Italy 43.84 N 13.26 E
	eSn	A	24 11	H = 22 21 17.0 h = normal MB = 4.4
				D = 6.91 Az = 351.3 (NEIC)
13.	e	A	23 55 53	Andreanof Islands, Aleutian Is.
				51.61 N 173.93 W
				H = 23 43 41.1 h = 26 km MB = 4.1
				D = 78.01 Az = 356.4 (NEIC)
14.	ePKP	AB	02 20 49	South of Java 10.51 S 113.92 E
	e	C	23 30	H = 02 02 17.6 h = normal MB = 5.4
	e	B	27 10	D = 105.85 Az = 319.7 (NEIC)

Day	Phase		h m s	Remarks
cont. 14.	e	B	02 35 46	LmH B 19s 0.3/um M = 4.9
	LmH	B	03 10.4	LmV B 18 0.3/um 4.8
	LmV	B	15.2	
14.	eP	A	13 11 10	Iran 30.08 N 50.85 E
				H = 13 04 11.8 h = 34 km MB = 4.4
				D = 35.71 Az = 316.4 (NEIC)
14.	iPg	A	15 05 20.5	Hilders/Rhön, Federal Rep. of Germany.
	iSg	A	05 35.5	50°32.53' N 10°02.31' E
				H = 15 05 00.95 yield 5.5 t (HAN)
				D = 1.0
14.	ePKP	A	22 19 39	Tonga Islands 16.08 S 173.75 W
	eX	A	19 48	H = 22 00 09.1 h = 80 km MB = 5.0
	epPKP	A	20 01	D = 145.25 Az = 354.0 (NEIC)
				h = 80 km
				PKPV A 1.4s 9.3nm
				XV A 1.5 35.2nm
				pPKPV A 1.5 35.2nm
14.	e(Sg)	A	22 42 46.5	
15.	eP	A	02 28 17.5	Southwestern Ryukyu Islands
	e	A	28 24.5	24.19 N 125.06 E
	LmV	B	03 11.6	H = 02 15 41.9 h = 29 km MB = 5.1
	LmH	B	11.7	D = 85.24 Az = 324.1 (NEIC)
				PV A 1.5s 25.1nm M = 5.2
				LmH B 17 0.3/um 4.7
				LmV B 18 0.4/um 4.9
15.	e(P)	A	07 02 37	North Atlantic Ridge 17.10 N 46.79 W
				H = 06 52 56.5 h = normal MB = 4.5
				D = 57.02 Az = 40.3 (NEIC)
15.	ePg	A	13 05 20	Bischhofsheim/Rhön, Federal Rep. of Germany
				50°25.31' N 10°01.62' E

Day	Phase		h m s	Remarks
cont. 15.	eSg	A	13 05 36	Explosion H = 13 05 00.12 yield 4.575 t (HAN)
				D = 1.1
15.	eP	A	16 29 45	Kurile Islands 48.86 N 151.94 E
				H = 16 18 15.2 h = 120 km MB = 4.6
				D = 75.20 Az = 335.1 (NEIC)
				PV A 1.7s 12.1nm M = 4.4
16.	ePKIKP	A	00 47 54.5	Fiji Islands Region 20.86 S 178.75 W
	iPKHKP	A	47 59.5	H = 00 29 16.6 h = 600 km MB = 5.4
	iPKP2	A	48 06	D = 149.17 Az = 347.1 (NEIC)
	epPKP	AB	50 20.5	PKIKPV A 1.5s 15.1nm
	pPKPm	A	50 25	PKHKPV A 1.8 142.0nm
	sPKP	C	51 16	PKP2V A 1.5 60.3nm
				pPKPmV A 1.5 70.4nm
16.	+iP	A	02 30 45	Tibet 32.50 N 95.89 E
	LmH	B	55.5	H = 02 20 23.6 h = normal MB = 5.2
				D = 62.23 Az = 314.3 (NEIC)
				PV A 1.8s 54.0nm M = 5.4
16.	eP1	A	02 52 17	Turkey 38.26 N 43.28 E
	eP2	A	52 20.5	H = 02 46 50.7 h = 40 km MB = 4.9
	eS	B	57 00	D = 25.56 Az = 309.2 (NEIC)
	LmH	B	03 03.6	P2V A 1.2s 16.3nm M = 4.5
	LmV	B	04.9	LmH B 17.5 2.0 µm 4.7
				LmV B 18 1.5 µm 4.7
16.	e	A	03 59 27	
16.	ePKIKP	A	04 41 35	South of Kermadec Islands
	ePKHKP	A	41 51	34.89 S 179.88 E
	ePKP2	A	42 21.5	H = 04 21 40.3 h = 51 km MB = 5.6
	epPKP2	A	42 38	D = 162.09 Az = 335.1 (NEIC)
	LmH	B	05 56.6	PKIKPV A 1.6s 35.7nm
	LmV	B	06 00.4	

Day	Phase		h m s	Remarks
16.	eP	A	05 47 38	Mid-Indian Rise 12.54 S 65.76 E
	e	A	47 44	H = 05 35 37.3 h = normal MB = 5.2
				D = 78.59 Az = 328.2 (NEIC)
				PV A 1.9s 26.5nm M = 4.9
16.	+eP	AB	17 22 13.5	Southern Sumatra 3.64 S 101.05 E
	e(sP)	A	22 27	H = 17 09 05.6 h = 28 km MB = 5.6
	ePP	B	25 54	D = 91.97 Az = 320.4 (NEIC)
	eSKS	B	32 46	PV A 1.6s 49.5nm M = 5.7
	eS	B	33 12	PV B 6 0.7 µm 6.2
	ePS	B	34 30	SH B 12 0.7 µm 5.8
	LmV	B	18 12.2	LmH B 17 1.3 µm 5.4
	LmH	B	14.5	LmV B 16 1.1 µm 5.4
16.	ePKHKP	A	18 28 48.5	Tonga Islands 21.24 S 175.72 W
				H = 18 09 00.6 h = normal MB = 5.3
				D = 150.09 Az = 350.6 (NEIC)
				PKHKPV A 1.3s 35.0nm
17.	LmH	B	05 01.6	Near Coast of Peru 14.43 N 76.20 W
	LmV	B	03.4	H = 04 02 57.2 h = 34 km MB = 5.0 (NEIC)
				D = 99.64
				LmH B 18s 0.3 µm M = 4.9
				LmV B 18 0.2 µm 4.7
17.	+eP	AB	08 40 11	Kamchatka 54.97 N 159.61 E
	eS	B	49 32	H = 08 28 52.0 h = normal MB = 5.3
	LmH	B	09 13.5	D = 71.41 Az = 339.1 (NEIC)
	LmV	B	16.3	PV A 1.8s 60.8nm M = 5.3
				LmH B 16 0.7 µm 5.0
				LmV B 14 0.3 µm 4.7
17.	eP	A	16 19 23	Ionian Sea 37.32 N 20.99 E
				H = 16 15 49 h = 6 km
				D = 14.91 Az = 336 (ISC)
17.				In the interval 17d 17h 27m - 17h 30m the short-period seismographs (type A) had been out of operation.

Day	Phase		h m s	Remarks
17.	LmH	B	17 50.3	Probably to: <u>South of Honshu, Japan</u>
	LmV	B	54.8	31.52 N 139.90 E
				H = 17 14 51.6 h = 42 km MB = 4.7 (NEIC)
				D = 86.31
				LmH B 15s 0.3 μm M = (4.8)
				LmV B 17 0.2 μm (4.7)
18.	LmV	B	00 55.4	<u>Off Coast of Mexico</u> 8.64 N 103.11 W
	LmH	B	55.7	H = 00 01 55.1 h = 33 km MB = 4.5 (NEIC)
				D = 98.52
				LmH B 24s 0.9 μm M = 5.2
				LmV B ·24 1.1 μm 5.3
18.	ePP	A	02 31 34.5	<u>Volcano Islands Region</u> 24.26 N 142.12 E
				H = 02 15 23.6 h = 577 km MB = 4.8 (NEIC)
				D = 93.57
				PPV A 1.1s 16.1nm M = 5.2
18.	ePn	A	04 45 25.5	
	ePg	A	45 44	
	eSg	A	46 45.5	
18.	e	A	13 49 02	<u>Greece-Bulgaria Border Region</u>
	e	A	50 16	41.63 N 23.80 E
	LmH	B	52.8	H = 13 45 48.3 h = normal MB = 4.0
	LmV	B	54.6	D = 12.34 Az = 321.1 (NEIC)
				LmH B 11.5s 0.5 μm M = 3.7
				LmV B 8 0.2 μm
18.	eP	A	14 06 33	<u>Hokkaido, Japan Region</u> 43.17 N 145.28 E
				H = 13 54 43.6 h = 108 km MB = 4.9
				D = 78.22 Az = 331.9 (NEIC)
				PV A 1.3s 8.7nm
18.	LmH	B	22 05.6	<u>Near Coast of Peru</u> 14.42 S 76.34 W
	LmV	B	07.0	H = 21 05 08 h = 13 km MB = 5.1 (NEIC)
				D = 99.72
				LmH B 18.5s 0.2 μm M = 4.7
				LmV B 18 0.2 μm 4.7

Day	Phase		h m s	Remarks
18.	eP	A	22 18 29.5	Hokkaido, Japan Region 42.60 N 144.63 E
	epP	A	18 45	H = 22 06 34.2 h = 66 km MB = 5.2
				D = 78.50 Az = 331.6 (NEIC)
				h̄ = 59 km
				PV A 1.3s 13.1nm M = 4.7
18.	e	A	22 29 36.5	Vancouver Island Region
	LmH	B	23 03.3	49.02 N 128.67 W
	LmV	B	05.0	H = 22 17 43.3 h = 33 km MB = 4.8 (NEIC)
				D = 75.04
				LmH B 16s 0.2 µm M = 4.5
				LmV B 16 0.2 µm 4.5
19.	ePKHKP	A	14 24 53.5	North of New Zealand 20.9 S 173.6 W
	e	A	25 09	H = 14 04 59.3 h = 0 km
				D = 150.05 Az = 353 (ISC)
20.	eP	A	17 28 18	Southern Nevada 37.22 N 116.18 W
				H = 17 16 00.2 h = 0 km MB = 5.0
				D = 81.21 Az = 30.6 (NEIC)
				Nuclear explosion "Diamond Skulls" (USAEC)
				PV A 1.2s 8.1nm M = 4.6
20.	LmH	B	20 37.8	Bismarck Sea 3.43 S 146.29 E
	LmV	B	40.9	H = 19 27 40.3 h = 33 km MB = 5.1 (NEIC)
				D = 119.51
21.	ePKP	A	01 01 18	Easter Island Cordillera
	ePKP2	A	01 25	50.07 N 114.0 W
	e	A	01 29	H = 00 41 37.8 h = 0 km MB = 4.8
	LmV	B	02 02.5	D = 145.97 Az = 68 (ISC)
	LmH	B	02.8	LmH B 18s 0.3 µm M = 5.1
				LmV B 19 0.4 µm 5.2
21.	ePg	A	04 30 16	Natters bei Innsbruck, Austria
	iSn	A	30 45	47.25 N 11.4 E
	eSg	A	31 00	H = 04 29 13 (Innsbruck, VIE)
				D = 3.37 (calculated under suggestion
				of the coordinates given by Innsbruck, VIE)

Day	Phase		h m s	Remarks
cont. 21.				Northern Italy 46.9 N 11.3 E H = 04 29 13 h = 5 km D = 3.80 Az = 4 (ISC)
21.	ePKIKP	A	08 56 46	Fiji Islands Region 21.92 S 176.05 W
	iPKHKP	A	56 51.7	H = 08 37 11.7 h = 118 km MB = 5.5
	ePKP2	A	57 01	D = 150.71 Az = 350.0 (NEIC)
	epPKHKP	A	57 23.5	h = 117 km
				PKHKPV A 1.5s 60.3nm
21.	eP	A	14 52 29	Mindanao, Philippine Islands 9.16 N 125.96 E H = 14 39 03.1 h = 104 km MB = 5.2 D = 97.88 Az = 324.2 (NEIC)
21.	LmV	C	17 03.7	Probably to:
	LmH	C	03.8	Taiwan Region H = 16 08 12.1 (NEIC) or: Szechwan Province, China H = 16 11 32.8 (NEIC)
22.	ePKIKP	A	03 35 30	Southwestern Atlantic Ocean
	ePP	A	36 29	60.70 S 19.40 W
	ePS	B	46 06	H = 03 16 56.1 h = normal MB = 5.2
	eSS	B	52 28	D = 113.76 Az = 21.0 (NEIC)
	LmV	B	04 19.4	PKIKPV A 1.2s 8.1nm
	LmH	B	19.8	LmH B 19 1.0/um M = 5.5
				LmV B 19 1.0/um 5.4
22.	eP	A	05 14 47.5	Crimea Region 44.93 N 36.91 E
	e	A	14 58	H = 05 10 39.5 h = normal MB = 4.6
				D = 17.89 Az = 297.7 (NEIC) PV A 1.6s 16.5nm M = 3.9
22.	ePKIKP	A	06 21 55	New Britain Region 6.20 S 149.85 E
	ePP	A	23 38	H = 06 03 01.1 h = 52 km MB = 4.9
	ePPP	A	26 18	D = 123.70 Az = 329.4 (NEIC)
	LmH	B	07 16.6	LmH B 21s 0.8/um
	LmV	B	16.9	LmV B 20 0.7/um

Day	Phase		h m s	Remarks
22.	eP	A	09 19 53	Kurile Islands 44.08 N 146.17 E
				H = 09 07 58.9 h = 41 km MB = 4.3
				D = 77.72 Az = 332.3 (NEIC)
22.	eP	AB	16 51 11.5	Tibet 31.43 N 91.49 E
	ePP	B	53 30	H = 16 41 04.0 h = normal MB = 5.5
	ePPP	B	54 52	D = 60.28 Az = 313.8 (NEIC)
	eS	B	59 26	PV A 2.0s 51.3nm M = 5.3
	eSS	B	17 03 20	SH B 13 0.7 µm 5.4
	eSSS	B	05 56	LmH B 18.5 6.5 µm 5.8
	LmH	B	17.1	LmV B 18 2.9 µm 5.5
	LmV	B	19.0	
23.	e	A	03 58 39.5	
23.	e	A	07 21 04	
23.	ePKP	A	10 10 00.5	Fiji Islands Region 15.72 S 179.15 W
				H = 09 51 19.6 h = 479 km MB = 4.6
				D = 144.10 Az = 348.3 (NEIC)
				PKPV A 1.2s 12.2nm
23.	ePg	A	10 17 32	Austria 47.6 N 15.8 E
	eSn	A	18 04.5	H = 10 16 04 (VIE)
	e	A	18 10.5	Austria 47.6 N 16.4 E
	eSg	A	18 23	H = 10 16 15 h = 5 km
				D = 4.40 Az = 316 (ISC)
23.	eP	A	11 04 09	Vancouver Island Region
	e	A	04 15.5	50.18 N 129.15 W
	LmH	B	39.2	H = 10 52 33.6 h = normal MB = 4.9
	LmV	B	39.9	D = 74.12 Az = 24.7 (NEIC)
				PV A 1.2s 8.1nm M = 4.6
				LmH B 18 0.4 µm 4.8
				LmV B 16 0.4 µm 4.9
23.	-eP	AB	19 24 43.5	Vancouver Island Region
	iX	AB	25 11	50.15 N 129.27 W

Day	Phase		h m s	Remarks
cont. 23.	ePP	B	19 27 29	H = 19 13 09.0 h = normal MB = 5.9
	ePP	A	27 31.5	D = 74.18 Az = 24.7 (NEIC)
	ePPP	B	29 14	PV A 2.1s 134.0nm M = 5.6
	iS	B	34 22	PV B 8 1.4/um 6.0
	ePS	B	34 44	XV B 6.5 3.3/um
	eSS	B	39 06	PPV A 2.0 85.5nm 4.9
	eSSS	B	42 46	PPV B 9 1.9/um 6.2
	LmH	B	57.7	SH B 11 5.2/um 6.7
	LmV	B	20 00.6	LmH B 17 21.7/um 6.5
				LmV B 16 17.3/um 6.5
23.	eP	A	20 29 08	<u>Vancouver Island Region</u>
				50.12 N 129.34 W
				H = 20 17 32.9 h = normal MB = 4.8 (NEIC)
				D = 74.23
				PV A 1.1s 10.1nm M = 4.7
23.	eP	A	21 54 43	<u>Vancouver Island Region</u>
				49.94 N 128.71 W
				H = 21 43 07.2 h = normal MB = 5.0
				D = 74.22 Az = 25.0 (NEIC)
				PV A 2.2s 27.3nm M = 4.9
24.	e	A	10 27 33.5	<u>Turkey</u> 39.40 N 40.14 E
				H = 10 22 22.72 h = 33 km MB = 4.4
				D = 22.94 Az = 308 (NEIC)
24.	eP	A	11 03 15	<u>Near Coast of Venezuela</u>
	e	A	03 30	10.80 N 65.60 W
	eS	B	12 40	H = 10 51 42.9 h = normal MB = 4.9
	LmH	B	27.6	D = 73.62 Az = 40.3 (NEIC)
	LmV	B	34.4	
24.	eP	A	15 07 12.5	<u>Kashmir-Tibet Border Region</u>
				35.81 N 80.57 E
				H = 14 58 13.9 h = normal MB = 4.8
				D = 50.58 Az = 309.7 (NEIC)

Day	Phase		h m s	Remarks
24.	eP	A	19 14 31.5	Fox Islands, Aleutian Is.
				53.18 N 166.72 W
				H = 19 02 45.2 h = 46 km MB = 4.6
				D = 76.54 Az = 1.1 (NEIC)
25.	eP1	A	01 59 20.5	Greece 38.73 N 21.35 E
	eP2	A	59 25	H = 01 56 07.3 h = 45 km MB = 4.5
	LmH	B	02 05.4	D = 13.76 Az = 333.1 (NEIC)
	LmV	B	05.4	P1V A 1.0s 7.9nm M = 4.4
				P2V A 1.0 19.7nm 4.8
25.	ePKHKP	A	07 53 30	Fiji Region 17.4 S 176.2 W
				H = 07 33 45.9 h = 0 km
				D = 146.27 Az = 351 (ISC)
25.	ePKIKP	A	08 49 26	Kermadec Islands Region
	ePKHKP	A	49 36	27.23 S 176.51 W
	ePKP2	A	49 54	H = 08 29 38.7 h = 62 km MB = 5.1
	e	A	50 00	D = 155.81 Az = 347.3 (NEIC)
	ePP	A	53 29	PKIKPV A 1.5s 10.0nm
				PPV A 1.5 15.1nm M = 4.9
25.	ePKIKP	AB	09 39 53	Fiji Islands Region 21.45 S 176.41 W
	ePKHKP	A	39 59	H = 09 20 28.3 h = 180 km MB = 5.4
	eiPKP2	A	40 05.5	D = 150.19 Az = 349.7 (NEIC)
	epPKP	A	40 48	PKIKPV A 1.5s 42.7nm
	e(SSP)	B	10 03 12	PKHKPV A 2.5 554.0nm
	ePSS	B	03 40	PKP2V A 1.2 146.3nm
	LmH	B	10.4	LmH B 17.5 0.2 µm
	LmV	B	10.8	LmV B 18 0.3 µm
25.	ePKP	A	10 02 52.5	Fiji Region 17.5 S 177.2 W
				H = 09 43 28 h = 167 km MB = 4.4
				D = 146.20 Az = 350 (ISC)
26.	LmH	C	08 20.0	Probably to:
	LmV	C	23.0	South of Bali Island 10.65 S 114.02 E
				H = 07 13 30.8 h = 50 km MB = 5.2 (NEIC)
				D = 106.02

Day	Phase		h m s	Remarks
26.	eP	A	18 56 29.5	Northern Yukon Territory, Canada
	e	A	56 41.5	66.54 N 135.99 W
				H = 18 46 20.6 h = normal MB = 4.8
				D = 60.57 Az = 23.1 (NEIC)
				PV A 1.0s 11.8nm M = 5.0
27.	ePn	A	00 10 30.5	Central Italy 43.7 N 13.5 E
	eSn	A	11 44	H = 00 08 48 (BCIS)
	eSg	A	12 37	D = 7.10
27.	eP	AB	00 32 40	Kurile Islands Region 49.98 N 159.06 E
	eX	A	32 46	H = 00 20 54.6 h = normal MB = 5.1
	ePcP	A	32 55	D = 75.96 Az = 339.3 (NEIC)
	eS	B	42 15	PV A 1.0s 19.7nm M = 5.1
	LmH	B	01 06.0	XV A 1.1 34.3nm
	LmV	B	11.0	PcPV A 1.7 69.7nm
				LmH B 17 0.4/um 4.8
				LmV B 19 0.3/um
27.	ePn	A	01 03 07.5	Central Italy 43.7 N 13.5 E
	eSn	A	04 19.5	H = 01 01 23 (BCIS)
	eSg	A	04 12	D = 7.10
27.	ePP	A	16 57 39	Ryukyu Islands Region 25.43 N 130.54 E
	LmH	B	17 30.6	H = 16 41 30.2 h = 33 km MB = 5.1 (NEIC)
	LmV	B	41.3	D = 87.08
				LmH B 18s 1.4/um M = 5.4
				LmV B 16 0.5/um 5.0
27.	e	A	20 16 40	Near Coast of Guatemala
				13.32 N 90.13 W
				H = 20 03 52.9 h = 78 km MB = 4.5
				D = 87.12 Az = 38.6 (NEIC)
28.	LmH	B	07 57.2	Taiwan Region 24.25 N 122.46 E
	LmV	B	57.3	H = 07 03 10.3 h = 65 km MB = 5.1 (NEIC)
				D = 83.79
				LmH B 15s 0.5/um
				LmV B 14 0.6/um

Day	Phase		h m s	Remarks
28.	eP	A	09 00 47.8	<u>Near Islands, Aleutian Is.</u>
	epP	A	00 58.5	52.57 N 173.21 E
	esP	A	01 02.5	H = 08 49 03.1 h = 46 km MB = 5.3
	e	B	10 55	D = 75.99 Az = 348.0 (NEIC)
	ePS	C	11 15	h = 41 km
	LmH	B	32.5	PV A 1.3s 21.8nm M = 4.9
	LmV	B	32.5	LmH B 24 0.6/um 4.8
				LmV B 24 0.7/um 5.0
28.	e	A	21 02 27	<u>North of New Zealand</u> 18.0 S 175.0 W
	e	A	02 47	H = 20 42 45 h = 0 km MB = 4.7
				D = 146.96 Az = 352 (ISC)
29.	+eP	AB	05 18 07	<u>Northern Celebes</u> 0.07 N 123.48 E
	epP	A	18 46	H = 05 04 22.7 h = 161 km MB = 5.8
	ePP	AB	22 25	D = 103.66 Az = 322.5 (NEIC)
	eSKS	B	28 35	h = 155 km
	eS	B	29 40	PV A 1.2s 20.3nm M = 5.9
	ePKKP	A	34 01	
29.	LmV	B	11 34.4	LmH B 11s 0.3/um
	LmH	B	34.8	
29.	ePKHKP	A	11 53 00	<u>North of New Zealand</u> 20.69 S 173.6 W
	e	A	53 11	H = 11 33 04.8 h = 0 km
				D = 149.82 Az = 353 (ISC)
29.	LmV	B	12 13.3	<u>South of Bali Island</u> 10.72 S 114.00 E
				H = 10 59 35.1 h = 33 km MB = 5.1 (NEIC)
				D = 106.06
				LmV B 16s 0.3/um M = 4.9
29.	eP	A	21 19 02	<u>Kurile Islands</u> 49.17 N 156.16 E
	e	A	19 19.5	H = 21 07 16.6 h = normal MB = 4.8
				D = 76.03 Az = 337.6 (NEIC)
				PV A 1.2s 20.3nm M = 5.0

Day	Phase		h m s	Remarks
29.	LmH	B	23 44.6	Bismarck Sea 3.34 S 146.45 E
	LmV	B	47.4	H = 22 34 33.3 h = 33 km MB = 4.8 (NEIC)
				D = 119.53
				LmH B 20s 1.5 μm M = 5.6
				LmV B 18 0.9 μm 5.4
30.	LmH	B	01 38.2	Aegean Sea 39.89 N 24.21 E
	LmV	B	39.2	H = 01 30 09.0 h = 33 km MB = 4.4 (NEIC)
				D = 13.92
				LmH B 9.5s 0.4 μm M = 3.8
				LmV B 9 0.3 μm
30.	+eiP	AB	03 12 50.5	Kurile Islands 49.22 N 156.15 E
	epP	A	13 04.5	H = 03 01 06.6 h = 45 km MB = 5.1
	LmH	B	50.6	D = 75.99 Az = 337.6 (NEIC)
	LmV	B	52.5	h = 50 km
				PV A 1.1s 96.7nm M = 5.7
				pPV A 1.9 53.0nm
				LmH B 15 0.7 μm 5.1
				LmV B 17 1.1 μm 5.3
30.	eP	A	05 01 17.5	Rat Islands, Aleutian Is.
				51.81 N 176.46 E
				H = 04 49 27.5 h = 46 km MB = 5.1
				D = 77.11 Az = 350.2 (NEIC)
				PV A 1.0s 15.8nm M = 5.0
30.	eP	A	05 11 56	Southern Sumatra 5.58 S 102.82 E
				H = 04 58 35.9 h = 46 km MB = 5.3
				D = 95.04 Az = 320.3 (NEIC)
				PV A 1.1s 12.1nm M = 5.2
30.	e	A	16 12 51	Taiwan Region 21.17 N 121.29 E
				H = 16 00 02.8 h = normal MB = 4.9
				D = 85.59 Az = 323.0 (NEIC)

Day	Phase		h m s	Remarks
30.	ePKIKP	A	17 59 07	New Britain Region 4.61 S 152.33 E
	eX	A	59 24	H = 17 40 22.9 h = 44 km MB = 5.3
				D = 123.56 Az = 331.1 (NEIC)
				XV A 1.5s 20.1nm
30.	LmH	B	18 22.7	LmH B 17s 1.1/um
	LmV	B	22.7	
30.	LmH	B	18 56.0	LmH B 19s 1.3/um
	LmV	B	56.0	LmV B 21 1.6/um
30.	ePKIKP	A	20 32 55.5	New Britain Region 4.64 S 152.30 E
	ePP	A	34 38	H = 20 13 56.1 h = 16 km MB = 5.5
				D = 123.57 Az = 331.0 (NEIC)
				PKIKPV A 1.2s 18.3nm
				PPV A 1.9 26.5nm M = 5.5
30.	+eP1	AB	21 56 24	Southeastern Alaska 56.82 N 135.69 W
	iP2	AB	56 29	H = 21 45 14.1 h = 25 km MB = 6.5
	iP3	A	56 32.5	D = 69.54 Az = 21.5 (NEIC)
	iP4	AB	56 36	P1V A 2.5s 430.0nm M = 6.1
	ePP	B	59 02	P2V A 1.5 266.0nm 6.1
	eiS	B	22 05 40	P3V A 1.6 434.0nm 6.3
	eiSS	B	10 08	P4V A 2.8 2420.0nm 6.8
	eiSSS	B	14 47	P4V B 9 25.5/um 7.3
	eP'P'	A	24 30	PPV B 6.5 20.7/um 7.5
	LmH	B	29.0	P'P'V A 1.8 179.0nm
	LmV	B	29.0	SH B 16 74.6/um 7.4
				LmH B 16 524.0/um 7.9
				LmV B 16 833.0/um 8.1
30.	ePKIKP	A	23 35 50.5	Banda Sea 5.68 S 130.47 E
	e	A	36 17.5	H = 23 17 19.5 h = 57 km MB = 5.9
	ePP	A	36 38	D = 112.44 Az = 322.9 (NEIC)
				PKIKPV A 2.0s 42.7nm
				PPV A 2.0 98.4nm M = 6.1

Day	Phase		h m s	Remarks
31.	+eP	AB	03 41 17	South of Honshu, Japan 30.68 N 141.65 E
	+ePP	AB	44 43	H = 03 28 30.2 h = 36 km MB = 5.7
	eS	B	51 58	D = 87.79 Az = 330.8 (NEIC)
	LmH	B	04 24.6	PV A 2.0s 137.0nm M = 5.9
	LmV	B	25.3	PV B 11 0.7/um 5.9
				PPV A 1.5 105.6nm 6.1
				PPV B 11 0.9/um 6.1
				SH B 11 1.6/um 6.0
				LmH B 16 2.6/um 5.7
				LmV B 17 2.7/um 5.8
31.	eP	A	06 51 44	Near East Coast of Kamchatka
				56.22 N 162.88 E
				H = 06 40 27.8 h = normal MB = 4.8
				D = 70.87 Az = 341.1 (NEIC)
				PV A 1.3s 10.9nm M = 4.8
31.	ePKHKP	A	08 13 29	Fiji Islands Region 20.23 S 178.69 W
				H = 07 54 50.0 h = 633 km MB = 4.5
				D = 148.57 Az = 347.4 (NEIC)
				ePKHKPV A 1.1s 14.1nm
31.	ePKP	A	11 15 20.5	New Hebrides Islands 18.41 S 168.87 E
				H = 10 56 14.2 h = 222 km MB = 4.7
				D = 143.08 Az = 335.8 (NEIC)
				PKPV A 1.2s 14.2nm
31.	eX	A	14 52 14	XV A 2.5s 76.7nm
	e	A	53 57	LmH B 16 0.4/um
	e	B	58 28	LmV B 16 0.3/um
	e	B	15 01 10	
	LmH	B	31.5	
	LmV	B	37.3	
31.	eP	AB	16 08 56.5	South of Honshu, Japan 30.74 N 141.57 E
	e	A	09 07	H = 15 56 08.5 h = normal MB=5.4 MS=5.6
	ePP	AB	12 22	D = 87.70 Az = 330.8 (NEIC)
	eS	B	19 36	PV A 2.4s 138.0nm M = 5.8

Day	Phase		h m s	Remarks
cont. 31.	eSS	C	16 25 25	PV B 10s 0.7 um M = 5.9
	LmH	B	50.2	PPV A 2.5 163.4nm 6.0
	LmV	B	52.7	PPV B 10 0.7 um 6.0
				SH B 10 1.5 um 6.0
				LmH B 15.5 3.3 um 5.9
				LmV B 17 2.6 um 5.7
31.	eiP	AB	23 59 49	Turkey 36.04 N 31.59 E
	eiX	A	59 52.	H = 23 55 11.66 h = 33 km MB = 4.5
	LmH	B	24 07.0	D = 20.50 Az = 321.6 (NEIC)
	LmV	B	09.0	PV A 1.1s 44.4nm M = 4.7
				LmV B 15 0.3 um 4.0

Day	Phase		h m s	Remarks
1.	eP	A	19 20 33	Molucca Passage 2.32 N 127.44 E
	epP	A	20 55	H = 19 06 33.8 h = 96 km MB = 5.9
	esP	A	21 05	D = 104.25 Az = 323.8 (NEIC)
	ePP	AB	24 53	PV A 1.6s 22.0nm M = 5.9
	epPP	B	25 21	PPV A 1.8 74.4 6.0
	eSKS	B	31 00	PPV B 11 0.87/um 6.3
	e(SKKKS)	BC	31 44	LmH B 17 4.4/um
	esS	B	32 45	LmV B 18 2.9/um
	ePS	B	35 00	
	eSS	B	40 00	
	LmH	B	20 06.6	
	LmV	B	10.6	
1.	eP	A	22 18 55	Norwegian Sea 72.4 N 6.3 E
				H = 22 13 59 h = 0 km
				D = 21.94 Az = 171 (ISC)
				PV A 1.3s 13.1nm M = 4.2
2.	ePKHKP	AB	01 30 43	Tonga Islands 18.89 S 173.34 W
	epPKP	AB	31 05	H = 01 11 02.0 h = 60 km MB = 4.9
	LmH	B	02 47.5	D = 148.07 Az = 354.0 (NEIC)
	LmV	B	47.5	PV A 1.9s 56.8nm
				LmH B 18 0.26/um M = 5.0
				LmV B 18 0.25/um (5.0)
2.	LmH	B	10 57.5	Solomon Islands 11.34 S 161.80 E
	LmV	B	11 03.8	H = 09 42 58.2 h = 59 km
				MB = 4.8 (NEIC)
				D = 133.82
				11.66 S 162.01 E
				H = 09 52 30.4 h = 44 km MB=4.7 (NEIC)
				D = 134.19
				LmH B 21s 0.53/um M = 5.2
				LmV B 18 0.76/um 5.4

Day	Phase		h m s	Remarks
2.	eP	A	11 24 27.5	Northwest of Kurile Islands
				47.43 N 147.25 E
				H = 11 13 22.2 h = 351 km MB = 5.0
				D = 75.10 Az = 332.6 (NEIC)
				PV A 1.0s 21.6nm M = 4.8
2.	eP	A	15 15 43	Turkey 37.71 N 32.56 E
	LmH	C	22.6	H = 15 11 09 h = 0 km MB = 4.3
	LmV	C	24.9	D = 19.73 Az = 318 (ISC)
				PV A 1.4s 14.0nm M = 3.9
				LmH C 15.5 0.31/um 3.7
				LmV C 9 0.38/um 4.2
2.	ePKIKP	AB	20 08 24	South of Fiji Islands 26.18 S 177.82 W
	ePKHKP	AB	08 34	H = 19 48 57.3 h = 224 km MB = 5.3
	ePKP2	AB	08 48.5	D = 154.52 Az = 346.0 (NEIC)
	e(pPKP)	A	09 19	h = (215) km
	e	A	09 34.5	PKIKPV A 1.4s 41.8nm
				PKHKPV A 1.3 44.0nm
				PKP2V A 1.4 65.0nm
2.	eP	A	21 40 45	Southern Iran 28.22 N 56.95 E
	epP	A	40 52	H = 21 33 06.3 h = 38 km MB = 4.7
				D = 40.74 Az = 316.1 (NEIC)
				h = 32 km
				PV A 1.3s 13.1nm M = 4.5
				pPV A 1.2 16.2nm
				BCIS gives:
				H = 21 33 20 28.6 N 56.4 E
2.	+iP	AB	21 50 08.1	Near East Coast of Kamchatka
	ePP	B	52 46	56.05 N 163.24 E
	eS	B	59 30	H = 21 38 50.4 h = normal MB=5.6 MS=5.7
	ePS	B	22 00 05	D = 71.09 Az = 341.3 (NEIC)
	eSS	B	04 05	PV A 1.8s 169.0nm M = 5.6
	LmH	B	27.3	PV B 7 0.65/um 5.8

Day	Phase		h m s	Remarks
cont. 2.	LmV	B	22 27.3	PPV B 12s 0.54 /um M = 5.6
				SH B 15 0.60 /um 5.5
				LmH B 18 2.5 /um 5.5
				LmV B 15 3.1 /um 5.7
2.	eP	A	23 11 09	Southern Iran 28.08 N 56.85 E
	epP	A	11 15	H = 23 03 28.90 h = 31 km MB = 5.0
				D = 40.78 Az = 316.2 (NEIC)
				h = 27 km
				PV A 1.2s 18.3nm M = 4.7
				BCIS gives:
				H = 23 03 37 28.2 N 56.7 E
3.	eP	A	02 08 55.5	Turkey 37.83 N 32.51 E
				H = 02 04 26.0 h = 28 km MB = 4.3
				D = 19.62 Az = 317.4 (NEIC)
				PV A 1.2s 14.2nm M = 4.1
3.	eP	A	02 37 15.5	Kurile Islands 46.86 N 152.63 E
				H = 02 25 23.1 h = normal MB = 4.5
				D = 77.21 Az = 335.8 (NEIC)
				PV A 1.3s 10.9nm M = 4.7
3.	+eP	AC	04 52 51.5	Andreanof Islands, Aleutian Is.
	Pm	A	52 57.5	51.20 N 178.12 W
	e	AB	53 10	H = 04 40 54.9 h = 49 km MB = 5.8
	eS	B	05 02 44	D = 78.20 Az = 353.7 (NEIC)
	eiSKS	B	03 12	PV A 1.5s 93.0nm M = 5.5
	eSS	B	08 00	PV C 12 3.57 /um 6.2
	LmH	B	32.8	PmV A 1.4 231.0nm 6.0
	LmV	B	33.8	LmH B 18.5 9.8 /um 6.2
				LmV B 18.5 8.1 /um 6.1
3.	eP	A	05 47 12.5	Andreanof Islands, Aleutian Is.
				51.21 N 177.87 W
				H = 05 35 16.4 h = 49 km MB = 4.8
				D = 78.20 Az = 353.8 (NEIC)
				PV A 1.3s 26.2nm M = 5.1

Day	Phase		h m s	Remarks
3.	+iP	AB	07 11 42.3	Andreanof Islands, Aleutian Is. 51.21 N 178.15 W H = 06 59 45.9 h = 45 km MB = 5.5 MS=5.4 D = 78.18 Az = 353.7 (NEIC) PV A 1.3s 109.2nm M = 5.7
3.	eP	AB	07 15 10.5	Andreanof Islands, Aleutian Is. 51.20 N 177.96 W H = 07 03 14.2 h = 48 km MB = 5.4 D = 78.21 Az = 353.8 (NEIC) PV A 1.1s 52.4nm M = 5.5
3.	+eP	A	12 47 43.5	Kamchatka 59.52 N 163.19 E H = 12 36 46.7 h = normal MB = 5.3 D = 67.79 Az = 340.9 (NEIC) PV A 1.4s 46.5nm M = 5.4
3.	e	A	18 45 00	
3.	e	A	19 52 03	
	e	A	52 28	
3.	eP	A	21 43 56	Turkey 37.71 N 32.71 E
	e	A	44 16	H = 21 39 26.0 h = 41 km MB = 4.5
	e	A	44 27	D = 19.82 Az = 317.5 (NEIC)
	eS	B	47 40	PV A 1.3s 30.6nm M = 4.5
	LmH	B	50.9	LmH B 15 0.8/um 4.2 LmV : defect
3.	eP	A	22 55 22.5	Southern Iran 28.21 N 56.99 E H = 22 47 45.5 h = 62 km MB = 4.8 D = 40.77 Az = 316.1 (NEIC)
3.	e	A	23 09 40	
	e	A	11 10	

Day	Phase		h m s	Remarks
4.	eP	A	05 34 31	Turkey 37.90 N 32.86 E
				H = 05 30 00.4 h = normal MB = 4.3
				D = 19.75 Az = 317.0 (NEIC)
				PV A 1.4s 23.2nm M = 4.3
4.	eP	A	08 10 26.5	Central Mid-Atlantic Ridge
				8.81 N 39.98 W
				H = 08 00 23.8 h = normal MB = 4.7
				D = 59.46 Az = 35.4 (NEIC)
4.	eP	A	09 59 24	Southeastern Alaska 56.22 N 135.53 W
	epP	A	59 36	H = 09 48 11.0 h = 18 km MB = 5.1
				D = 70.07 Az = 21.6 (NEIC)
				PV A 1.3s 17.5nm M = 5.0
				pPV A 1.5 23.8nm
4.	+eP	A	11 22 09.5	Iceland Region 61.67 N 26.88 W
				H = 11 16 58.7 h = normal MB = 4.3
				D = 23.73 Az = 100.0 (NEIC)
				PV A 1.8s 84.5nm M = 5.0
4.	-eP	A	11 49 21	Southeastern Alaska 56.21 N 135.34 W
	e(pP)	A	49 27	H = 11 38 08.3 h = 20 km MB = 5.6
				D = 70.05 Az = 21.7 (NEIC)
				PV A 1.4s 55.8nm M = 5.5
4.	eP	A	17 04 35	Andreanof Islands, Aleutian Is.
				51.17 N 177.99 W
				H = 16 52 39.1 h = 46 km MB = 4.2
				D = 78.23 Az = 353.8 (NEIC)
				PV A 1.8s 33.8nm M = 5.1
4.	+eP1	AB	18 02 56	Kurile Islands 49.16 N 156.07 E
	+iP2	A	03 02.5	H = 17 51 12.9 h = 54 km MB = 5.7
	ePP	B	05 52	D = 76.02 Az = 337.6 (NEIC)
	ePPP	B	07 40	P1V A 1.7s 254.2nm M = 5.9
	eS	B	12 35	P2V A 2.2 654.0nm 6.2
	ePS	B	13 27	PV B 10 3.02/um 6.2

Day	Phase		h m s	Remarks
cont. 4.	eSS	B	18 17 35	PPV B 16s 1.17 µm M = 5.8
	e	B	21 30	SH B 17 2.56 µm 6.1
	LmH	B	41.5	LmH B 16 28.5 µm 6.7
	LmV	B	41.5	LmV B 16.5 30.0 µm 6.7
4.	eP	A	18 37 57	<u>Kurile Islands</u> 48.96 N 156.21 E H = 18 26 11.3 h = normal MB = 4.4 D = 76.23 Az = 337.7 (NEIC) PV A 0.8s 9.6nm M = 4.9
4.	e(SKP)	A	20 24 23	<u>Solomon Islands</u> 11.24 S 162.05 E
	e	A	24 29	H = 20 01 37.7 h = 34 km MB = 5.2 (NEIC) D = 133.84 SKPV A 1.8s 20.3nm (disturbance ?)
4.	ePKIKP	AB	22 49 52.5	<u>Solomon Islands</u> 11.11 S 162.03 E
	ePP	AB	52 17	H = 22 30 36.3 h = 36 km MB = 5.5
	eSKP	AB	53 23	D = 133.71 Az = 334.2 (NEIC)
	eSS	B	23 10 00	PKIKPV A 2.0s 59.9nm
	LmV	B	51.5	LmH B 19.5 3.5 µm M = 6.1
	LmH	B	51.9	LmV B 20 4.3 µm 6.2
5.	ePg	A	01 15 45	<u>Switzerland</u> 46.1 N 7.6 E
	eSg	A	16 57	H = 01 14 02 (BCIS) D = 5.3
5.	ePKIKP	A	04 02 33	<u>Fiji Islands Region</u> 17.94 S 178.57 W
	ePKHKP	A	02 36	H = 03 44 01.8 h = 619 km MB = 5.2 D = 146.37 Az = 348.3 (NEIC) PKIKPV A 1.2s 18.3nm PKHKPV A 1.4 51.2nm
5.	eP	A	04 28 17	<u>Kurile Islands</u> 49.25 N 156.0 E H = 04 16 27.8 h = 0 km MB = 4.5 D = 75.92 Az = 338 (ISC) PV A 0.9s 9.7nm M = 4.9

Day	Phase		h m s	Remarks
5.	eP	A	05 03 48	Kurile Islands 49.02 N 156.31 E
				H = 04 52 02.7 h = normal MB = 4.7
				D = 76.20 Az = 337.7 (NEIC)
				PV A 1.1s 10.1nm M = 4.7
5.	eP	A	05 50 47	Solomon Islands 11.28 S 162.15 E
	e	A	51 02	H = 05 31 29 h = 10 km MB = 5.0
				D = 133.91 Az = 334 (ISC)
5.	+eP	A	05 58 13.5	Kurile Islands 49.04 N 156.22 E
	ePcP	A	58 21	H = 05 46 29.2 h = 46 km MB = 4.9
	eSP	A	58 34	D = 76.17 Az = 337.7 (NEIC)
	LmH	B	06 36.0	PV A 1.4s 41.8nm M = 5.2
	LmV	B	36.0	LmH B 17 0.50/um 4.9
				LmV B 17 0.49/um 4.9
5.	ePKIKP	A	06 50 16	Solomon Islands 11.18 S 162.13 E
	eX	A	53 30	H = 06 30 58.8 h = normal MB = 5.4
	eSKP	A	53 43.5	D = 133.82 Az = 334.3 (NEIC)
	e	A	54 33	PKIKPV A 2.5s 61.5nm
	eSS	B	07 10 24	XV A 1.5 15.1nm
	LmH	B	55.8	SKPV A 1.6 33.0nm
	LmV	B	55.8	LmH B 18 2.27/um M = 5.9
				LmV B 18 2.16/um 5.9
5.	LmH	B	06 52.3	LmH B 19s 0.65/um
	LmV	B	52.4	LmV B 19 0.68/um
5.	ePKP2	A	14 10 48	Kermadec Islands 29.76 S 177.21 W
	epPKP2	A	10 54.5	H = 13 50 24.3 h = 63 km MB = 4.5 (NEIC)
	LmH	B	15 32	D = 158.11
	LmV	B	33	LmH B 18s 0.57/um M = 5.3
				LmV B 17 0.55/um 5.4
6.	eP	A	00 56 46	Black Sea 44.68 N 32.64 E
	e	A	01 00 41.5	H = 00 53 12.3 h = normal MB = 4.5
				D = 15.34 Az = 300.3 (NEIC)
				PV A 0.7s 9.6nm M = 4.1

Day	Phase		h m s	Remarks
6.	eP	AB	01 21 10.5	Southern Iran 25.07 N 61.23 E
	epP	A	21 18	H = 01 12 50.4 h = 33 km MB = 5.5 (NEIC)
	eS	B	27 52	D = 45.65
	eSS	C	30 55	PV A 1.3s 37.1nm M = 5.0
	LmH	B	44.5	LmH B 20 0.9 µm 4.7
	LmV	B	49.7	LmV B 14 0.8 µm 4.9
6.	e	A	07 07 44.5	Western Iran 31.73 N 49.98 E
				H = 07 00 56 h = 26 km MB = 5.01 (ISC)
				D = 34
6.	ePKIKP	AB	07 34 00	Solomon Islands 11.09 S 162.03 E
	esPKP	A	34 03.5	H = 07 14 46.6 h = 56 km MB = 5.9
	ePP	AB	36 30	D = 133.70 Az = 334.2 (NEIC)
	ePKS	B	37 28	PKIKPV A (2.5)s 76.7nm
	eSS	B	54 05	PKIKPV B 5 0.66 µm
	LmH	B	08 39.5	sPKPV A 3.0 197.0nm
	LmV	B	39.5	PPV A 3.0 290.0nm M = 5.8
				PPV B 5 1.48 µm 6.3
				LmH B 18.5 6.9 µm 6.4
				LmV B 18.5 6.2 µm 6.4
6.	ePKHKP	A	08 12 55	Loyalty Islands Region 22.25 S 174.06 E
				H = 07 53 13.9 h = 60 km MB = 5.0
				D = 148.51 Az = 338.4 (NEIC)
				PKHKPV A 1.2s 12.2nm
6.	eP	A	10 11 17	Dodecanese Islands 35.77 N 27.19 E
				H = 10 06 57.1 h = 8 km
				D = 18.60 Az = 328 (ISC)
				PV A 1.0s 11.8nm M = 4.0
6.	ePKIKP	A	10 34 51.5	Solomon Islands 11.22 S 162.01 E
	e	A	35 24	H = 10 15 40.4 h = 67 km MB = 5.1 (NEIC)
	ePP	A	37 24	D = 133.80
	eSKP	A	38 22	

Day	Phase		h m s	Remarks
6.	eP	A	11 19 16.5	Kurile Islands 49.02 N 156.41 E H = 11 07 29.8 h = normal MB = 4.2 D = 76.23 Az = 337.8 (NEIC)
6.	eP	A	15 45 54	Arabian Sea 14.64 N 55.84 E H = 15 36 55.4 h = normal MB = 4.7 D = 50.56 Az = 324.9 (NEIC) PV A 2.0s 34.2nm M = 5.0
6.	eP	A	16 02 51.5 H	Arabian Sea 14.71 N 55.62 E H = 15 53 55.6 h = normal MB = 5.2 D = 50.38 Az = 325.0 (NEIC) PV A 3.0s 105.0nm M = 5.3
6.	eP LmH LmV	A B B	16 15 05 38.5 38.5	Arabian Sea 14.76 N 55.59 E H = 16 06 09.4 h = normal MB = 5.3 D = 50.33 Az = 324.9 (NEIC) PV A 2.2s 38.1nm M = 5.0 LmH B 20 0.50/um 4.5 LmV B 20 0.52/um 4.6
6.	eP e	A A	16 53 35.5 53 45	Arabian Sea 14.41 N 55.61 E H = 16 44 37.6 h = normal MB = 5.1 D = 50.62 Az = 325.1 (NEIC) PV A 2.2s 21.8nm M = 4.8
7.	e	A	01 22 39	Poland, Rock-burst
7.	e e e	A A A	01 38 58 40 07.5 41 04	Southern Italy 41 1/4 N 15.8 E H = 01 36 14.4 (BCIS) D = 9.84
7.	+ePKIKP1 eiPKIKP2 eiSPKP2 eSKSP	AB A AB A	09 43 51 43 54.5 44 04 57 29	Samoa Islands Region 16.65 S 172.10 W H = 09 24 14.7 h = 33 km MB = 5.8 D = 145.95 Az = 355.8 (NEIC) PKIKP1V A 1.4s 270.0nm

Day	Phase		h m s	Remarks
cont. 7.	LmH	B	10 52	PKIKPV B 5.5s 4.08 µm
	LmV	B	53.7	PKIKP2V A 1.4 354.0nm
				SPKP2V A 1.3 655.0nm
				LmH B 19 2.0 µm M = 5.9
				LmV B 18.5 2.9 µm 6.1
7.	+ePKP	A	23 35 51.5	New Hebrides Islands 19.08 S 169.14 E
	epPKP	A	36 21	H = 23 16 39.0 h = 184 km MB = 5.5
	epPKP2	A	36 37	D = 143.79 Az = 335.7 (NEIC)
				h = 115 km
				PKPV A 1.3s 152.5nm
8.	eP	A	09 57 42	Kurile Islands Region 48.07 N 157.11 E
	eX	A	57 48.5	H = 09 45 49.3 h = normal MB = 5.1
	LmH	B	10 38	D = 77.29 Az = 338.3 (NEIC)
	LmV	B	38	PV A 1.2s 10.2nm M = 4.7
				XV A 1.1 16.1nm
				LmH B 15 0.23 µm 4.6
				LmV B 15 0.29 µm 4.8
8.	ePKP	A	10 08 41	New Zealand 20.61 S 174.5 W
				H = 09 48 47.4 h = 0 km MB = 4.7
				D = 149.64 Az = 352 (ISC)
8.	iPg	A	11 55 23.8	
	iSg	A	55 44.5	
8.	ePg	A	13 20 38	Northern Italy 44.24 N 8.01 E
	eSn	A	21 20	H = 13 18 24.8 h = 0 km (ISC)
	eSg	A	22 17	D = ca. 6.85
8.	ePKP2	A	18 03 09	South of Fiji Islands 26.39 S 179.25 E
				H = 17 43 53.4 h = 533 km MB = 4.4 (NEIC)
				D = 153.99
8.	eP	AB	19 17 51.5	Southern Iran 25.03 N 61.13 E
	esP	A	18 05	H = 19 09 33.5 h = 41 km MB = 5.5
	eS	C	24 35	D = 45.62 Az = 317.3 (NEIC)

Day	Phase		h m s	Remarks
cont. 8.	e(SS)	C	19 27 45	PV A 1.3s 43.7nm M = 5.2
	LmH	B	41.4	LmH B 19 0.7/um 4.6
	LmV	B	46.3	LmV B 14 0.6/um 4.7
8.	eP	A	24 09 22	Andreanof Islands, Aleutian Is.
				51.26 N 177.95 W
				H = 23 57 27.7 h = 67 km MB = 4.9
				D = 78.15 Az = 353.8 (NEIC)
				PV A 1.4s 23.3nm M = 4.9
9.	ePKHKP	A	00 57 15	South of Fiji Islands 23.25 S 178.33 W
	ePKP2	A	57 27.5	H = 00 37 57.9 h = 323 km MB = 5.0
				D = 151.58 Az = 346.6 (NEIC)
				PKHKPV A 1.1s 16.1nm
9.	eP	A	04 05 07	Greece - Albania Border Region
	LmH	B	11.3	39.48 N 20.68 E
	LmV	B	11.3	H = 04 02 05.0 h = 42 km MB = 4.5
				D = 12.86 Az = 333.2 (NEIC)
				PV A 0.7s 19.1nm M = 5.2
				LmH B 10 0.4/um 3.7
				LmV B 11 0.4/um
9.	eP1	A	04 29 41	Nicobar Islands Region 7.93 N 94.09 E
	eP2	AB	29 45.5	H = 04 17 37.3 h = normal MB = 5.1
	eS	B	39 40	D = 79.15 Az = 320.0 (NEIC)
	LmV	C	05 01.6	P1V A 1.9s 53.0nm M = 5.2
	LmH	C	04.3	P2V A 1.3 65.5nm 5.5
				LmH C 26 2.36/um 5.4
				LmV C 36 3.1/um 5.4
9.	eP	A	08 27 40	Chiapas, Mexico 17.08 N 94.84 W
				H = 08 15 05.4 h = 124 km MB = 5.0
				D = 86.99 Az = 37.7 (NEIC)
				PV A 1.6s 19.2nm M = 4.8

Day	Phase		h m s	Remarks
9.	eP	A	13 00 44.5	Nicobar Islands Region 7.97 N 94.15 E
	LmH	C	42	H = 12 48 41.4 h = normal MB = 5.0
	LmV	C	42	D = 79.16 Az = 320.0 (NEIC)
				PV A 1.4s 18.6nm M = 4.9
				LmV C 16 0.37/um 4.8
9.	eP1	AB	15 47 04	Bonin Islands Region 26.40 N 140.53 E
	eP2	A	47 17	H = 15 34 49.6 h = 470 km MB = 5.1
	epP	A	48 48	D = 91.01 Az = 330.3 (NEIC)
	ePP1	AB	50 47	P1V A 1.9s 75.8nm M = 5.3
	ePP2	A	51 01	P2V A 1.4 37.2nm 5.1
	eSP	B	58 25	PP1V A 1.6 33.0nm 5.2
	eSS	B	16 03 35	PP2V A 1.6 30.0nm 5.2
	LmH	B	31	LmH B 16 0.2/um
	LmV	B	33.5	LmV B 14 0.4/um
9.	eP1	A	19 51 47	Lake Baikal Region 52.96 N 107.54 E
	eP2	A	51 49.5	H = 19 42 17.3 h = normal MB = 5.1
	eP3	A	51 52.5	D = 54.95 Az = 309.3 (NEIC)
	eS	C	59 25	P2V A 1.1s 20.2nm M = 5.1
	eSS	C	20 03 25	P3V A 1.1 20.2nm 5.1
	LmH	B	18.6	LmH B 13 1.8/um 5.3
	LmV	B	18.6	LmV B 13 2.4/um 5.5
9.	eP1	A	21 01 58.5	Eastern Russia 56.83 N 127.16 E
	eP2	A	02 05	H = 20 51 51.0 h = 36 km MB=4.8 MS=4.8
	LmH	B	28.4	D = 60.44 Az = 318.7 (NEIC)
	LmV	B	32.6	P2V A 1.0s 23.6nm M = 5.3
				LmH B 14 0.6/um 4.9
				LmV B 14 0.4/um 4.8
10.	LmV	B	03 47.8	LmH B 12s 0.14/um
	LmH	B	49.5	LmV B 12 0.22/um
10.	ePKIKP	A	15 51 09.5	Tonga Islands 21.47 S 174.56 W
	ePKHKP	AB	51 15	H = 15 31 24.7 h = normal MB = 5.7
	LmH	B	17 09	D = 150.48 Az = 352.0 (NEIC)
	LmV	B	09	PKIKPV A 1.2s 12.2nm
				PKHKPV A 1.6 90.6nm

Day	Phase		h m s	Remarks
cont. 10.				LmH B 16.5s 0.7/um M = 5.4
				LmV B 17 0.7/um 5.5
10.	e	A	18 54 11.5	Poland, Rock-burst 50.35 N 18.90 E
	eSg	A	54 31.5	H = 18 51 54.9 M = 2.9
				D = 4.66 Az = 276 (ISC)
10.	eP	A	21 16 51	Tibet 32.42 N 93.47 E
	ePcP	A	17 36	H = 21 06 40.1 h = normal MB = 5.2
	LmH	B	40.5	D = 60.81 Az = 313.8 (NEIC)
	LmV	B	44.8	PV A 1.1s 16.1nm M = 5.1
				LmH B 20 0.7/um 4.8
				LmV B 14 0.4/um 4.8
11.	eP1	A	02 32 02	Mongolia 44.74 N 102.04 E
	eP2	A	32 07.5	H = 02 22 14.2 h = normal MB = 5.0
	LmH	B	57.3	D = 57.48 Az = 311.0 (NEIC)
	LmV	B	58.8	P1V A 1.2s 8.5nm M = 4.7
				P2V A 1.6 44.0nm
				LmH B 15 1.1/um 5.1
				LmV B 12 1.1/um 5.2
11.	eP	A	06 51 30.5	Southern Italy 40.60 N 15.68 E
	e	A	51 33	H = 06 48 55.6 h = 0 km
	LmH	B	55.3	D = 10.45 Az = 346 (ISC)
	LmV	B	56.2	
11.	ePKIKP	A	07 50 35.5	South of Fiji Islands 25.13 S 179.19 W
	ePKHKP	A	50 43.5	H = 07 31 31.0 h = 390 km MB = 4.9
	ePKP2	A	50 57.5	D = 153.19 Az = 344.7 (NEIC)
	epPKP2	A	52 28	h = 373 km
				PKIKPV A 1.2s 18.3nm
11.	eP	A	09 17 38	Hindu Kush Region 36.64 N 70.74 E
				H = 09 09 56.0 h = 271 km MB = 4.7
				D = 43.86 Az = 307.9 (NEIC)
				PV A 1.8s 23.7nm M = 4.2

Day	Phase		h m s	Remarks
11.	+iP	AB	13 36 07	Near East Coast of Kamchatka
	epP	A	36 17	54.62 N 161.56 E
	esP	A	36 22	H = 13 24 44.0 h = 33 km MB = 5.3
	LmV	B	13.6	D = 72.14 Az = 340.4 (NEIC)
	LmH	B	13.7	h = 41 km
				PV A 1.0s 92.5nm M = 5.7
				LmH B 15 0.7/um 5.1
				LmV B 16 0.7/um 5.1
11.	ePKP	A	16 49 31	Fiji Islands Region 18.38 S 176.70 E
	eX	A	49 36.5	H = 16 29 55.1 h = 47 km MB = 4.7
	LmH	C	17 50.5	D = 145.68 Az = 343.1 (NEIC)
	LmV	C	50.5	PKPV A traces
				XV A 1.6s 60.5nm
				LmH C 27 0.52/um M = 5.1
				LmV C 27 0.70/um 5.3
11.	e(Pn)	A	20 40 09.5	
	ei(Sg)	A	40 49	
12.	+iP	AB	09 54 02	Andreanof Islands, Aleutian Is.
	e(PcP)	A	54 11	51.38 N 179.32 W
	epP	A	54 20	H = 09 42 05.2 h = 29 km MB = 5.9
	esP	A	54 28	D = 77.93 Az = 352.9 (NEIC)
	ePP	AC	57 09.5	h = 71 km
	epPP	A	57 29	PV A 1.5s 158.0nm M = 5.8
	eS	B	10 03 50	PV B 6 0.87/um 6.5
	e(ScS)	B	04 20	PPV A 1.5 45.3nm 5.4
	eSS	B	09 15	LmH B 20 3.1/um 5.6
	LmH	B	29.4	LmV B 17 2.3/um 5.6
	LmV	B	36.6	
12.	eP	A	12 46 06.5	Unimak Island Region 53.81 N 164.79 W
	e	A	46 29	H = 12 34 20.4 h = normal MB = 5.0
	LmH	B	13 23	D = 75.88 Az = 2.4 (NEIC)
	LmV	B	26.3	PV A 0.8s 15.4nm M = 5.1
				LmH B 22 0.74/um 4.9
				LmV B 18 0.45/um 4.9

Day	Phase		h m s	Remarks
12.	eP	AB	13 28 41	South of Panama 5.01 N 82.58 W
	epP	A	28 53	H = 13 15 48.1 h = normal MB = 5.7
	ePP	AB	32 10	D = 88.83 Az = 39.4 (NEIC)
	eS	B	39 18	PV A 2.0s 85.5nm M = 5.7
	eSS	B	45 38	PV B 5 0.44 μm 6.0
	LmH	B	14 01.1	PPV A 2.8 86.0nm 5.7
	LmV	B	01.1	SH B 15 0.84 μm 5.7
				LmH B 22 3.0 μm 5.7
				LmV B 24 2.8 μm 5.6
12.	eP	A	23 50 55	Yugoslavia 41.13 N 22.75 E
	e	A	51 06	H = 23 47 57.3 h = 12 km MB = 4.9
	e	A	51 16	D = 12.27 Az = 324.6 (NEIC)
	LmH	B	55.2	PV A 1.3s 15.3nm M = 5.1
	LmV	B	56	LmH B 11 5.7 μm 4.8
				LmV B 10 2.0 μm
13.	eP	A	02 41 37.5	Gulf of Alaska 59.55 N 144.48 W
	e	A	41 42.5	H = 02 30 37.4 h = normal MB = 4.6
	LmH	B	03 18.5	D = 68.46 Az = 16.1 (NEIC)
	LmV	B	18.5	PV A 1.2s 10.2nm M = 4.8
				LmH B 14 0.14 μm 4.3
				LmV B 14 0.28 μm 4.7
13.	LmH	B	05 35.5	Solomon Islands 11.10 S 161.65 E
	LmV	B	45.5	H = 04 22 28.7 h = 33 km MB = 4.9 (NEIC)
				D = 133.54
				LmH B 17s 0.33 μm M = 5.1
				LmV B 18 0.32 μm 5.1
13.	eP	A	08 18 18.5	Sea of Okhotsk 47.16 N 144.67 E
	epP	A	19 55	H = 08 07 22.6 h = 417 km MB = 4.8
				D = 74.52 Az = 331.1 (NEIC)
				h = 460 km
				PV A 1.4s 23.3nm M = 4.7

Day	Phase		h m s	Remarks
13.	ePKP2	A	10 14 39	Kermadec Islands 29.56 S 177.11 W H = 09 54 12.6 h = 39 km MB = 4.7 (NEIC) D = 157.94
13.	ePKIKP	A	12 06 45	Near Coast of Southern Chile
	LmH	C	59.5	51.85 S 73.91 W
	LmV	C	13 00	H = 11 47 48.1 h = normal MB = 5.6 D = 125.03 Az = 50.8 (NEIC) PKIKPV A 1.2s 30.5nm LmH B 20 0.36 /um M = 5.0 LmV B 20 0.53 /um 5.2
13.	ePn	A	17 17 34	Northern Italy 44.8 N 10.4 E
	ePg	A	18 03.5	H = 17 16 06 (BCIS)
	eSn	A	18 37.5	D = 5.94
	i	A	18 39.0	
	eSg	A	19 18	
14.	eP	A	07 24 21.5	Afghanistan - USSR Border Region 36.52 N 71.54 E H = 07 16 21.8 h = 130 km MB = 4.6 D = 44.44 Az = 308.1 (NEIC) PV A 1.8s 27.0nm M = 4.7
14.	e(P)	A	10 57 37	Caribbean Sea 14.28 N 68.55 W H = 10 45 56.0 h = 20 km MB = 5.0 D = 72.88 Az = 41.0 (NEIC)
14.	ePKIKP	A	12 05 22	Fiji Islands Region 17.93 S 178.58 W
	ePKHKP	AB	05 24.5	H = 11 46 50.6 h = 622 km MB = 4.8 D = 146.36 Az = 348.3 (NEIC) PKIKPV A 1.2s 12.2nm PKHKPV A 1.3 30.6nm
14.	eP	A	15 47 18	Molucca Passage 1.70 N 126.39 E
	LmH	C	16 39	H = 15 33 15.3 h = normal MB = 5.5
	LmV	C	40.2	D = 104.12 Az = 323.4 (NEIC) LmH C 25s 0.28 /um M = 4.7

Day	Phase		h m s	Remarks
14.	eP	A	18 41 12	South of Honshu, Japan 29.02 N 139.95 E
				H = 18 29 19.6 h = 560 km MB = 4.1
				D = 88.49 Az = 330.0 (NEIC)
				traces
14.	ePKIKP	A	22 48 17	New Guinea 6.31 S 144.42 E
	epPKIKP	A	48 24.5	H = 22 29 27.6 h = 44 km MB = 5.6
	ePP	A	49 43	D = 120.95 Az = 327.0 (NEIC)
	ePP	B	49 48	h = 28 km
	e	A	49 53	PKIKPV A 1.4s 30.2nm
	eSP	C	59 30	LmH B 19.5 1.3/um M = 5.6
	LmV	B	23 43.2	LmV B 19 1.5/um 5.7
	LmH	B	43.3	
15.	ePKIKP	A	04 09 14	Fiji Islands Region 17.77 S 178.76 W
	ePKHKP	A	09 16.5	H = 03 50 38.3 h = 573 km MB = 4.9
				D = 146.17 Az = 348.1 (NEIC)
				PKIKPV A 1.3s 13.1nm
				PKHKPV A 1.6 49.5nm
15.	ePKHKP	A	10 41 35	Tonga Islands 17.66 S 173.05 W
	eX	A	41 43.5	H = 10 21 53.2 h = normal MB = 4.8
				D = 146.88 Az = 354.6 (NEIC)
				PKHKPV A 1.5s 35.2nm
				XV A 1.8 74.4nm
15.	-eP	A	11 07 24.5	Southeastern Alaska 56.25 N 135.50 W
	LmH	B	39.5	H = 10 56 12.8 h = 21 km MB = 5.6
	LmV	B	41.5	D = 70.03 Az = 21.6 (NEIC)
				PV A 1.2s 52.8nm M = 5.5
				LmH B 15 0.5/um 4.9
				LmV B 15 0.4/um 4.8
15.	e	A	21 49 39	Alaska 64.15 N 148.75 W
				H = 21 39 04.2 h = 20 km MB = 4.3
				D = 63.55 Az = 13.8 (NEIC)

Day	Phase		h m s	Remarks
16.	+iP	A	03 24 46.8	<u>Eastern Kazakh SSR</u> 49.76 N 78.15 E
	ePn	A	26 19	H = 03 16 57.2 h = 0 km MB = 5.2
				D = 41.29 Az = 297.7 (NEIC)
				PV A 0.7s 44.0nm M = 5.3
				underground explosion
16.	ePKHKP	A	11 45 36.5	<u>South of Australia</u> 51.91 S 139.55 E
	ePKP2	AB	45 41	H = 11 25 53.9 h = normal MB = 5.1
	LmH	B	12 56	D = 147.98 Az = 288.7 (NEIC)
	LmV	B	56.1	PKHKPV A 0.8s 7.7nm
				PKP2V A 2.4 124.5nm
				LmH B 20 0.8 µm M = 5.5
				LmV B 20 1.3 µm 5.7
16.	e(Pn)	A	23 24 39	<u>Northern Italy</u> 44.26 N 8.06 E
	ePg	A	24 58	H = 23 22 44.5 h = 4 km
	eSn	A	25 41	D = 6.83 Az = 19.4 (NEIC)
	eSg	A	26 30	
17.	eP	A	06 32 47	<u>Southern Sinkiang Prov., China</u>
				40.74 N 78.34 E
				H = 06 24 23.7 h = normal MB = 4.6
				D = 46.17 Az = 305.8 (NEIC)
				PV A 1.0s 23.6nm M = 5.1
17.	LmH	B	15 47.5	<u>Peru</u> 13.96 S 74.62 W
	LmV	B	47.5	H = 14 50 51.0 h = 14 km MB = 5.4 (NEIC)
				D = 98.30
				LmH B 20s 0.8 µm M = 5.2
				LmV B 20 1.03 µm 5.3
17.	eP	A	17 12 06	<u>Dodecanese Islands</u> 35.71 N 27.18 E
				H = 17 07 41.0 h = 22 km
				D = 18.70 Az = 328 (ISC)
17.	e	A	17 48 45	<u>South of Sumbawa Island</u>
	LmH	C	18 37	11.09 S 117.25 E

Day	Phase		h m s	Remarks
cont. 17.	LmV	C	18 37.3	H = 17 29 52.1 h = normal MB = 5.4
				D = 108.41 Az = 319.7 (NEIC)
				LmH C 22s 0.31/um M = 4.8
				LmV C 20 0.53/um 5.1
17.	eP diff	B	23 59 44	New Britain Region 5.96 S 152.90 E
	e	A	59 50.5	H = 23 44 05.9 h = 10 km MB=6.4 MS=7.1
	ePKIKP	AB	00 03 08	(NEIC)
	PKIKPm	A	03 17	D = 125.07 Az = 331 (ISC)
	eiPP	AB	04 58	P diff V B 8s 0.71/um
	e	B	05 19	PKIKPV A 2.5 660.0nm
	ePS	B	15 00	PKIKPV B 12 2.11/um
	LmH	B	58.5	PKIKPmV A 2.0 556.0nm
	LmV	B	01 04.2	PPV A 2.0 478.0nm M = 6.7
				PPV B 11 7.67/um 7.1
				LmH B 19 26.4/um 6.9
				LmV B 16 23.2/um 6.9
18.	e(Sg)	A	00 31 27	Switzerland 46.0 N 7.5 E
				H = 00 28 18 (BCIS)
				D = 5.1
18.	ePKIKP	A	00 42 50	New Britain Region 6.02 S 152.64 E
	esPKIKP	A	43 00	H = 00 23 48.2 h = 8 km MB = 6.0
	ePP	A	44 30	D = 124.93 Az = 330.8 (NEIC)
	e	A	44 52	PKIKPV A 1.3s 21.8nm
				PPV A 0.8 9.6nm M = 5.4
18.	ePKP	A	00 46 29.5	New Britain Region 5.22 S 152.90 E
				H = 00 27 28.9 h = 0 km
				D = 124.36 Az = 331 (ISC)
18.	ePKIKP	A	01 31 03	New Britain Region 6.03 S 152.77 E
				H = 01 12 02.5 h = 35 km MB = 5.6
				D = 125.00 Az = 330.8 (NEIC)

Day	Phase		h m s	Remarks
18.	+eP	A	19 13 31	Near East Coast of Kamchatka
	epP	A	13 49	52.96 N 159.87 E
	esP	A	13 58	H = 19 02 00.8 h = normal MB = 5.1
	LmV	C	39	D = 73.35 Az = 339.5 (NEIC)
				PV A 1.0s 35.4nm M = 5.3
19.	e(Pn)	A	10 06 13	
	eSg	A	06 50	
19.	eSn	A	16 29 29.5	Austria 47.84 N 15.92 E
	iSg	A	29 47	H = 16 27 40.2 h = 7 km
				D = 3.96 Az = 332 (ISC)
19.	eP	A	23 32 49	Kurile Islands Region 43.50 N 148.37 E
	epP	A	32 56.5	H = 23 20 48.1 h = 39 km MB = 4.9
	esP	A	33 00.5	D = 78.96 Az = 333.6 (NEIC)
	LmH	C	00 05	h = 30 km
				PV A 1.2s 24.4nm M = 5.1
				LmH C : NS : 20s 0.29/um
				EW : disturbance
20.	+eP1	AB	03 05 09.3	Western Kazakh SSR 49.46 N 48.18 E
	eiP2	AB	05 13.5	H = 02 59 57.9 h = 0 km MB = 5.7
	LmV	B	14.0	D = 23.36 Az = 287.0 (NEIC)
	LmH	B	14.1	P1V A 1.1s 335.0nm M = 5.8
				P1V B 1.5 0.49/um 6.1
				P2V A 1.4 790.0nm 5.8
				P2V B 2.0 1.04/um 6.0
				LmH B 5 0.8/um 4.8
				LmV B 4.5 0.8/um 5.0
				Probably underground explosion
20.	ePP	A	05 28 43.5	Halmahera 1.34 N 126.98 E
				H = 05 10 23.2 h = 57 km MB = 5.2 (NEIC)
				D = 104.76
				PPV A 1.4s 18.6nm M = 5.5

Day	Phase		h m s	Remarks
20.	eP	A	08 21 48.5	Off East Coast of Kamchatka
	LmH	B	53	51.34 N 161.59 E
	LmV	B	09 01.5	H = 08 10 08.1 h = normal MB = 5.2
				D = 75.24 Az = 340.8 (NEIC)
				PV A 1.3s 39.3nm M = 5.2
				LmH B 24 0.7 /um 4.9
				LmV B 14 0.6 /um 5.1
20.	eP	A	10 21 20	Honshu, Japan 38.71 N 139.97 E
	ePcP	A	21 23.5	H = 10 09 10.8 h = 25 km MB = 5.5
	LmH	B	54.5	D = 80.14 Az = 329.6 (NEIC)
	LmV	B	58.4	PV A 1.2s 28.5nm M = 5.2
				PcP A 1.6 55.0nm
				LmH B 16 0.6 /um 5.1
				LmV B 14 0.3 /um 4.9
21.	-iP	AB	06 34 22.5	Sea of Okhotsk 49.51 N 147.04 E
	epP	AB	36 22	H = 06 23 48.9 h = 578 km MB = 5.9
	ipP	AB	36 26	D = 73.19 Az = 332.2 (NEIC)
	isP	B	37 16	h = 580 km
	e	B	38 16	PV A 1.5s 1673.0nm M = 6.3
	e	B	38 52	PV B 5 4.07 /um 6.2
	eiS	AB	43 03	
	eiSKS	B	43 32	
	esS	B	46 38	
	eisSS	B	51 05	
	eP'P'	A	07 01 43.5	
21.	ePKIKP	A	09 06 04	New Britain Region 5.75 S 152.96 E
				H = 08 47 05.4 h = 35 km MB = 5.3
				D = 124.86 Az = 331.0 (NEIC)
				PKIKPV A 1.6s 22.0nm
21.	eP	A	10 27 00	Near East Coast of Kamchatka
				55.71 N 161.33 E
				H = 10 15 43.5 h = normal MB = 4.5
				D = 71.06 Az = 340.2 (NEIC)
				PV A 1.0s 15.8nm M = 5.0

Day	Phase		h m s	Remarks
21.	e	A	16 40 15	
21.	e(P)	A	22 14 56	PV A 1.5s 20.1nm
	eX	A	15 05	XV A 2.0 34.2nm
21.	e(P)	A	23 05 39.5	Mediterranean Sea 34.4 N 22.6 E
				H = 23 01 12 h = 50 km
				D = 18.10 Az = 337 (ISC)
22.	LmH	B	06 26.9	Volcano Islands Region 23.23 N 144.21 E
	LmV	B	31.7	H = 05 29 37.9 h = 33 km MB = 4.9 (NEIC)
				D = 95.39
				LmH B 15s 0.51 µm M = 5.1
				LmV B 15 0.57 µm 5.2
22.			07 10	Between 07^h 10^m and 12^h 58^m no records
			12 58	by short period seismographs.
22.	e	A	12 56 39	
22.	eP	A	14 31 56	Kurile Islands 50.22 N 156.74 E
	LmH	B	53.4	H = 14 20 19.4 h = 63 km MB = 5.2
	LmV	B	54.7	D = 75.20 Az = 337.9 (NEIC)
				PV A 1.4s 23.3nm M = 4.9
22.	LmH	C	15 06	West New Guinea 3.78 S 136.20 E
	LmV	C	10	H = 14 11 38.4 h = 36 km MB = 4.6 (NEIC)
				D = 114.28
				LmH C 25s 0.4 µm M = 4.9
				LmV C 20 0.5 µm 5.1
22.	e	A	16 19 23	
	eSg	A	19 26	
22.	LmH	C	22 35	LmV C 15s 0.35 µm
	LmV	C	37	

Day	Phase		h m s	Remarks
22.	LmH	C	23 40	<u>South of Honshu, Japan</u>
	LmV	C	45	31.66 N 139.41 E
				H = 22 47 51.0 h = 32 km MB = 4.5 (NEIC)
				D = 85.98
				LmV C 15s 0.35/um M = 4.9
23.	+iP	AB	08 58 28	<u>Kodiak Island Region</u> 58.25 N 153.58 W
	e	A	58 33.5	H = 08 47 16.1 h = 61 km MB = 5.5
	eS	B	09 07 39	D = 70.79 Az = 9.9 (NEIC)
				PV A 1.0s 197.0nm M = 6.0
23.	ePP	A	12 18 10	<u>Halmahera</u> 1.82 N 127.26 E
				H = 11 59 52.4 h = 106 km MB = 5.3 (NEIC)
				D = 104.54
				PPV A 1.5s 32.7nm M = 5.7
23.	eP	A	18 03 09.5	<u>Rumania</u> 45.85 N 26.77 E
	LmH	C	08	H = 18 00 31.3 h = 82 km MB = 4.7
	LmV	C	08	D = 11.18 Az = 300.9 (NEIC)
				PV A 1.3s 6.5nm M = 4.4
24.	LmH	B	21 24.8	LmV B 9s 0.4/um
	LmV	B	24.8	
24.	LmH	C	15 17.4	<u>New Britain Region</u> 6.36 S 153.16 E
	LmV	C	24.5	H = 14 10 58.5 h = 10 km MB = 4.6 (NEIC)
				D = 125.47
				LmH C 22.5s 0.3/um M = 4.9
				LmV C 18 0.3/um 5.0
24.	ePn	A	19 15 20.5	<u>Yugoslavia</u> 44.9 N 19.0 E
	e	A	16 53.5	H = 19 13 20 h = 0 km
				D = 7.64 Az = 322 (ISC)
25.	eP	A	09 00(01)	<u>Jan Mayen Island Region</u> 71.44 N 12.41 W
	eS	B	04 20	H = 08 54 51.6 h = normal MB = 4.2
	LmH	B	09.9	D = 23.54 Az = 139.5 (NEIC)
	LmV	B	10.9	LmH B 17s 0.4/um M = 4.0
				LmV B 14 0.3/um 4.0

Day	Phase		h m s	Remarks
25.	ePKP2	A	14 28 31	South of Kermadec Islands 33.39 S 179.12 W H = 14 07 57.1 h = 111 km (NEIC) D = 161.03 PKP2V A 1.6s 13.7nm
25.	LmH	B	16 20.2	LmH B 12.5s 0.3 μm
	LmV	B	20.3	LmV B 12 0.3 μm
25.	LmH	B	23 10.5	New Britain Region 6.14 S 152.59 E H = 22 10 23.4 h = 33 km MB = 5.1 (NEIC) D = 125.01 LmH B 17s 0.4 μm M = 5.2
	LmV	B	10.5	LmV B 18 0.5 μm 5.2
26.	e	A	02 11 02	
26.	e	A	02 21 57	
26.	LmH	B	03 42	South Sandwich Islands Region 59.56 S 26.11 W H = 02 39 02.2 h = 33 km MB = 4.9 (NEIC) D = 114.03
	LmV	B	42	
26.	+iP	A	03 54 43.5	Eastern Kazakh SSR 49.99 N 77.58 E H = 03 46 56.9 h = 0 km MB = 5.5 D = 40.97 Az = 297.3 (NEIC) PV A 0.7s 49.8nm M = 5.3 underground explosion
	ePn	A	56 15	
26.	LmH	B	05 50.6	West New Guinea 3.14 S 135.97 E H = 04 39 15.2 h = 33 km MB = 5.3 (NEIC) D = 113.62 LmH B 20s 0.3 μm M = 4.9
	LmV	B	51.4	LmV B 18 0.3 μm 4.9

Day	Phase		h m s	Remarks
26.	eP	A	11 49 59.5	Fox Islands, Aleutian Is. 52.55 N 170.56 W H = 11 38 09.8 h = 52 km MB = 4.7 D = 77.17 Az = 358.6 (NEIC) PV A 1.1s 14.1nm M = 4.9
26.	ePKHKP LmV	A C	13 00 11 14 18	Tonga Islands 21.48 S 174.00 W H = 12 40 22.8 h = normal MB = 4.4 D = 150.56 Az = 352.7 (NEIC) PKHKPV A 1.3s 10.9nm
26.	ePKP	A	13 45 31	Loyalty Islands Region 22.14 S 169.89 E H = 13 25 48.2 h = 42 km D = 146.85 Az = 334.5 (NEIC)
26.	e	A	17 31 01.5	
26.	LmH LmV	B B	18 53.3 54.3	Off Coast of Oregon 43.22 N 127.39 W H = 18 05 01.4 h = 33 km MB = 4.9 (NEIC) D = 79.93 LmH B 19s 0.12/um M = 4.3
27.	iSn eSg	A A	06 14 58 15 22	Switzerland 47.30 N 7.4 E H = 06 13 07 h = 0 km (ISC) D ca. 4.1
27.	eP LmH LmV	A B B	15 00 55.5 29.7 36.2	Burma - China Border Region 22.63 N 100.69 E H = 14 49 31.9 h = normal MB = 4.8 D = 72.27 Az = 318.1 (NEIC) PV A 1.4s 14.0nm M = 4.8 LmH B 18.5 0.6/um 4.9 LmV B 16 0.5/um 4.9
27.	e	A	18 21 03	
27	eP ePcP	A A	19 48 08 48 52	North Atlantic Ridge 10.60 N 42.51 W H = 19 38 05.2 h = normal MB = 4.8

Day	Phase		h m s	Remarks
cont. 27.	eSP	C	19 56 28	D = 59.49 Az = 36.8 (NEIC)
	eSS	C	20 00 10	PV A traces
	LmH	B	11.8	PcPV A 1.6s 24.8nm
	LmV	B	11.9	LmH B 18.5 0.4/um M = 4.5
				LmV B 18.5 0.4/um 4.6
28.	eP	A	02 38 07.5	Northern Territory, Australia
				24.93 S 136.32 E
				H = 02 19 00.5 h = 59 km MB = 5.6
				D = 130.71 Az = 316.3 (NEIC)
28.	+iP	AB	06 06 02.1	Novaya Zemlya 73.34 N 55.09 E
	ePn	A	06 28	H = 05 59 56.5 h = 0 km MB = 6.3
	ePcP	B	09 07.5	D = 29.35 Az = 243.3 (NEIC)
	eS	B	10 58	PV A 1.4s 503.0nm M = 6.2
	eSn	B	11 40	PV B 4 1.21/um 6.1
	LmH	B	17.8	LmH B 14 10.4/um 5.6
	LmV	B	21.5	LmV B 7.5 5.9/um 5.8
	ePKPPKP	A	38 29	underground explosion
	e	A	39 20.5	
28.	ePKHKP	A	14 08 18.5	Fiji Islands Region 18.15 S 176.55 W
				H = 13 48 37.9 h = normal MB = 5.1
				D = 146.93 Az = 350.5 (NEIC)
				PKHKPV A \ 1.4s 14.0nm
28.	eP	A	14 37 47	Mindoro, Philippine Islands
	ePP	A	41 26	13.27 N 120.56 E
	LmH	B	15 25.4	H = 14 24 46.1 h = 62 km MB = 5.2
	LmV	B	26.7	D = 91.45 Az = 323.0 (NEIC)
				PV A 1.0s 17.7nm M = 5.4
				LmH B 18 0.26/um 4.7
				LmV B 16 0.24/um 4.7
28.	eP	A	15 32 56	Andreanof Islands, Aleutian Is.
	LmH	B	08.6	51.37 N 179.22 W
	LmV	B	10.6	H = 15 21 01.8 h = 45 km MB = 5.5

Day	Phase		h m s	Remarks
cont. 28.				D = 77.95 Az = 353.0 (NEIC)
				PV A 1.1s 38.3nm M = 5.3
				LmH B 22 0.4/um 4.7
				LmV B 19 0.3/um 4.6
28.	ePKIKP	AB	19 17 48.5	Tonga Islands 21.26 S 174.53 W
	ePKHKP	A	17(52)	H = 18 58 03.2 h = normal MB = 5.6
	ePKP2	AB	18 01	D = 150.28 Az = 352.1 (NEIC)
	e	B	20 27	PKIKPV A 2.6s 86.6nm
	ePP	B	21 40	PKHKPV A 2.4 362.0nm
	eSKKS	B	28 17	LmH B 19 2.7/um M = 6.0
	eSS	B	40 52	LmV B 20 2.8/um 6.0
	LmV	B	20 26.3	
	LmH	B	30.9	
28.	ePn	A	20 53 42	Polkowice / Lubin, Poland
	ePg	A	53 49.5	51°25' N 16°11' E
	e	A	54 20	D = 2.99 rockburst
	e	A	54 22	
	eSg	A	54 29	
28.	e(P)	A	21 59 11.5	
	e	A	59 17.5	
29.	LmH	C	02 19.8	Gulf of California 25.39 N 109.77 W
	LmV	C	25.7	H = 01 35 27.2 h = 33 km MB = 4.8 (NEIC)
				D = 88.35
				LmH C 28s 0.4/um M = 4.7
				LmV C 22 0.5/um 4.9
29.	e(P)	A	02 53 06	Turkey 37.00 N 29.14 E
				H = 02 48 36.9 h = 0 km MB = 4.4
				D = 18.54 Az = 323 (ISC)
				PV A 1.1s 8.1nm M = 3.8
29.	eP	A	03 40 38	Windward Islands 11.02 N 62.28 W
				H = 03 29 23.2 h = 70 km MB = 5.3
				D = 71.35 Az = 40.2 (NEIC)

Day	Phase		h m s	Remarks
29.	ePKIKP	A	06 18 28	Tonga Islands 19.99 S 175.28 W
	ePKP2	AB	18 32.5	H = 05 59 01.7 h = 154 km MB = 5.4
				D = 148.92 Az = 351.5 (NEIC)
29.	LmH	C	07 54	Banda Sea 6.54 S 129.84 E
	LmV	C	59	H = 06 53 28.9 h = 33 km MB = 5.4 (NEIC)
				D = 112.73
29.	eP	AB	19 25 20	Central Mid - Atlantic Ridge
	e	A	25 26	8.54 N 39.40 W
	e	A	25 32	H = 19 15 17.8 h = normal MB = 4.7
	eS	B	33 30	D = 59.34 Az = 35.1 (NEIC)
	LmV	B	48.8	PV A 2.0s 47.0nm M = 5.3
	LmH	B	49.7	LmH B 19.5 0.5/um 4.6
				LmV B 20 0.6/um 4.8
29.	eP	A	22 04 00	Crete 34.4 N 26.1 E
				H = 21 59 30 h = 45 km MB = 3.9
				D = 19.43 Az = 331 (ISC)
29.	eP	A	22 15 29	Central Mid - Atlantic Ridge
	LmV	C	41.1	8.46 N 39.47 W
	LmH	C	41.3	H = 22 05 27.1 h = normal MB = 4.9
				D = 59.45 Az = 35.1 (NEIC)
				PV A 1.3s 17.5nm M = 5.0
				LmH C 18 0.16/um 4.2
				LmV C 18 0.22/um 4.4
30.	ePn	A	00 10 10	Yugoslavia 44.05 N 16.20 E
	e	A	10 26	H = 00 08 23.1 h = normal MB = 4.5
	eSn	A	11 30.5	D = 7.30 Az = 336.4 (NEIC)
	eSg	A	12 12	
30.	ePKIKP	A	05 39 55.5	Fiji Islands Region 19.80 S 177.74 W
	ePKHKP	A	39 59.5	H = 05 21 15.6 h = 562 km MB = 4.9
	epPKP2	A	42 12.5	D = 148.34 Az = 348.6 (NEIC)

Day	Phase		h m s	Remarks
30.	ePKIKP	A	10 48 41.5	Near North Coast of New Guinea
	ePP	B	50 00	3.48 S 144.93 E
	ePPP	B	52 32	H = 10 29 51.1 h = 13 km MB = 5.6
	ePS	B	59 40	D = 118.85 Az = 328.1 (NEIC)
	eSS	B	11 06 20	PKIKPV A 1.2s 8.1nm
	eSSS	B	10 36	PPV B 9 2.0 µm M = 6.8
	LmH	B	39.5	LmH B 22.5 21.1 µm 6.7
	LmV	B	40.6	LmV B 21 24.4 µm 6.8
30.	eP	AB	15 24 12.5	Tsinghai Province, China
	esP	A	24 20	36.72 N 96.47 E
	eS	C	32 25	H = 15 14 09.9 h = normal MB = 5.5
	LmH	B	47.4	D = 59.67 Az = 312.7 (NEIC)
	LmV	B	51.4	PV A 1.4s 93.2nm M = 5.7
				sPV A 1.6 68.6nm
				LmH B 20 2.3 µm 5.3
				LmV B 15 2.1 µm 5.5
30.	+eP	AB	18 57 46.5	Tsinghai Province, China
	esP	A	57 52.5	36.60 N 96.42 E
	LmH	B	19 20.9	H = 18 47 42.6 h = normal MB = 5.5
	LmV	B	25	D = 59.73 Az = 312.8 (NEIC)
				PV A 1.5s 85.4nm M = 5.7
				sPV A 1.6 115.3nm
				LmH B 20 1.5 µm 5.1
				LmV B 15 1.03 µm 5.2
30.	eP	AB	19 01 37.5	Tsinghai Province, China
	esP	A	01 44	36.69 N 96.34 E
	LmV	B	30.3	H = 18 51 34.8 h = normal MB = 5.5
				D = 59.62 Az = 312.7 (NEIC)
				PV A 1.5s 82.9nm M = 5.6
				sPV A 1.4 51.1nm
				LmV B 14.5 1.5 µm 5.3
30.	eP	A	20 52 35	Chinghai Province 37.1 N 95.5 E
				H = 20 42 44 h = 97 km
				D = 58.82 Az = 313 (ISC)
				PV A 1.5s 10.1nm M = 4.6

Day	Phase		h m s	Remarks
31.	eP	A	02 36 46	El Salvador 13.27 N 88.51 W
	epP	A	37 05	H = 02 24 09.6 h = 50 km MB = 5.4
	esP	A	37 16	D = 86.17 Az = 38.9 (NEIC)
	e	A	37 28	h = 75 km
	ePP	A	40 29.5	PV A 1.5s 20.1nm M = 5.1
	eS	B	47 14	PPV A 2.0 42.8nm 5.5
	LmH	B	03 14.7	LmH B 20 0.6/um 5.0
	LmV	B	16.8	LmV B 18 0.5/um 5.0
31.	ePKHKP	A	05 53 28	Fiji Islands Region 21.18 S 179.18 W
				H = 05 34 47.5 h = 625 km MB = 4.8
				D = 149.39 Az = 346.5 (NEIC)
				PKHKPV A 1.1s 16.1nm
31.	eP	AB	08 19 37.5	Southern Honshu, Japan 35.87 N 136.72 E
	ePP	B	22 42	H = 08 07 22.5 h = 24 km MB = 5.5
	ePPPP	B	26 00	D = 81.23 Az = 328.2 (NEIC)
	eS	B	29 48	PV A 2.0s 42.7nm M = 5.1
	eSS	C	35 00	PV B 6 0.44/um 5.7
	LmH	B	54.6	LmH B 15 3.9/um 5.9
	LmV	B	09 00.6	LmV B 13 2.0/um 5.7
31.	+iP1	AB	14 12 04.2	Central Russia 52.32 N 95.37 E
	-iP2	AB	12 07.7	H = 14 03 16.4 h = normal MB = 5.5
	eS	B	19 16	D = 49.36 Az = 303.5 (NEIC)
	LmH	B	35.2	P1V A 1.4s 132.2nm M = 5.8
	LmV	B	35.2	P2V A 1.4 163.0nm 5.9
				LmH B 14 2.9/um 5.4
				LmV B 13 3.3/um 5.6

Day	Phase		h m s	Remarks
1.	ePKIKP	A	13 06 19	New Hebrides Islands 14.98 S 167.31 E
	e	A	06 23	H = 12 47 14.2 h = 130 km MB = 5.6
	e	A	06 27	D = 139.36 Az = 336.3 (NEIC)
	LmH	C	53.0	PKIKPV A 1.4s 11.6nm
				LmH C 35 0.5 µm
1.	eP	A	14 42 44	Andreanof Islands, Aleutian Is.
				51.38 N 178.13 W
				H = 14 30 50.6 h = 63 km MB = 5.2
				D = 78.02 Az = 353.7 (NEIC)
				PV A 1.0s 15.7nm M = 4.9
1.	iPg	A	16 05 20.4	Hilders/Rhön, Federal Rep. of Germany
	iSg	A	05 34.8	50°32.56'N 10°02.38'E
				Explosion yield 12.2 t (Hannover)
				H = 16 05 01.25
				D = 1.0
2.	+eP	AB	02 02 04	Ryukyu Islands 29.37 N 130.54 E
	epP	A	02 16	H = 01 49 36.9 h = 41 km MB = 6.1
	ePP	AB	05 15	D = 83.83 Az = 325.9 (NEIC)
	iS	B	12 22	h = 42 km
	eSP	C	13 05	PV A 1.8s 324.0nm M = 6.1
	ePS	C	13 10	PV B 8 1.6 µm 6.2
	eSS	C	18 20	PPV A 1.8 95.4nm 5.9
	ePKKP	A	20 21	SH B 9 8.5 µm 6.9
	e	A	20 31	LmH B 18.5 19.0 µm 6.5
	eP'P'	A	28 25	LmV B 18 21.0 µm 6.6
	LmH	B	43.6	
	LmV	B	43.7	
2.	eP	A	02 54 08.5	Kurile Islands Region 48.78 N 156.49 E
	epP	A	54 21.5	H = 02 42 21.7 h = normal MB = 5.1
				D = 76.47 Az = 337.9 (NEIC)
				h = 47 km
				PV A 1.3s 52.4nm M = 5.4
				pPV A 1.5 47.7nm

Day	Phase		h m s	Remarks
2.	eP	A	03 54 43	Kurile Islands Region 48.75 N 156.52 E
	epP	A	54 54	H = 03 42 55.0 h = normal MB = 4.4
				D = 76.51 Az = 337.9 (NEIC)
				h = 43 km
				PV A 1.2s 14.2nm M = 4.9
				pPV A 1.8 33.8nm
2.	eP	A	04 00 51.5	Kurile Islands Region 48.91 N 156.22 E
	epP	A	01 04.5	H = 03 49 06.0 h = 47 km MB = 4.9
				D = 76.28 Az = 337.7 (NEIC)
				h = 47 km
				PV A 1.0s 33.5nm M = 5.3
2.	ePKHKP	A	04 34 33	South of Fiji Islands 22.13 S 179.67 W
	ePKP2	A	34 42	H = 04 15 47.7 h = 591 km MB = 4.9
	ePKP	A	36 51.5	D = 150.20 Az = 345.5 (NEIC)
				PKHKPV A 1.3s 43.7nm
2.	ePKP	A	05 46 55.5	Loyalty Islands Region 22.01 S 169.97 E
				H = 05 27 15.4 h = 34 km MB = 4.8
				D = 146.77 Az = 334.6 (NEIC)
				PKPV A 1.1s 22.2nm
2.	eP	A	09 04 44	Eastern Kazakh SSR 49.96 N 77.73 E
				H = 08 56 57.6 h = 0 km MB = 5.1
				D = 40.95 Az = 297.4 (NEIC)
				Probably artificial (BCIS)
				traces
2.	eP	A	10 43 57	Turkmen SSR 39.88 N 53.65 E
	LmH	C	59.0	H = 10 37 39.4 h = normal MB = 4.9
	LmV	C	59.0	D = 31.09 Az = 304.4 (NEIC)
				PV A 1.5s 25.1nm M = 4.8
				LmH C 15 0.2 μm 3.8
				LmV C 15 0.4 μm 4.3
2.	eP1	AB	14 58 18	Near Coast of Libya 31.49 N 16.30 E
	iP2	AB	58 24	H = 14 53 51.4 h = normal MB = 5.0

Day	Phase		h m s	Remarks
cont. 2.	eS1	B	15 01 51	D = 19.45 Az = 351.0 (NEIC)
	eiS2	B	01 55	P1V A 1.6s 99.0nm M = 4.8
	LmH	B	05.9	P1V B 3 0.8 /um 5.4
	LmV	B	09.2	P2V A 1.4 209.0nm 5.2
				P2V B 6.5 1.3 /um 5.3
				S2H B 12 8.4 /um 5.7
				LmH B 18 19.0 /um 5.5
				LmV B 14 8.4 /um 5.4
2.	eP	A	18 41 31.5	Kurile Islands Region 48.75 N 156.41 E
	epP	A	41 43.5	H = 18 29 43.3 h = normal MB = 4.6
				D = 76.48 Az = 337.8 (NEIC)
				h = 45 km
				PV A 1.0s 13.8nm M = 4.9
2.	eP	A	18 44 19.5	Kurile Islands Region 48.89 N 156.36 E
	epP	A	44 31.5	H = 18 32 32.1 h = normal MB = 4.6
				D = 76.34 Az = 337.8 (NEIC)
				h = 45 km
				PV A 1.1s 20.2nm M = 5.0
2.	eP	A	19 02 24	Kurile Islands 48.95 N 156.34 E
				H = 18 50 37.5 h = normal MB = 4.2
				D = 76.28 Az = 337.8 (NEIC)
2.	eP	A	19 24 49	Near East Coast of Honshu, Japan
				35.54 N 140.88 E
				H = 19 12 26.8 h = 54 km MB = 4.7
				D = 83.25 Az = 330.2 (NEIC)
2.	LmH	C	20 02.6	LmH C 19s 0.4 /um
	LmV	C	04.5	LmV C 20 0.3 /um
3.	eSn	A	00 44 24	Northern Italy 44.33 N 11.23 E
	eSg	A	45 04	H = 00 41 31.6 h = 0 km (ISC)
				D = 6.35

Day	Phase		h m s	Remarks
3.	ePKIKP	A	02 13 12.5	New Britain Region 6.21 S 152.80 E
				H = 01 54 13.0 h = normal MB = 4.7
				D = 125.18 Az = 330.8 (NEIC)
				traces
3.	ePKIKP	A	06 50 18	New Britain Region 5.55 S 151.28 E
	e	A	52 08	H = 06 31 21.6 h = 45 km MB = 5.4
	ePP	B	52 10	D = 123.86 Az = 330.2 (NEIC)
	ePP	A	52 13	PPV A 2.5s 123.0nm M = 5.8
	ePS	C	07 01 55	PPV B 11 0.8/um 5.9
	eSS	C	08 45	LmH B 22.5 6.6/um 6.2
	LmH	B	35.5	LmV B 20 5.3/um 6.2
	LmV	B	42.7	
3.	ePKIKP	A	08 14 55	New Britain Region 5.62 S 151.23 E
	e	A	16 45	H = 07 55 59.1 h = 49 km MB = 5.2
	ePP	AB	16 52	D = 123.90 Az = 330.2 (NEIC)
	eSS	C	33 40	PPV A 3.0s 105.3nm M = 5.6
3.	eP1	A	08 42 35	Turkey 39.15 N 28.13 E
	eiP2	AB	42 38.5	H = 08 38 43.2 h = 16 km MB = 4.6
	LmH	B	50.4	D = 16.36 Az = 320.0 (NEIC)
	LmV	B	51.2	P1V A 1.6s 38.5nm M = 4.3
				P2V A 1.6 209.0nm 5.0
				P2V B 2 0.6/um 5.3
				LmH B 14 1.6/um 4.4
				LmV B 15 1.5/um 4.5
3.	LmH	C	13 07.0	New Britain Region 5.52 S 151.15 E
	LmV	C	07.0	H = 11 55 45.2 h = 33 km MB = 4.5 (NEIC)
				D = 123.77
3.	ePKIKP	A	16 23 38	New Britain Region 5.96 S 152.82 E
	e	A	23 50	H = 16 04 40.1 h = 48 km MB = 5.3
	e	A	25 22	D = 124.96 Az = 330.9 (NEIC)
	ePP	B	25 25	PKIKPV A 1.6s 27.4nm
	ePP	A	25 28	PPV A 1.8 27.0nm M = 5.4
	eSS	C	42 28	PPV B 7 0.3/um 5.6

Day	Phase		h m s	Remarks
3.	+iP1	AB	16 56 49.9	<u>Northwestern Kashmir</u> 35.98 N 73.42 E
	iP2	AB	56 51.5	H = 16 48 28.8 h = 36 km MB = 6.3
	iP3	B	56 53.5	D = 45.97 Az = 308.7 (NEIC)
	iP3	A	56 54.5	P1V A 1.4s 326.0nm M = 6.1
	PmV	B	56 56	P2V A 1.4 614.0nm 6.3
	ePcP	AB	58 25	P3V A 1.7 1030.0nm 6.5
	ePP	B	58 39	PmV B 6 4.4 μm 6.6
	ePPPP	B	59 40	PPV B 9 2.4 μm 6.1
	eS	C	17 03 30	LmH B 14 14.1 μm 6.1
	eSS	C	06 50	LmV B 12 17.9 μm 6.3
	LmH	B	15.1	P'P'V A 1.6 22.0nm
	LmV	B	18.5	
	eP'P'	A	28 22	
3.	eP	A	17 17 40.5	<u>Northwestern Kashmir</u> 35.96 N 73.35 E
				H = 17 09 18.9 h = normal MB = 5.0
				D = 45.94 Az = 308.7 (NEIC)
3.	eP	A	17 54 37	<u>Northwestern Kashmir</u> 36.05 N 73.26 E
				H = 17 46 17.2 h = normal MB = 5.1
				D = 45.83 Az = 308.6 (NEIC)
				traces
3.	+iP	AB	23 12 13.8	<u>Northwestern Kashmir</u> 35.91 N 73.26 E
	LmH	B	33.4	H = 23 03 52.1 h = normal MB = 5.6
	LmV	B	35.6	D = 45.91 Az = 308.7 (NEIC)
				PV A 1.4s 93.0nm M = 5.5
				PV B 3 0.6 μm 6.4
				LmH B 16 0.6 μm 4.6
				LmV B 12 0.9 μm 5.0
4.	eP	A	00 17 46.5	<u>Hokkaido, Japan Region</u> 41.56 N 141.87 E
				H = 00 05 51.4 h = 76 km MB = 5.3
				D = 78.41 Az = 330.3 (NEIC)
				PV A 0.8s 15.4nm M = 5.0
4.	eP	A	00 22 29	<u>Northwestern Kashmir</u> 35.92 N 73.37 E
				H = 00 14 10.0 h = 55 km MB = 5.5
				D = 45.97 Az = 308.7 (NEIC)

Day	Phase		h m s	Remarks
4.	LmH	B	00 55.4	LmH B 20s 0.5/um
	LmV	B	55.5	LmV B 17 0.5/um
4.	+iP	A	00 58 44.0	Northwestern Kashmir 35.99 N 73.27 E H = 00 50 26.7 h = 68 km MB = 5.3 D = 45.87 Az = 308.7 (NEIC) PV A 1.0s 35.4nm M = 5.2
4.	eP	A	01 32 12.5	Northwestern Kashmir 35.80 N 73.31 E H = 01 23 49.5 h = normal MB = 5.4 (NEIC) D = 46.01 PV A 1.3s 52.4nm M = 5.3
4.	eP	A	02 44 39	Northwestern Kashmir 35.94 N 73.35 E H = 02 36 17.1 h = 30 km MB = 5.2 D = 45.95 Az = 308.7 (NEIC) PV A 1.7s 60.6nm M = 5.3
4.	eP	A	03 59 42.5	Northwestern Kashmir 35.93 N 73.46 E H = 03 51 20.9 h = 35 km MB = 5.3 D = 46.03 Az = 308.7 (NEIC) PV A 1.2s 60.9nm M = 5.4
4.	eP	A	07 04 41.5	Western Russia 67.69 N 33.45 E H = 07 00 03.6 h = 7 km MB = 4.6
	e	A	14 00	D = 20.22 Az = 223.2 (NEIC) Underground explosion (UPP) PV A 0.9s 16.1nm M = 4.4
4.	eP	A	10 44 06.5	Northwestern Kashmir 35.67 N 73.42 E H = 10 35 45.6 h = 61 km MB = 4.9 D = 46.16 Az = 308.9 (NEIC) PV A 1.4s 16.3nm M = 4.8
4.	eP	A	13 46 13.5	Northwestern Kashmir 35.80 N 73.30 E H = 13 37 51.4 h = normal MB = 5.2 D = 46.01 Az = 308.8 (NEIC) PV A 1.9s 41.6nm M = 5.0

Day	Phase		h m s	Remarks
4.	+iP	AB	13 50 40.5	Northwestern Kashmir 35.92 N 73.41 E
	eSS	B	14 00 52	H = 13 42 18.1 h = normal MB = 5.8
	LmH	B	09.3	D = 46.00 Az = 308.7 (NEIC)
	LmV	B	12.3	PV A 1.3s 175.0nm M = 5.8
				PV B 4 0.8/um 6.0
				LmH B 10.5 0.6/um 4.8
				LmV B 12 1.1/um 5.1
4.	eP diff	C	18 27 35	Santa Cruz Islands 11.68 S 166.24 E
	ePKIKP	C	30 12	H = 18 11 12.0 h = 62 km MB = 6.0
	ePKIKP	A	30 28.5	D = 135.94 Az = 336.9 (NEIC)
	ePP	AB	33 03	PKIKPV A 1.4s 18.6nm
	ePS	C	43 15	PKIKPV B 9 3.4/um
	ePPS	C	45 15	PPV B 9 7.0/um M = 6.9
	eSS	C	51 12	LmH B 21.5 26.0/um
	eSSS	C	56 07	LmV B 20 27.3/um
	LmH	B	19 33.7	
	LmV	B	33.7	
4.	ePKIKP	A	20 17 32.5	New Britain Region 5.98 S 152.43 E
				H = 19 58 36.9 h = 53 km MB = 5.5
				D = 124.79 Az = 330.7 (NEIC)
				traces
4.	ePKIKP	A	21 43 03.5	Fiji Islands Region 17.88 S 178.43 W
	-iPKHKP	A	43 06	H = 21 24 30.6 h = 607 km MB = 5.3
	iPKP2	A	43 08.8	D = 146.34 Az = 348.4 (NEIC)
4.	ePKP	AB	21 47 33	Tonga Islands 17.65 S 173.01 W
	LmH	B	22 58.6	H = 21 27 51.5 h = normal MB = 4.9
	LmV	B	58.8	D = 146.87 Az = 354.6 (NEIC)
				PKPV A 2.0s 85.5nm
				PKPV B 7 0.5/um
				LmH B 18 0.4/um M = 5.1
				LmV B 16 0.4/um 5.2
4.	+eP	AB	23 45 39	Peru-Ecuador Border Region
	eSKS	B	56 12	3.77 S 80.61 W

Day	Phase		h m s	Remarks
cont. 4.	eS	C	23 56 48	H = 23 32 24.3 h = 63 km MB = 5.4
	LmH	B	24 25.3	D = 94.30 Az = 39.6 (NEIC)
	LmV	B	25.4	PV A 1.4s 51.2nm M = 5.8
				LmH B 20 0.5 µm
				LmV B 19 0.4 µm
5.	+ePKP	AB	02 59 22.5	<u>Loyalty Islands Region</u> 21.64 S 169.74 E
	LmH	C	04 13.0	H = 02 39 46.1 h = 55 km MB = 4.9
	LmV	C	13.5	D = 146.35 Az = 334.7 (NEIC)
				PKPV A 1.1s 46.4nm
				PKPV B 9 0.7 µm
5.	eP	A	03 16 23	<u>Northwestern Kashmir</u> 35.85 N 73.52 E
				H = 03 08 00.7 h = 45 km MB = 5.2
				D = 46.11 Az = 308.8 (NEIC)
5.	eP	A	04 15 50	<u>Northwestern Kashmir</u> 35.87 N 73.41 E
				H = 04 07 27.0 h = normal MB = 5.2
				D = 46.04 Az = 308.8 (NEIC)
5.	eP diff	A	05 37 38	<u>Banda Sea</u> 6.92 S 129.69 E
	ePKIKP	A	41 31	H = 05 23 02.1 h = 86 km MB = 5.8
	epP	A	41 54	D = 112.94 Az = 322.4 (NEIC)
	e	A	42 16	h = 86 km
	ePP	B	42 20	PKIKPV A 1.3s 65.5nm
	e	A	42 23	PPV A 6 0.8 µm M = 6.3
	eSP	B	51 38	LmH B 16.5 0.9 µm
	e	B	53 40	LmV B 17 0.6 µm
	LmH	B	06 28.3	
	LmV	B	35.5	
5.	eP	A	09 22 19.5	<u>Northwestern Kashmir</u> 35.82 N 73.41 E
	e	A	22 27	H = 09 13 57.3 h = normal MB = 5.1
	ePcP	A	23 57	D = 46.07 Az = 308.8 (NEIC)
				PV A 1.1s 16.1nm M = 4.9
5.	ePKP	A	16 43 47.5	<u>New Hebrides Islands</u> 20.88 S 169.72 E
				H = 16 24 17.8 h = 95 km MB = 4.3
				D = 145.65 Az = 335.1 (NEIC)

Day	Phase		h m s	Remarks
5.	+eP	AB	17 32 21	<u>Halmahera</u> 1.88 N 128.17 E
	epP	B	32 51	H = 17 18 27.1 h = 132 km MB = 6.1
	esP	C	34(10)	D = 105.03 Az = 323.9 (NEIC)
	ePP	AB	36 43.5	h = 115 km
	epPP	B	37 12	PV A 1.8s 87.7nm M = 6.5
	e	B	37 25	PPV A 2.0 299.0nm 6.5
	ePPP	B	38 58	PPV B 7 2.0 um 6.8
	e	B	39 38	PKKPV A 1.4 16.3nm
	eS	C	43 45	
	eSPP	C	46 26	
	ePKKP	A	48 04.5	
5.	+iPKP	AB	17 50 06.5	<u>New Hebrides Islands</u> 19.88 S 169.00 E
	ipPKP	AB	50 19.5	H = 17 30 35.7 h = 52 km MB = 6.1
	e	B	51 18	D = 144.47 Az = 335.1 (NEIC)
	ePP	B	53 26	h = 46 km
	eipPP	B	53 43	PKPV A 1.3s 874.0nm
	LmH	B	18 53.7	PKPV B 9.5 3.6 um
	LmV	B	57.1	LmH B 22 4.8 um
				LmV B 22 5.3 um
5.	+ePKP	AB	18 23 03	<u>New Hebrides Islands</u> 19.85 S 168.95 E
	+ipPKP	AB	23 16.5	H = 18 03 35.1 h = 74 km MB = 5.2
				D = 144.42 Az = 335.1 (NEIC)
				h = 48 km
				PKPV A 1.6s 99.0nm
				pPKPV A 1.5 252.0nm
5.	eP	A	19 55 08.5	<u>Fox Islands, Aleutian Is.</u>
	epP	A	55 19	52.41 N 169.50 W
				H = 19 43 15.7 h = 37 km MB = 4.4
				D = 77.32 Az = 359.3 (NEIC)
				h = 39 km
				PV A 1.0s 15.7nm M = 5.0
6.	eP	A	03 00 34.5	<u>Kashmir-Tibet Border Region</u>
	LmH	B	21.0	32.50 N 78.31 E

Day	Phase		h m s	Remarks
cont. 6.	LmV	B	03 24.5	H = 02 51 32.1 h = 44 km MB = 5.3 D = 51.30 Az = 311.5 (NEIC) PV A 1.5s 35.2nm M = 5.1 LmH B 15 0.2 µm 4.3 LmV B 12 0.3 µm 4.6
6.	LmH	B	05 10.0	LmH B 17.5s 0.3 µm
	LmV	B	10.1	LmV B 18 0.4 µm
6.	ePKP	A	06 15 04.5	Fiji Islands 15.98 S 177.96 E
	LmH	C	07 26.5	H = 05 55 28.5 h = normal MB = 5.2
	LmV	C	28.5	D = 143.71 Az = 345.3 (NEIC) PKPV A 2.4s 55.2nm LmH C 18 0.4 µm M = 5.2
6.	eP	A	07 28 58.5	Crete 34.75 N 23.89 E H = 07 24 46 h = 46 km MB = 4.3 D = 18.23 Az = 334 (ISC) PV A 0.6s 11.5nm
6.	ePn	A	08 52 25	Western Mediterranean Sea
	eSn	A	54 12	41.34 N 4.10 E
	eSg	A	55 29	H = 08 49 47.4 h = normal MB = 4.9 D = 10.67 Az = 26.7 (NEIC)
6.	eP	A	11 54 39	Kyushu, Japan 32.87 N 130.30 E
	LmH	B	12 28.1	H = 11 42 24.0 h = 15 km MB = 4.8
	LmV	B	34.3	D = 80.83 Az = 325.5 (NEIC) PV A 2.6s 52.0nm M = 5.1 LmH B 16 0.8 µm 5.2 LmV B 17 1.1 µm 5.3
6.	eP	A	18 16 34.5	Crete 35.51 N 25.51 E H = 18 12 26.4 h = 70 km MB = 4.6 D = 18.17 Az = 330.6 (NEIC) PV A 0.6s 9.6nm
6.	e	A	23 29 35.5	

Day	Phase		h m s	Remarks
7.	ePKHKP	A	00 56 34	South of Fiji Islands 22.23 S 179.65 W
				H = 00 37 42.8 h = 520 km MB = 4.0
				D = 150.29 Az = 345.5 (NEIC)
7.	+eP	AB	03 06 15	Carlsberg Ridge 2.05 S 68.01 E
	ePP	AC	08 49.5	H = 02 54 58.3 h = normal MB = 5.8
	eS	B	15 30	D = 71.05 Az = 325.9 (NEIC)
	eSS	C	20 00	PV A 2.0s 150.0nm M = 5.7
	LmH	B	42.6	PV B 4.5 0.6 µm 5.9
	LmV	B	46.6	PPV A 2.1 63.7nm 5.5
				PPV C 10 0.6 µm 5.7
				LmH B 19 0.6 µm 4.9
				LmV B 18 0.4 µm 4.8
7.	eP	A	04 32 36.5	Northwestern Kashmir 35.96 N 73.56 E
				H = 04 24 15.0 h = normal MB = 4.8
				D = 46.08 Az = 308.7 (NEIC)
7.	eP	A	05 29 40.5	Hokkaido, Japan Region 40.99 N 141.97 E
				H = 05 17 43.0 h = 67 km MB = 4.5
				D = 78.95 Az = 330.4 (NEIC)
7.	eSg	A	19 16 02	Poland 50.23 N 18.91 E
				H = 19 13 27.3 M = 2.8 (WAR)
				D = 4.7
7.	ePKIKP	A	20 19 21	New Britain Region 6.35 S 148.41 E
	epPKIKP	A	19 30	H = 20 00 26.92 h = 51 km MB = 5.3
	ePP	A	21 16.5	D = 123.10 Az = 328.6 (NEIC)
	epPP	A	21 28	h = 32 km
	LmH	B	21 14.8	PKIKPV A 0.8s 7.7nm
	LmV	B	14.8	
7.	eP1n	AB	22 29 17	France 46.03 N 1.06 W
	eP2n	A	29 18.4	H = 22 26 57.9 h = normal MB = 5.3
	eP3n	AB	29 20.6	D = 9.62 Az = 56.7 (NEIC)
	eiPg	A	30 08	P1nV A 0.6s 34.4nm M = 5.8
	e	B	31 16	P2nV A 0.5 71.2nm 6.2

Day	Phase		h m s	Remarks
cont. 7.	e	B	22 31 39	P3nV A 1.0s 140.0nm M = 6.2
	ei	A	32 02	P3nV B 4.5 0.5 μm 6.0
	eiSg	AB	32 17	LmH B 10 45.6 μm 5.6
	LmH	B	32.7	LmV B 4 15.4 μm
	LmV	B	32.8	
8.	eSg	A	01 57 05	<u>France</u> 46.0 N 1.5 W
				H = 01 51 48 (BCIS)
				D = 9.92
8.	e(PP)	A	03 32 11	<u>Banda Sea</u> 5.77 S 130.48 E
				H = 03 12 57.4 h = 64 km MB = 5.6 (ISC)
				D = 112.3
8.	eP	A	06 42 28	<u>Hokkaido, Japan Region</u> 43.24 N 145.47 E
	LmH	C	07 13.5	H = 06 30 34.0 h = 69 km MB = 5.2
				D = 78.23 Az = 332.0 (NEIC)
				PV A 1.0s 25.6nm M = 5.1
8.	eSg	A	08 06 46	<u>France</u> 46.0 N 1.5 W
	e	A	06 57	H = 08 01 23 (BCIS)
				D = 9.92
8.	+iP1	AB	11 39 50.3	<u>Jan Mayen Island Region</u> 71.59 N 10.03 W
	+iP2	AB	39 52.3	H = 11 34 46.2 h = normal MB = 5.7
	eiP3	AB	39 55	D = 23.19 Az = 143.4 (NEIC)
	iS	B	43 59	P1V A 1.5s 136.0nm M = 5.2
	LmH	B	50.8	P2V A 2.0 1990.0nm 6.2
	LmV	B	53.8	P3V A 2.0 2780.0nm 6.4
				SH B 13 49.2nm 6.7
				LmH B 14 75.6 μm 6.3
				LmV B 12 55.1 μm 6.4
8.	eP	A	12 42 31	<u>Kurile Islands</u> 46.04 N 151.81 E
				H = 12 30 37.8 h = 47 km MB = 4.6
				D = 77.73 Az = 335.3 (NEIC)
				PV A 1.2s 20.3nm M = 5.0

Day	Phase		h m s	Remarks
8.	ePKP	A	18 11 15	Tonga Islands 15.30 S 175.54 W
	e	A	11 21	H = 17 51 36.0 h = 33 km MB = 5.2 (NEIC)
	LmH	C	19 12.0	D = 144.27
	LmV	C	12.0	
9.	eP	A	01 49 43.5	Southern Iran 28.50 N 52.67 E
				H = 01 42 29.0 h = 48 km MB = 5.1
				D = 37.94 Az = 317.2 (NEIC)
9.	eSg	A	02 53 32	France 46.0 N 1.5 W
				H = 02 48 01 (BCIS)
				D = 9.92
9.	ePKP	AC	03 03 36	Tonga Islands 15.37 S 175.67 W
	ePP	B	07 00	H = 02 44 03.1 h = 35 km MB = 5.9
	eSS	B	25 40	D = 144.32 Az = 352.0 (NEIC)
	e	B	26 14	PKPV A 2.4s 248.0nm
	LQ	C	44 00	LmH B 25 6.3 µm M = 6.2
	LR	C	51 35	LmV B 23 5.6 µm 6.2
	LmH	B	58.0	
	LmV	B	04 05.4	
9.	LmH	B	07 03.0	LmH B 18s 0.3 µm
	LmV	B	02.6	LmV B 18 0.5 µm
10.	eP	A	02 58 25.5	Kodiak Island Region 56.46 N 153.56 W
	e	A	58 36	H = 02 46 59.1 h = normal MB = 4.9
	e	A	58 50	D = 72.56 Az = 9.8 (NEIC)
				PV A 1.2s 36.6nm M = 5.3
10.	eP	A	03 05 19.5	Kodiak Island Region 56.49 N 153.48 W
				H = 02 53 53.2 h = normal MB = 4.2
				D = 72.52 Az = 9.9 (NEIC)
10.	e(P)	A	03 10 52	Kodiak Island Region 56.47 N 153.54 W
				H = 02 59 24.3 h = normal MB = 4.0
				D = 72.55 Az = 9.8 (NEIC)

Day	Phase		h m s	Remarks
10.	+ePKP	AB	06 22 59	Loyalty Islands 20.20 S 168.80 E
	eSS	C	45 00	H = 06 03 25.7 h = 39 km MB = 5.4
	LmH	B	07 35.4	D = 144.68 Az = 334.7 (NEIC)
	LmV	B	36	PKPV A 1.5s 100.0nm
				PKPV B 10 4.5 μm
				LmH B 19 3.4 μm M = 6.1
				LmV B 19 4.0 μm 6.2
10.	ePKP	A	09 00 10	Fiji Islands Region 18.77 S 177.63 W
				H = 08 41 27.4 h = 541 km MB = 4.5
				D = 147.36 Az = 349.1 (NEIC)
				PKPV A 1.8s 47.3nm
10.	ePKP	A	09 36 08	Loyalty Islands 20.32 S 168.59 E
	e	A	36 17	H = 09 16 35.6 h = normal MB = 4.7
				D = 144.70 Az = 334.5 (NEIC)
				PKPV A 1.5s 20.1nm
10.	ePKP	A	10 26 47	New Hebrides Islands 20.38 S 168.96 E
				H = 10 07 12.6 h = 35 km MB = 4.7
				D = 144.90 Az = 334.8 (NEIC)
				PKPV A 1.5s 32.7nm
10.	ePKP	A	11 08 03	Loyalty Islands 20.28 S 168.89 E
				H = 10 48 28.2 h = 41 km MB = 4.7
				D = 144.79 Az = 334.7 (NEIC)
				PKPV A 1.4s 23.3nm
10.	+iPKP	AB	11 27 02.4	Loyalty Islands 20.06 S 168.80 E
	epPKP	A	27 14	H = 11 07 31.2 h = 58 km MB = 5.2
	LmH	B	12 40.0	D = 144.55 Az = 334.8 (NEIC)
	LmV	B	40.3	h = 41 km
				PKPV A 1.0s 114.0nm
				PKPV B 5 0.8 μm
				LmH B 18 0.6 μm
				LmV B 19 0.6 μm

Day	Phase		h m s	Remarks
10.	eP	A	21 06 41.5	Southern Sinkiang Prov., China
	epP	A	06 55	39.19 N 81.28 E
				H = 20 57 57.1 h = normal MB = 5.1
				D = 48.91 Az = 307.6 (NEIC)
				h = 58 km
				PV A 0.8s 19.2nm M = 5.2
10.	ePKP	A	22 08 04	Loyalty Islands 20.47 S 168.77 E
	e	A	08 14	H = 21 48 27.2 h = 20 km MB = 4.9
	e	A	08 28.5	D = 144.91 Az = 334.5 (NEIC)
				PKPV A 1.3s 21.8nm
11.	eP	A	03 14 54	Kenai-Peninsula, Alaska 59.63 N 148.94 W
	epP	A	15 00	H = 03 03 50.1 h = 27 km MB = 5.1
				D = 68.95 Az = 13.1 (NEIC)
				h = 23 km
				PV A 1.0s 35.4nm M = 5.4
				pPV A 1.4 53.5nm
11.	ePn	A	07 50 56.5	France 46.02 N 1.05 W
	ePg	A	51 45.5	H = 07 48 37.5 h = normal MB = 4.9
	eSg	A	54 00	D = 9.62 Az = 56.7 (NEIC)
				PnV A 0.8s 9.6nm M = 5.1
11.	ePP	B	10 57 22	Bismarck Sea 3.59 S 149.83 E
	LmH	B	12 19.8	H = 10 36 56.9 h = 33 km MB = 5.2 (NEIC)
	LmV	B	23.0	D = 121.46
				LmH B 20s 1.4/um M = 5.6
				LmV B 18 1.4/um 5.7
11.	ePKP	A	12 10 01	Loyalty Islands 20.54 S 168.84 E
				H = 11 50 25.9 h = 44 km MB = 4.2
				D = 145.00 Az = 334.6 (NEIC)
				PKPV A 1.4s 9.3nm
11.	ePKP	A	13 54(25)	Ceram 3.24 S 130.76 E
	ePP	A	54 58.5	H = 13 35 48.6 h = normal MB = 5.5
	e	A	55 08.5	D = 110.67 Az = 323.5 (NEIC)

Day	Phase		h m s	Remarks
cont. 11.	eS diff	B	14 02 32	LmH B 19s 3.6/um M = 6.0
	ePS	B	04 50	LmV B 17.5 3.4/um 6.0
	LmH	B	38.4	
	LmV	B	52.4	
11.	LmV	B	18 39.5	<u>Bismarck Sea</u> 3.65 S 149.91 E
	LmH	B	40.5	H = 17 24 57.1 h = 33 km MB = 5.3 (NEIC)
				D = 121.54
				LmH B 18.5s 0.4/um M = 5.1
				LmV B 17 0.5/um 5.2
12.	ePKP	A	02 36 55.5	<u>Loyalty Islands</u> 20.49 S 168.91 E
	e	A	37 02	H = 02 17 21.6 h = 30 km MB = 4.8
				D = 144.98 Az = 334.6 (NEIC)
				PKPV A 1.2s 30.5nm
12.	ePP	B	04 18 32	<u>Off Coast of Mexico</u> 8.42 N 103.01 S
	eS	B	26 09	H = 04 00 53.3 h = 33 km MB = 5.3 (NEIC)
	ePS	B	27 16	D = 98.65
	eSS	B	32 52	LmH B 20.5s 1.0/um M = 5.3
	LmH	B	56.4	LmV B 16 0.9/um 5.4
	LmV	B	05 05.2	
12.	ePn	A	14 42 35.5	<u>Yugoslavia</u> 44.03 N 17.11 E
	eSg	A	44 45	H = 14 40 44.0 h = normal MB = 4.4
	LmH	B	45.1	D = 7.60 Az = 332.6 (NEIC)
				PnV A 1.5s 22.6nm M = 4.7
				LmH B 9 1.0/um 3.8
13.	-eP1	AB	04 16 43.5	<u>Southern Greece</u> 37.93 N 22.39 E
	+iP2	AB	16 46.2	H = 04 13 20.5 h = 83 km MB = 6.0
	eP3	A	16 51.5	D = 14.85 Az = 332.3 (NEIC)
	eiS	B	19 28	P1V A 1.6s 57.7nm M = 4.6
	LmH	B	23.4	P2V A 1.2 232.0nm 5.3
	LmV	B	23.4	P2V B 4 3.8/um 6.0
				P3V A 1.3 1832.0nm 6.2
				(SH) B 10 13.6/um 6.7

Day	Phase		h m s	Remarks
13.	ePKP	A	06 41 55	Loyalty Islands 20.34 S 168.72 E
	LmH	B	07 55.0	H = 06 22 22.5 h = 48 km MB = 5.1
	LmV	B	55.0	D = 144.77 Az = 334.6 (NEIC)
				PKPV A 1.5s 126.0nm
				LmH B 19 0.2/um M = 4.9
				LmV B 19 0.3/um 4.9
13.	ePKP	AB	13 11 51.5	Loyalty Islands 20.38 S 168.72 E
	e	B	12 17	H = 12 52 17.6 h = 32 km MB = 5.0
	eSS	C	34 10	D = 144.81 Az = 334.5 (NEIC)
	LmH	B	14 25.4	PKPV A 2.6s 251.0nm
	LmV	B	25.4	PKPV B 8.5 2.0/um
				LmH B 18 1.0/um M = 5.6
				LmV B 18 0.9/um 5.6
13.	ePKP	A	13 18 24.5	Loyalty Islands 20.51 S 168.79 E
				H = 12 58 56.1 h = 88 km MB = 4.9
				D = 144.95 Az = 334.5 (NEIC)
				PKPV A 1.3s 45.9nm
13.	ePKP	A	13 26 25	Loyalty Islands 20.28 S 168.78 E
				H = 13 06 55.9 h = 75 km MB = 4.9
				D = 144.74 Az = 334.7 (NEIC)
				PKPV A 1.5s 90.5nm
13.	ePKP	A	13 52 07	Loyalty Islands 20.54 S 168.81 E
				H = 13 32 33.6 h = 34 km MB = 4.7
				D = 144.98 Az = 334.5 (NEIC)
				PKPV A 1.2s 24.4nm
13.	ePKP	AB	14 11 30	Loyalty Islands 20.21 S 168.80 E
				H = 13 51 59.0 h = 58 km MB = 4.9
				D = 144.69 Az = 334.7 (NEIC)
				PKPV A 2.0s 128.0nm
				PKPV B 8 1.4/um

Day	Phase		h m s	Remarks
13.	+ePKP	AB	14 53 11	Loyalty Islands 20.37 S 168.84 E
	ePP	B	56 36	H = 14 33 38.0 h = 35 km MB = 5.9
	eSKSP	B	15 06 44	D = 144.85 Az = 334.7 (NEIC)
	LmH	B	16 06.5	PKPV A 1.4s 906.0nm
	LmV	B	06.5	PKPV B 8.5 12.2 /um
				LmH B 19 10.3 /um M = 6.6
				LmV B 18.5 11.1 /um 6.6
13.	+eP	AB	15 15 18	Loyalty Islands 20.41 S 168.85 E
				H = 14 55 46.1 h = 53 km MB = 5.6
				D = 144.89 Az = 334.6 (NEIC)
				PKPV A 1.5s 344.0nm
13.	eP	A	18 49 32	Near East Coast of Kamchatka
				51.79 N 158.02 E
				H = 18 37 57.2 h = normal MB = 4.7
				D = 74.04 Az = 338.5 (NEIC)
				PV A 0.6s 13.4nm M = 5.1
14.	+ePKP	AB	09 43 47	Loyalty Islands 20.54 S 168.90 E
	e	A	43 55	H = 09 24 12.1 h = 33 km MB = 5.3
	LmH	B	10 56.0	D = 145.02 Az = 334.6 (NEIC)
	LmV	B	56.5	PKPV A 1.5s 211.0nm
				PKPV B 8 0.8 /um
				LmH B 19 0.4 /um M = 5.1
				LmV B 18 0.4 /um 5.2
14.	ePKP	A	12 57 20.5	New Hebrides Islands 18.64 S 169.05 E
				H = 12 38 12.5 h = 211 km MB = 4.7
				D = 143.36 Az = 335.8 (NEIC)
				PV A 1.3s 24.0nm
14.	ePg	A	22 19 20	Yugoslavia 42 3/4 N 18.0 E
	e	A	21 14	H = 22 17 03 (BCIS)
	LmH	B	22.1	D = 9.05
15.	ePKP	A	03 31 48	Fiji Islands Region 17.89 S 178.60 W
				H = 03 13 09.9 h = 600 km MB = 5.0
				D = 146.31 Az = 348.2 (NEIC)

Day	Phase		h m s	Remarks
15.	ePKP	AB	13 38 19	Loyalty Islands 20.05 S 168.76 E
	LmH	B	14 51.5	H = 13 18 45.4 h = 21 km MB = 5.2
	LmV	B	51.5	D = 144.53 Az = 334.8 (NEIC)
				PKPV A 1.3s 56.7nm
				PKPV B 9 0.8 µm
				LmH B 19 0.5 µm M = 5.3
				LmV B 19 0.7 µm 5.4
15.	e	A	16 11 04.5	
15.	ePKP	AB	18 06 26	Loyalty Islands 20.44 S 168.68 E
	LmH	B	19 15.0	H = 17 46 50.8 h = 25 km MB = 5.4
	LmV	B	19.8	D = 144.84 Az = 334.5 (NEIC)
				PKPV A 2.2s 180.0nm
				PKPV B 5.5 1.1 µm
				LmH B 18 0.6 µm M = 5.4
				LmV B 18 0.6 µm 5.4
15.	eP	A	21 38 09	Southwestern Ryukyu Islands
				24.94 N 123.70 E
				H = 21 25 49.7 h = 117 km MB = 5.0
				D = 83.92 Az = 323.6 (NEIC)
				PV A 1.5s 25.1nm M = 4.9
15.	ePKP	A	22 02 13.5	Fiji Islands Region 17.22 S 178.90 W
	e	A	05 04.5	H = 21 43 34.1 h = 539 km MB = 5.3
				D = 145.61 Az = 348.1 (NEIC)
				PKPV A 2.0s 85.5nm
15.	e	A	23 43 00	New Hebrides Islands 19.65 S 168.95 E
				H = 23 23 13.7 h = 51 km
				D = 144.24 Az = 335.2 (NEIC)
16.	eP	AB	03 56 14.5	Albania 40.33 N 19.70 E
	Pm	A	56 17.5	H = 03 53 26.5 h = 15 km MB = 5.1
	LmH	B	04 01.8	D = 11.77 Az = 334.0 (NEIC)
	LmV	B	02.1	PmV A 1.0s 88.5nm M = 6.0
				LmH B 10.5 11.3nm 5.1

Day	Phase		h m s	Remarks
16.	LmH	B	06 32.5	West Chile Rise 41.21 S 85.79 W
	LmV	B	32.5	H = 05 23 09.2 h = 33 km MB = 4.3 (NEIC)
				D = 124.61
				LmH B 20s 0.3/um M = 4.9
				LmV B 20 0.4/um 4.9
16.	eP	AB	09 27 29	Near Coast of Oaxaca, Mexico
	ePP	AB	31 00	15.24 N 96.23 W
	eSKS	B	37 57	H = 09 14 34.6 h = normal MB = 6.0
	eS	B	38 20	D = 89.25 Az = 37.3 (NEIC)
	eSS	B	44 07	PV A 1.5s 75.4nm M = 5.8
	LmH	B	10 03.6	PV B 10 1.6/um 6.3
	LmV	B	12.1	PPV A 2.5 338.0nm 6.3
				PPV B 8 2.2/um 6.6
				SH B 12 3.2/um 6.1
				LmH B 22.5 3.2/um 5.7
				LmV B 16 2.2/um 5.7
16.	ePKP2	A	11 03(28)	Tonga Islands 20.65 S 174.81 W
				H = 10 43 32.7 h = normal MB = 4.6
				D = 149.63 Az = 351.9 (NEIC)
16.	LmV	B	13 31.8	Solomon Islands 10.61 S 161.20 E
	LmH	B	35.8	H = 12 11 26.6 h = 56 km MB = 5.5 (NEIC)
				D = 132.91
				LmH B 17.5s 0.9/um
				LmV B 20 1.0/um
16.	eP	A	14 03 05	North of Ascension Island
	LmH	B	28.7	0.79 S 13.05 W
	LmV	B	28.7	H = 13 53 32.0 h = normal MB = 5.2
				D = 55.39 Az = 18.8 (NEIC)
				PV A 1.5s 25.1nm M = 5.0
				LmH B 17 0.5/um 4.7
				LmV B 16 0.8/um 4.9
16.	eP	A	14 09 12	Albania 41.35 N 20.73 E
	e	B	12 26	H = 14 06 26.1 h = 6 km MB = 5.2

Day	Phase		h m s	Remarks
cont. 16	LmH	B	14 13.9	D = 11.25 Az = 328.8 (NEIC)
	LmV	B	14.0	LmH B 11s 1.0/um M = 4.0
				LmV B 11 1.5/um
16.	LmV	B	17 14.6	Solomon Islands 10.61 S 161.19 E
	LmH	B	17.4	H = 15 53 18.3 h = 52 km MB = 5.4 (NEIC)
				D = 132.90
				LmH B 18s 0.4/um
				LmV B 18.5 0.4/um
17.	ePKHKP	A	09 37 29.5	Fiji Islands Region 20.45 S 178.78 W
	eP'KP2	A	37 35	H = 09 18 49.7 h = 625 km MB = 4.4
				D = 148.77 Az = 347.2 (NEIC)
				PKHKPV A 1.5s 22.6nm
17.	eP	AB	14 10 26.5	Greece 38.28 N 20.34 E
	eS	C	12 48	H = 14 07 15.6 h = normal MB = 5.6
	LmH	B	15.6	D = 13.83 Az = 336.2 (NEIC)
	LmV	B	16.9	LmH B 11s 204.0/um M = 6.4
				LmV B 14.5 172.0/um
17.	eP	A	14 29 48	Greece 38.0 N 20.3 E
				H = 14 26 27 h = 0 km
				D = 14.02 Az = 337 (ISC)
17.	eP	A	14 47 23	Greece 38.47 N 20.34 E
				H = 14 44 10.3 h = normal MB = 4.8
				D = 13.66 Az = 335.9 (NEIC)
				PV A 1.0s 23.6nm M = 5.0
17.	eP	A	15 40 57.5	Greece 38.06 N 20.16 E
				H = 15 37 37.0
				D = 13.97 Az = 337 (ISC)
				PV A 1.0s 11.8nm
17.	eP	A	16 30 10.5	South of Atlantic Ridge 24.39 S 13.02 W
				H = 16 18 13.6 h = normal MB = 4.8
				D = 77.82 Az = 15.8 (NEIC)
				PV A 1.0s 11.8nm M = 4.9

Day	Phase		h m s	Remarks
17.	eP	A	17 02 19	Greece 38.3 N 20.2 E H = 16 59 04 h = 48 km D = 13.71 Az = 336 (ISC)
17.	+iP	AB	17 46 10.0	Northwestern Kashmir 35.94 N 73.26 E H = 17 37 48.5 h = normal MB = 5.4 D = 45.89 Az = 308.7 (NEIC) PV A 1.7s 109.0nm M = 5.5 PV B 3.5 0.4 µm 5.8
17.	eP	A	19 53 50	Greece 38.24 N 20.06 E H = 19 50 30.5 h = 0 km D = 13.78 Az = 337 (ISC)
17.	eP	A	21 09 28.5	Near West Coast of Colombia 5.74 N 77.63 W H = 20 56 52.3 h = 22 km MB = 4.9 D = 85.13 Az = 39.7 (NEIC) PV A 1.6s 24.7nm M = 5.2
17.	eP	A	21 23 03.5	Near West Coast of Colombia 5.80 N 77.52 W H = 21 10 30.5 h = 48 km MB = 4.8 D = 85.01 Az = 39.7 (NEIC)
17.	eP	AC	21 30 15	Near West Coast of Colombia
	epP	A	30 24.5	5.74 N 77.57 W
	eS	B	40 48	H = 21 17 39.2 h = 22 km MB = 5.4
	eSS	C	46 36	D = 85.09 Az = 39.7 (NEIC)
	LmV	C	22 05.4	h = 34 km
	LmH	C	06.5	PV A 2.1s 129.0nm M = 5.8
				PV C 11 0.9 µm 5.9
				SH B 14 1.0 µm 5.7
				LmH C 20.5 0.6 µm 4.9
				LmV C 21 1.1 µm 5.3
17.	eP	A	23 38 12	South of Honshu, Japan
	LmH	B	24 18.9	31.80 N 141.64 E

Day	Phase		h m s	Remarks
cont. 17.	LmV	C	24 23.4	H = 23 25 26.4 h = 16 km MB = 4.8
				D = 86.81 Az = 330.8 (NEIC)
				LmH B 17s 0.4/um M = 4.9
				LmV C 16 0.6/um 5.1
18.	eP	A	01 00 54.5	Greece 38.30 N 20.30 E
				H = 00 57 40.0 h = 40 km MB = 4.3
				D = 13.80 Az = 336 (ISC)
				PV A 1.0s 11.8nm
18.	ePKHKP	A	01 09 34.5	South of Fiji Islands 24.75 S 179.85 E
	e	A	09 47.5	H = 00 50 34.6 h = 499 km MB = 4.7
				D = 152.59 Az = 343.6 (NEIC)
18.	eP	A	02 21 58	South Atlantic Ridge 24.48 S 13.26 W
				H = 02 10 01.9 h = normal MB = 4.6
				D = 77.96 Az = 15.9 (NEIC)
18.	eP	A	05 20 11.5	Greece 38.19 N 20.0 E
				H = 05 16 54 h = 13 km
				D = 13.82 Az = 337 (ISC)
				traces
18.	eP	A	05 48 07	Dominican Republic Region
	LmH	B	06 13.0	19.54 N 70.12 W
	LmV	B	13.0	H = 05 36 53.3 h = 13 km MB = 4.8
				D = 69.96 Az = 42.1 (NEIC)
				PV A 1.9s 30.3nm M = 5.1
				LmH B 22 1.1/um 5.0
				LmV B 21 0.9/um 5.0
18.	ePKP	A	07 33 57	Loyalty Islands 20.37 S 168.24 E
	e	A	34 07	H = 07 14 18.8 h = 10 km MB = 4.7
	e	A	34 19.5	D = 144.60 Az = 334.1 (NEIC)
	e	A	34 32	PKPV A 1.2s 16.2nm

Day	Phase		h m s	Remarks
18.	eP	A	08 23 40	Greece 38.25 N 20.24 E
	LmH	B	28.8	H = 08 20 24.8 h = 14 km MB = 4.6
	LmV	B	30.3	D = 13.83 Az = 336.5 (NEIC)
				PV A 1.1s 34.2nm M = 5.2
				LmH B 16 2.4 μm 4.4
				LmV B 12 1.3 μm
18.	ePKP	AB	11 02 26	Loyalty Islands 20.56 S 168.52 E
	e	A	02 37	H = 10 42 49.8 h = 21 KM MB = 4.8
	e	A	02 47	D = 144.88 Az = 334.3 (NEIC)
				PKPV A 1.3s 39.3nm
18.	eP	A	12 26 12	Near East Coast of Honshu, Japan
				40.74 N 141.23 E
				H = 12 14 18.7 h = 105 km MB = 4.9
				D = 78.89 Az = 330.0 (NEIC)
				PV A 0.8s 11.5nm M = 4.7
18.	eP	A	14 40 31.5	Greece 38.69 N 20.42 E
				H = 14 37 22 h = 59 km
				D = 13.48 Az = 335 (ISC)
				PV A 1.1s 16.1nm M = 4.7
18.	LmH	C	14 47.0	Halmahera 2.61 N 129.17 E
	LmV	C	47.2	H = 13 38 27.6 h = 88 km (NEIC)
				D = 105.03
				LmH C 12s 0.3 μm
				LmV C 13 0.5 μm
18.	eP	A	15 43 32.5	Greece 38.2 N 20.3 E
	LmH	C	50.0	H = 15 40 14 h = 2 km
	LmV	C	50.0	D = 13.87 Az = 336 (ISC)
				LmH C 14s 0.6 μm M = 3.8
				LmV C 14 0.9 μm
18.	ePKHKP	A	18 22 51	North of New Zealand 20.9 S 174.3 W
	eX	A	23 03	H = 18 03 08 h = 85 km
				D = 149.90 Az = 352 (ISC)

Day	Phase		h m s	Remarks
cont. 18.				PKHKPV A 1.5s 15.1nm
				XV A 1.5 20.1nm
18.	ePKIKP	A	20 54 05	Sumba Island Region 9.97 S 119.55 E
	ePP	AB	54 27	H = 20 35 38.2 h = 46 km MB = 5.7
	PPm	A	54 35	D = 109.01 Az = 320.2 (NEIC)
	LmH	C	21 40.5	PKIKPV A 2.0s 29.9nm
	LmV	C	40.5	PPV A 2.0 42.7nm M = 5.8
				PPmV A 1.8 54.0nm 6.0
				LmH C 26 0.6/um 5.0
				LmV C 28 0.4/um 5.1
18.	eP	A	21 50 22	Northwestern Kashmir 35.85 N 73.50 E
				H = 21 42 00.4 h = 35 km MB - 4.8
				D = 46.10 Az = 308.8 (NEIC)
18.	eP	A	22 18 20	Greece 38.26 N 20.13 E
	e	A	18 26	H = 22 15 02 h = 0 km
				D = 13.78 Az = 337 (ISC)
				PV A 0.7s 11.5nm
19.	eP1	AB	01 48 02.5	Dominican Republic Region
	eiP2	AB	48 06	19.54 N 70.15 W
	ePP	B	50 37	H = 01 36 52.4 h = normal MB = 5.8
	ei	B	50 53	D = 69.98 Az = 42.1 (NEIC)
	eS	B	57 14	P2V A 2.3s 280.0nm M = 5.9
	ePS	B	57 50	P2V B 8 1.4/um 6.1
	eSS	B	02 01 50	SH B 11 2.7/um 6.3
	eSSS	B	05 00	LmH B 22 13.2/um 6.1
	LmH	B	12.7	LmV B 20 11.2/um 6.1
	LmV	B	14.2	P'P'V A 2.0 51.3nm
	eP'P'	A	16 05.5	
19.	ePKIKP	A	15 26 54	New Britain Region 5.67 S 148.09 E
	e	A	27 14	H = 15 08 19.1 h = 202 km MB = 5.4
	e	A	28 05.5	D = 122.35 Az = 328.7 (NEIC)
	ePP	AB	28 29.5	PKIKPV A 1.0s 11.8nm
	e(PS)	C	38 20	LmH B 16.5 0.6/um

Day	Phase		h m s	Remarks
cont. 19.	e	B	15 46 24	LmV B 16s 0.4/um
	LmH	B	16 14.3	
	LmV	B	18.0	
19.	e	A	16 45 28.5	
19.	ePKIKP	A	17 59 52.5	Tonga Islands 19.20 S 173.78 W
	ePKHKP	AB	59 56.5	H = 17 40 13.6 h = 55 km MB = 5.4
	ePKP2	A	59 59.5	D = 148.33 Az = 353.5 (NEIC)
				PKIKPV A 1.1s 12.1nm
				PKHKPV A 1.7 84.9nm
				PKP2V A 1.6 55.0nm
19.	e	A	18 19 23	
19.	eP	A	19 53 10	Albania 40.2 N 19.9 E
	e	A	56 02	H = 19 50 22 (BCIS)
				D = 11.98
20.	e	A	02 18 22.5	
20.	ePKIKP	A	08 04 57.5	Tonga Islands 19.33 S 173.83 W
	+ePKHKP	AB	05 03.5	H = 07 45 17.0 h = normal MB = 5.4
	LmH	B	09 16.0	D = 148.46 Az = 353.4 (NEIC)
	LmV	B	16.3	PKIKPV A 1.6s 24.7nm
				PKHKPV A 1.7 121.0nm
				PKHKPV B 10 1.4/um
				LmH B 20 1.0/um M = 5.5
				LmV B 20 0.7/um 5.5
20.	eP	A	17 15 56	Southern Sumatra 4.16 S 101.58 E
	LmH	B	18 03.0	H = 17 02 47.1 h = 66 km MB = 5.2
	LmV	B	08.0	D = 93.17 Az = 320.4 (NEIC)
				PV A 1.5s 20.1nm M = 5.3
				LmH B 19 0.6/um
				LmV B 16 0.4/um

Day	Phase		h m s	Remarks
20.	LmH	B	20 01.5	Dominican Republic Region
	LmV	B	01.5	19.67 N 70.12 W
				H = 19 14 13.4 h = 33 km MB = 4.9 (NEIC)
				D = 69.87
				LmH B 20s 0.5/um M = 4.7
				LmV B 24 0.6/um 4.8
20.	eP1	A	20 53 54.5	Arabian Sea 14.42 N 56.64 E
	eP2	AB	53 56	H = 20 44 53.2 h = normal MB = 5.3
	ePP	A	55 58	D = 51.18 Az = 324.7 (NEIC)
	eS	B	21 01 15	P1V A 1.0s 13.8nm M = 4.8
	LmH	B	19.4	P2V A 2.0 85.5nm 5.3
	LmV	B	20.4	PPV A 2.2 81.6nm 5.3
				LmH B 19 0.7/um 4.7
				LmV B 17 0.7/um 4.9
20.	eP	A	21 01 39.5	Arabian Sea 14.25 N 56.52 E
	eS	B	08 55	H = 20 52 36.3 h = normal MB = 4.9
				D = 51.25 Az = 324.8 (NEIC)
20.	eP	AB	22 22 07	Near Islands, Aleutian Islands
	LmH	B	10.0	51.79 N 174.02 E
	LmV	B	10.0	H = 22 10 15.4 h = 26 km MB = 5.0
				D = 76.86 Az = 348.6 (NEIC)
				PV A 1.5s 25.1nm M = 5.0
				LmH B 14 1.0/um 5.3
				LmV B 14 1.0/um 5.3
21.	eP	A	00 22 43	Northern Sumatra 2.98 N 96.15 E
	LmH	B	01 05.6	H = 00 10 13.3 h = normal MB = 5.2
	LmV	B	10.5	D = 84.24 Az = 320.4 (NEIC)
				PV A 1.5s 35.2nm M = 5.3
				LmH B 24 0.9/um 5.1
				LmV B 16 0.4/um 4.9
21.	eP	A	00 59 10	Northern Sumatra 2.96 N 96.16 E
				H = 00 46 40.4 h = normal MB = 5.0
				D = 84.26 Az = 320.4 (NEIC)
				PV A 1.6s 22.0nm M = 5.1

Day	Phase		h m s	Remarks
21.	eP	A	02 25 05	Greece 37.97 N 20.28 E
	e	A	25 12	H = 02 21 48.5 h = normal MB = 4.0
	e	A	25 15	D = 14.09 Az = 336.8 (NEIC)
21.	ePn	A	05 47 35.5	Yugoslavia 43 3/4 N 18.0 E
	eSn	A	49 03	H = 05 45 29 (BCIS)
	eSg	A	49 58	D = 8.2
21.	eP	A	09 05 24	Western Russia 52.13 N 51.99 E
	e	A	05 33	H = 09 00 01.2 h = 28 km MB = 5.1
				D = 25.02 Az = 282.8 (NEIC)
				PV A 1.2s 20.3nm M = 4.6
21.	eP	AB	15 42 18.5	Southern Nevada 37.08 N 116.04 W
	e	A	45 12	H = 15 30 00.2 h = 0 km MB = 5.7
	ePP	A	45 23	D = 81.26 Az = 30.7 (NEIC)
				Nuclear Explosion OSCURO (USAEC)
				PV A 1.5s 65.3nm M = 5.5
				PPV A 1.6 27.5nm 5.3
21.	ePKP2	A	21 19 27	North of New Zealand 20.91 S 174.7 W
				H = 20 59 26.7 h = 0 km MB = 4.7
				D = 149.91 Az = 352 (ISC)
22.	ePKHKP	A	08 21 37	Fiji Islands Region 20.24 S 178.02 W
	ePKP2	A	21 42	H = 08 02 55.9 h = 608 km MB = 4.5
				D = 148.72 Az = 348.2 (NEIC)
				PKHKPV A 1.3s 19.6nm
22.	-eiPKP	AB	12 04 29	Tonga Islands 16.47 S 174.60 W
	+epPKP	AB	05 20	H = 11 45 14.8 h = 219 km MB = 5.5
				D = 145.54 Az = 353.0 (NEIC)
				h = 197 km
				PKPV A 1.5s 350.0nm
				PKPV B 3 1.0/um

Day	Phase		h m s	Remarks
22.	eP	A	14 21 38	South of Honshu, Japan 33.29 N 140.59 E
				H = 14 09 06.6 h = 59 km MB = 5.0
				D = 85.08 Az = 330.2 (NEIC)
				PV A 2.0s 29.9nm M = 5.0
22.	-eP	AB	20 09 59	Taiwan Region 22.33 N 121.20 E
	ePP	AC	13 16	H = 19 57 27.4 h = normal MB = 5.7
	eSKS	B	20 20	D = 84.62 Az = 323.0 (NEIC)
	eS	B	20 40	PV A 2.3s 256.0nm M = 6.0
	eSP	B	21 14	PV B 15 2.9 μm 6.3
	eSS	C	25 33	PPV A 2.0 102.5nm 5.9
	LmH	B	52.7	PPV C 15 2.5 μm 6.4
	LmV	B	52.7	LmH B 17.5 37.6 μm 6.8
				LmV B 17 51.6 μm 7.0
22.	e(P)	A	21 46 47	Northeast of Taiwan 25.05 N 123.94 E
				H = 21 34 15.0 h = normal MB = 4.2
				D = 83.95 Az = 323.7 (NEIC)
22.	ePKHKP	A	23 11 57.5	Fiji Islands Region 20.23 S 178.44 W
				H = 22 53 13.2 h = 578 km MB = 4.2
				D = 148.62 Az = 347.7 (NEIC)
				PKHKPV A 1.4s 18.6nm
23.	-ePKP	A	01 29 54	Loyalty Islands 20.98 S 168.47 E
	+ipPKP	A	30 04.8	H = 01 10 18.1 h = 33 km MB = 4.9
				D = 145.24 Az = 334.0 (NEIC)
				h = 39 km
				PKPV A 1.4s 88.4nm
				pPKPV A 1.0 74.8nm
23.	eP	A	01 56 17	Bulgaria 42.12 N 25.22 E
	e	A	56 21.5	H = 01 53 16.7 h = normal MB = 4.7
	e	A	58 55	D = 12.67 Az = 316.9 (NEIC)
23.	eP	AB	02 26 58.5	Taiwan Region 22.32 N 121.36 E
	e	A	27 08.5	H = 02 14 26.8 h = normal MB = 5.4
	ePP	AB	30 14.5	D = 84.72 Az = 323.0 (NEIC)

Day	Phase		h m s	Remarks
cont. 23.	eSKS	B	02 37 20	PV A 3.0s 447.0nm M = 6.4
	ePS	B	38 14	PV B 4 1.1/um 6.4
	eSS	C	42 55	PPV A 3.0 302.3nm 6.2
	LmH	B	03 10.1	LmH B 16 23.2/um 6.7
	LmV	B	10.1	LmV B 16.5 24.3/um 6.7
23.	ePKP	A	03 19 09	Loyalty Islands 20.88 S 168.54 E
	epPKP	A	19 18.5	H = 02 59 33.0 h = normal MB = 4.7
				D = 145.18 Az = 334.1 (NEIC)
				h = 34 km
23.	e	A	03 38 34.5	Ionian Sea 38.2 N 19.9 E
				H = 03 35 03 h = 0 km
				D = 13.77 Az = 337 (ISC)
23.	eP	A	07 32 46	West Pakistan 25.39 N 66.73 E
				H = 07 23 58.6 h = 10 km MB = 4.4
				D = 48.84 Az = 316.1 (NEIC)
23.	eP	A	10 52 10	Andreanof Islands, Aleutian Is.
				51.23 N 175.01 W
				H = 10 40 12.9 h = 41 km MB = 4.8
				D = 78.35 Az = 355.7 (NEIC)
23.	eP	A	11 17 58	North of Svalbard 85.5 N 7.0 W
				H = 11 11 36 h = 0 km
				D = 31.52 Az = 157 (ISC)
23.	ePKHKP	A	11 39 01.5	South of Fiji Islands 22.13 S 179.70 W
				H = 11 20 19.6 h = 640 km MB = 4.6
				D = 150.19 Az = 345.4 (NEIC)
23.	eP	A	22 09 12	Austria 47.7 N 15.8 E
				H = 22 08.0 M = 1.7 (VIE)
				D = 4.03 Az = 319 (ISC)

Day	Phase		h m s	Remarks
24.	e(pP)	A	13 22 13	Taiwan Region 22.53 N 121.28 E
	LmH	B	14 02.4	H = 13 09 32.0 h = 53 km MB = 4.2 (NEIC)
	LmV	B	04.7	D = 84.51
				LmH B 16s 0.6 /um
				LmV B 16 0.7 /um
24.	ePKP	A	14 43 00.5	Loyalty Islands 20.71 S 168.41 E
	epPKP	A	43 10.5	H = 14 23 21.2 h = 11 km
				D = 144.97 Az = 334.1 (NEIC)
				h = 34 km
24.	ePKP	A	15 19 24.5	Loyalty Islands 21.67 S 167.94 E
	epPKP	A	19 35	H = 14 59 58.4 h = 85 km
				D = 145.64 Az = 333.1 (NEIC)
				h = 38.5 km
				PKPV A 1.3s 15.6nm
24.	eP diff	AB	20 24 20.5	Tanimbar Islands Region
	ePKIKP	AB	28 15	6.28 S 131.22 E
	e	C	28 50	H = 20 09 35.6 h = normal MB = 6.1
	ePP	AB	29 03.5	D = 113.36 Az = 322.9 (NEIC)
	ePP	C	29 10	PKIKPV A 1.4s 14.0nm
	eSKS2	C	34 58	PPV A 2.0 179.3nm M = 6.5
	ePS	C	38 40	PPV B 4.5 1.0 /um 6.9
	+iSPP	C	39 55	PPV C 20 4.0 /um 6.8
	eSS	C	45 12	LmH B 21 16.6 /um 6.6
	LmH	B	21 13.1	LmV B 21 18.7 /um 6.7
	LmV	B	26.2	
24.	eX	A	22 48 49	XV A 1.4s 18.6nm
24.	ePP	A	22 58 40	Tanimbar Islands Region
				6.19 S 131.20 E
				H = 22 39 15.1 h = normal MB = 5.5
				D = 113.27 Az = 322.9 (NEIC)
24.	ePP	A	23 39 47	Tanimbar Islands Region 6.39 S 131.21 E
				H = 23 20 14.6 h = 33 km MB = 5.5 (NEIC)
				D = 113.43

Day	Phase		h m s	Remarks
24.	ePP	A	24 18 37	Tanimbar Islands Region 6.35 S 131.14 E
	e	A	19 26	H = 23 59 04.8 h = 33 km MB = 5.8 (NEIC)
	e(SKS)	BZ	24 45	D = 113.34
	LmH	B	25 09.2	PPV A 1.9s 19.0nm M = 5.5
	LmV	B	09.8	LmH B 16 0.4 μm 5.1
				LmV B 20 0.4 μm 5.1
25.	eP	A	09 39 43.5	Near East Coast of Honshu, Japan
	epP	A	40 13	38.49 N 141.81 E
	LmV	B	10 19.6	H = 09 27 33.5 h = 58 km MB = 5.2
	LmH	B	19.7	D = 81.06 Az = 330.5 (NEIC)
				h = 118 km
				PV A 1.7s 36.4nm M = 5.0
				pPV A 1.5 42.7nm
				LmH B 16 0.5 μm
				LmV B 16 0.6 μm
25.	-eP	A	11 42 11.5	Nicobar Islands Region 9.83 N 93.94 E
				H = 11 30 17.8 h = normal MB = 5.3
				D = 77.61 Az = 319.8 (NEIC)
				PV A 1.6s 33.0nm M = 5.1
25.	eP	A	16 12 30.5	Honshu, Japan 35.87 N 138.44 E
				H = 16 00 30.5 h = 186 km MB = 4.9
				D = 81.96 Az = 329.0 (NEIC)
				PV A 1.2s 12.2nm M = 4.5
25.	eP	A	18 09 31	Dodecanese Islands 36.63 N 26.83 E
				H = 18 05 30.0 h = 152 km MB = 4.2
				D = 17.77 Az = 326.8 (NEIC)
25.	ePKP2	A	19 13 07	Loyalty Islands Region 21.96 S 170.19 E
				H = 18 53 21.8 h = normal
				D = 146.81 Az = 334.9 (NEIC)
26.	ePKP	A	08 39 20	Samoa Islands Region 16.31 S 172.31 W
				H = 08 19 40.6 h = normal MB = 4.5
				D = 145.60 Az = 355.6 (NEIC)
				PKPV A 1.7s 24.2nm

Day	Phase		h m s	Remarks
26.	eP1	AB	12 21 26	<u>Crete</u> 34.37 N 26.38 E
	iP2	A	21 33.5	H = 12 17 00.9 h = normal MB = 5.1
	LmH	B	31.0	D = 19.51 Az = 330.9 (NEIC)
	LmV	B	31.0	P1V A 0.8s 17.3nm M = 4.4
				P2V A 0.9 46.7nm
				LmH B 15 0.7 µm 4.1
				LmV B 13 0.7 µm 4.3
26.	eP	A	14 42 16.5	<u>Southern Nevada</u> 37.12 N 116.09 W
				H = 14 30 00.2 h = 0 km MB = 4.4
				D = 81.25 Az = 30.6 (NEIC)
				Nuclear explosion DELPHINIUM (USAEC)
				traces
26.	eP diff	A	21 20 11.5	<u>San Juan Province, Argentina</u>
	ePKIKP	A	24 12.5	30.91 S 68.10 W
	ePP1	AB	24 27	H = 21 05 44.4 h = 16 km MB = 5.9
	ePP2	A	24 34	D = 107.25 Az = 41.0 (NEIC)
	ePS	B	33 44	PKIKPV A 2.3s 61.0nm
	eSS	B	39 36	PP1V A 2.2 147.3nm M = 6.3
	eSSS	C	43 20	PP2V A 2.2 266.2nm 6.5
	LmV	B	22 10.0	LmH B 18.5 3.6 µm 6.0
	LmH	B	10.3	LmV B 18.5 3.6 µm 6.0
27.	eP	A	00 19 23	<u>Szechwan Province, China</u>
	e	A	19 37.5	30.31 N 101.67 E
	LmH	B	45.6	H = 00 08 29.9 h = normal MB = 5.0
	LmV	B	51.2	D = 67.25 Az = 316.3 (NEIC)
				PV A 1.6s 22.0nm M = 5.0
				LmH B 18.5 2.8 µm 5.5
				LmV B 13 1.3 µm 5.4
27.	eP	A	02 12 06.5	<u>West Pakistan</u> 33.92 N 72.72 E
	LmV	C	37.7	H = 02 03 39.1 h = 46 km MB = 4.9
	LmH	C	35.8	D = 46.84 Az = 310.2 (NEIC)
				PV A 1.4s 18.6nm M = 4.9
				LmH C 18 0.4 µm 4.4
				LmV C 18 0.6 µm 4.6

Day	Phase		h m s	Remarks
27.	+ePKP	A	05 07 58.5	Samoa Islands Region 16.47 S 172.22 W
				H = 04 48 21.0 h = 33 km MB = 4.9 (NEIC)
				D = 145.77
				PKPV A 1.6s 102.0nm
27.	-ePKP	AB	09 21 20.5	Samoa Islands Region 16.47 S 172.17 W
	i	A	21 25	H = 09 01 43.8 h = normal MB = 5.9 (NEIC)
	e	B	21 58	D = 145.77
	e	C	22 21	PKPV A 1.3s 476.0nm
	e(SKS)	BZ	25 28	PKPV B 5 2.4 µm
	LmH	B	10 29.5	LmH B 18.5 2.8 µm M = 6.0
	LmV	B	35.5	LmV B 17.5 2.6 µm 6.1
27.	ePKP	A	09 49 48.8	Samoa Islands Region 16.26 S 172.32 W
	e	A	49 53	H = 09 30 11.5 h = 33 km MB = 5.2 (NEIC)
	e	A	50 10	D = 145.55
	e	A	50 23	PKPV A 1.2s 44.7nm
27.	e	A	17 28 40	
27.	eP	AB	18 07 19	Taiwan Region 22.37 N 121.19 E
	eS	B	17 40	H = 17 54 49.7 h = 54 km MB = 5.0 (NEIC)
	LmV	B	50.1	D = 84.58
	LmH	B	50.7	PV A 2.0s 94.0nm M = 5.5
				LmH B 16 2.8 µm
				LmV B 13.5 2.5 µm
27.	eP	A	18 40 25	Taiwan Region 22.10 N 121.01 E
				H = 18 27 50.4 h = 20 km MB = 4.7 (NEIC)
				D = 84.70
				PV A 1.5s 15.1nm M = 5.0
27.	eP	A	20 33 17	Northwestern Kashmir 35.17 N 73.04 E
				H = 20 24 55.6 h = 39 km MB = 5.0 (NEIC)
				D = 46.24
				traces

Day	Phase		h m s	Remarks
27.	ePKHKP	A	21 53 55	Tonga Islands Region 23.03 S 175.90 W
	ePKP2	A	54 07	H = 21 34 03.2 h = normal MB = 5.0 (NEIC)
				D = 151.81
				PKHKPV A 1.3s 24.0nm
28.	eP	A	06 06 31	Near East Coast of Kamchatka
				53.46 N 159.96 E
				H = 05 55 08.0 h = 70 km MB = 5.1 (NEIC)
				D = 72.91
				PV A 1.0s 9.8nm M = 4.7
28.	ePKHKP	A	07 17 19	South Pacific Ocean 47.79 S 119.70 W
	ePKP2	A	17 23	H = 06 57 35.2 h = normal MB = 5.1 (NEIC)
				D = 148.53
29.	eP	A	01 20 16	Greece 38.2 N 20.6 E
	LmH	B	25.2	H = 01 16 59 h = 34 km MB = 4.0
	LmV	B	27	D = 14.02 Az = 336 (ISC)
				LmH B 15s 0.4/um M = 3.6
29.	eP	A	02 37 02	Ionian Sea 38.0 N 20.1 E
	LmH	C	41.9	H = 02 33 38 h = 0 km MB = 4.0
				D = 14.02 Az = 337 (ISC)
				LmH C 16s 0.5/um M = 3.7
29.	eP	A	06 30 06.5	Central Russia 52.38 N 95.27 E
				H = 06 21 18.8 h = normal MB = 4.7 (NEIC)
				D = 49.27
				PV A 1.6s 19.2nm M = 4.9
29.	ePKIKP	AB	06 57 19.5	Tonga Islands 21.13 S 174.63 W
	ePKHKP	AB	57 25.5	H = 06 37 35.3 h = normal MB = 5.8 (NEIC)
	ei(PKP2)	AB	57 35	D = 150.13
	LmH	B	08 06.4	PKIKPV A 2.0s 51.3nm
	LmV	B	14.2	PKHKPV A 1.9 190.0nm
				PKP2V A 1.9 257.9nm
				LmH B 18.5 1.7/um M = 5.8
				LmV B 18 1.7/um 5.9

Day	Phase		h m s	Remarks
29.	eP	A	09 11 32	Southeastern Alaska 60.20 N 141.04 W
				H = 09 00 36.3 h = 15 km MB = 4.6 (NEIC)
				D = 67.32
				PV A 1.0s 9.8nm M = 5.0
29.	eP	A	13 04 12	Kurile Islands 44.91 N 147.76 E
				H = 12 52 17.6 h = normal MB = 4.6 (NEIC)
				D = 77.51
				PV A 1.2s 18.3nm M = 5.0
29.	eP	A	14 05 24.5	Southern Sinkiang Prov., China
				39.73 N 77.75 E
				H = 13 56 59.5 h = normal MB = 5.1 (NEIC)
				D = 46.40
29.	eP	A	16 32 35	Szechwan Province, China
	ePcP	A	33 05	30.40 N 101.52 E
	eS	C	41 24	H = 16 21 38.0 h = 33 km MB = 5.1 (NEIC)
	eSa	C	49 00	D = 67.10
	LmH	B	58.7	PV A 1.1s 8.1nm M = 4.7
	LmV	B	17 04.3	LmH B 20 3.7/um 5.6
				LmV B 12.5 1.6/um 5.5
29.	eP	A	20 35 36	Szechwan Province, China
	ePcP	A	36 10.5	30.28 N 101.67 E
	eS	C	44.5	H = 20 24 42.0 h = 33 km MB = 5.1 (NEIC)
	eSa	C	52 00	D = 67.28
	LmH	B	21 01.8	PV A 2.0s 34.2nm M = 5.1
	LmV	B	07.3	LmH B 20 2.9/um 5.5
				LmV B 12 1.3/um 5.4
30.	ePKHKP	A	03 29 21.5	Tonga Islands 21.09 S 174.40 W
	ePKP2	A	29 27.5	H = 03 09 32.5 h = normal MB = 5.3 (NEIC)
				D = 150.12
				PKHKPV A 1.4s 30.2nm
30.	eP	A	07 07 47	Hindu Kush Region 36.4 N 70.6 E
	epP	A	08 31.5	H = 06 59 59.8 h = 204 km MB=4.9 (NEIC)

Day	Phase		h m s	Remarks
cont. 30.				D = 43.88 h = 211 km
				PV A 1.4s 11.6nm M = 4.2
30.	-iP	A	10 37 02.5	Afghanistan-USSR Border Region
				36.60 N 71.42 E
				H = 10 29 07.8 h = 176 km MB = 4.9 (NEIC)
				D = 44.31
30.	e	A	22 00 23	
	e	A	00 29	
30.	LmH	C	22 22.7	LmH C 28s 0.3/um

Day	Phase		h m s	Remarks
1.	eP	A	00 59 02.5	Rumania 45.80 N 26.18 E
	e	A	59 24	H = 00 56 25.6 h = 155 km MB = 4.6 (NEIC)
	e	A	01 00 35	D = 10.85
				PV A 1.4s 18.6nm M = 4.5
1.	eP	A	04 34 26	Yugoslavia 43.48 N 21.46 E
	e	A	34 57	H = 04 32 00.2 h = 3 km
	eS	A	37 18	MB = 4.9 MS = 4.3 (NEIC)
	LmH	C	37.7	D = 9.82
	LmV	C	38.7	PV A 1.2s 28.4nm M = 5.4
				LmH C 13.5 3.4/um 4.3
				LmV C 11 4.6/um
1.	eP	A	06 40 42	Nicaragua 11.1 N 84.03 W
				H = 06 28 11 h = 60 km MB = 4.1
				D = 85.08 Az = 39 (ISC)
				PV A 1.0s 11.8nm M = 4.9
1.	eP	A	10 19 29	Central Alaska 62.74 N 149.08 W
	epP	A	19 47.5	H = 10 08 49.7 h = 76 km MB = 4.7
				D = 65.93 Az = 13.3 (NEIC)
				h = 73 km
				PV A traces
				pPV A 0.8s 19.2nm
1.	eP	A	19 54 20.5	Utah 40.58 N 111.30 W
				H = 19 42 29.2 h = 5 km MB = 4.7
				D = 76.34 Az = 33.4 (NEIC)
1.	-iP	AB	24 02 09	Mindanao, Philippine Islands
	e	A	04 29	7.47 N 123.75 E
	ePP	A	06 16.5	H = 23 49 36.2 h = 613 km MB = 6.0
	eS	B	12 43	D = 97.95 Az = 323.5 (NEIC)
	eSP	B	14 08	PV A 1.4s 214.0nm M = 6.3
	ePKKP	A	18 41	PKKPV A 1.2 16.3nm
	eSSS	C	23 10	LmH C 22 0.45/um
	LmH	C	50.8	
	LmV	B	51.5	

Day	Phase		h m s	Remarks
2.	eP	A	02 34 27.5	Kurile Islands 47.94 N 153.43 E H = 02 22 39.1 h = normal MB = 4.9 D = 76.45 Az = 336.1 (NEIC)
2.	e	A	13 56 06.5	
2.	eX	A	14 16 52.5	XV A 0.6s 7.7nm
2.	e	A	14 35 40.5	
2.	ePKIKP ePKHKP	A A	15 49 16 49 19	Fiji Islands Region 18.03 S 178.39 W H = 15 30 43.7 h = 609 km MB = 4.8 D = 146.49 Az = 348.4 (NEIC) PKHKPV A 1.2s 22.4nm
2.	LmH	C	21 39.0	Chile-Argentina Border Region 33.90 S 70.83 W H = 20 35 45.9 h = 81 km MB = 5.5 (NEIC) D = 111.00 LmH C 22s 0.4/um
2.	eP	A	22 03 45.5	Kurile Islands 43.11 N 146.37 E H = 21 51 45.4 h = normal MB = 4.4 D = 78.65 Az = 332.5 (NEIC)
2.	epP	A	23 30 50.5	South of Honshu, Japan 32.09 N 141.74 E H = 23 17 56.8 h = 34 km MB = 4.6 D = 86.59 Az = 330.8 (NEIC)
3.	ePKP	A	00 25 11	Samoa Region 16.08 S 172.10 W H = 00 05 29.8 D = 145.39 Az = 356 (ISC) PV A 1.1s 16.1nm
3.	+eP1 eiP2	AB A	09 04 57 05 02	Southwestern Russia 46.85 N 45.01 E H = 08 59 57.8 h = 0 km MB = 5.8 D = 22.23 Az = 292.2 (NEIC)

Day	Phase		h m s	Remarks
cont. 3.				underground explosion, MB = 6.2 (UPP)
				P1V A 1.4s 209.0nm M = 5.4
				P2V A 1.3 284.0nm 5.6
3.	eP	A	10 26 13	Kurile Islands 47.27 N 153.99 E
	e	A	26 19	H = 10 14 19.8 h = normal MB = 5.3
	e	A	26 24.5	D = 77.21 Az = 336.5 (NEIC)
	LmH	B	11 00.3	PV A 1.4s 83.6nm M = 5.6
	LmV	B	06.8	LmH B 18 0.5/um 4.9
				LmV B 16 0.4/um 4.9
3.	ePKIKP	A	18 24 29	Tonga Islands 21.53 S 174.43 W
	ePKHKP	A	24 34.5	H = 18 04 41.9 h = 61 km MB = 5.1
	ePKP2	A	24 42	D = 150.56 Az = 352.2 (NEIC)
	LmH	B	19 39.8	PKIKPV A 1.3s 17.5nm
	LmV	B	39.8	LmH B 18 0.35/um
				LmV B 18 0.3/um
3.	e	A	22 14 58	
4.	eP	A	05 35 02	Kurile Islands 46.52 N 153.58 E
				H = 05 23 06.5 h = normal MB = 4.3
				D = 77.80 Az = 336.3 (NEIC)
4.	eP	A	06 18 24.5	Turkey 39.12 N 29.36 E
				H = 06 14 25.1 h = normal MB = 4.4
				D = 17.00 Az = 318.4 (NEIC)
				PV A 1.7s 24.2nm M = 4.1
4.	eP	A	08 52 02.5	Greece 38.32 N 20.22 E
	LmH	B	57.1	H = 08 48 45.9 h = normal MB = 4.8
	LmV	B	58.5	D = 13.75 Az = 336.4 (NEIC)
				PV A 1.0s 27.6nm M = 5.0
				LmH B 16 3.1/um 4.8
				LmV B 13 2.7/um

Day	Phase		h m s	Remarks
5.	LmH	B	05 00.1	Kyushu, Japan 30.08 N 130.27 E
	LmV	B	01.1	H = 04 06 23.6 h = 24 km MB = 5.0 (NEIC)
				D = 83.11
				LmH B 18s 1.2μm M = 5.3
				LmV B 13 1.0μm 5.4
5.	eP	A	11 00 57.5	Near Coast of Guatemala 13.84 N 91.07 W
	epP	A	01 16.5	H = 10 48 19.1 h = 89 km MB = 5.4
	LmV	B	39.8	D = 87.28 Az = 38.4 (NEIC)
	LmH	B	40.5	h = 74 km
				PV A 1.4s 25.6nm M = 5.1
				LmH B 18.5 2.2μm
				LmV B 18 1.8μm
5.	ePn	A	18 42 41	Northern Italy 44.49 N 11.02 E
	e	A	42 56.5	H = 18 41 08.5 h = 19 km MB = 4.2
	ePg	A	43 07.5	D = 6.17 Az = 3.6 (NEIC)
	eSn	A	43 47	
	eSg	A	44 28.5	
6.	ePn	A	04 36 02.5	Switzerland 46.2 N 7.2 E
	e	A	36 16	H = 04 34 41 (BCIS)
	iPg	A	36 19	D = 5.35
	eSg	A	37 27	
6.	eX	A	05 32 09	Fiji Region 17.7 S 176.6 W
				H = 05 12 17.2 h = 0 km
				D = 146.48 Az = 350 (ISC)
				XV A 1.9s 19.0nm
6.	ePg	A	05 43 12	Northern Italy 44.5 N 11.0 E
	eSn	A	43 46.5	H = 05 41 08 (BCIS)
	eSg	A	44 30	D = 6.20
6.	eP	A	11 43 27	Near S. Coast of Honshu, Japan
	e(pP)	A	43 35	34.35 N 138.41 E
	LmH	B	12 18.6	H = 11 31 04.3 h = 51 km MB = 5.2

Day	Phase		h m s	Remarks
cont. 6.	LmV	B	12 26.1	D = 83.25 Az = 329.1 (NEIC)
				PV A 1.3s 13.1nm M = 4.8
				LmH B 18.5 1.7/um
				LmV B 13.5 1.0/um
6.	ePKIKP	A	16 54 03	New Britain Region 6.13 S 149.88 E
	e(pPKIKP)	A	54 13	H = 16 35 10.9 h = 61 km MB = 5.4
				D = 123.66 Az = 329.4 (NEIC)
				PKIKPV A 1.1s 14.1nm
6.	ePKIKP	A	17 01 47.5	New Britain Region 6.06 S 149.84 E
	e	A	02 08	H = 16 42 54.4 h = 59 km MB = 5.5
	e	A	02 25.5	D = 123.58 Az = 329.4 (NEIC)
	e	A	02 32	PPV B 16s 1.42/um M = 5.9
	ePP	B	03 32	LmH B 21 13.0/um
	ePS	B	13 20	LmV B 21 13.3/um
	ePPS	C	14 48	
	eSS	C	20 48	
	eSSS	C	25 26	
	LmV	B	56.4	
	LmH	B	56.5	
6.	eP	A	17 11 49	North-Western Kashmir 36.29 N 73.4 E
				H = 17 03 15 h = 33 km (ISC)
				D = 46.5
6.	eP	A	20 34 28	Northern Sumatra 3.27 N 98.52 E
	epP	A	35 04	H = 20 22 03.2 h = 124 km MB = 5.3
				D = 85.52 Az = 320.4 (NEIC)
				h = 142 km
				PV A 1.2s 20.3nm M = 4.9
6.	ePKP	A	23 33 30	Fiji Region 16.4 S 178.68 W
				H = 23 13 55.5 h = 0 km MB = 4.9
				D = 144.86 Az = 349 (ISC)
				PKPV A 1.7s 27.2nm

Day	Phase		h m s	Remarks
7.	eP	A	01 42 34.5	Laptev Sea 73.39 N 134.49 E
				H = 01 33 37.9 h = normal MB = 4.7
				D = 50.27 Az = 316.0 (NEIC)
7.	ePKIKP	A	05 52 11.5	Fiji Islands Region 21.44 S 177.80 W
	ePKHKP	A	52 17	H = 05 33 08.2 h = 365 km MB = 5.0
	ePKP2	A	52 25	D = 149.93 Az = 348.0 (NEIC)
	epPKIKP	A	53 40	h = 368 km
	epPKHKP	A	53 46	
	epPKP2	A	53 51	
7.	ePn	A	08 34 20.5	Yugoslavia 44.8 N 14.8 E
	eSn	A	35 40	H = 08 32 35 h = 50 km
				D = 7.45 Az = 324 (ISC)
7.	ePKIKP	A	12 53 47	New Ireland Region 4.49 S 153.10 E
	epPKIKP	A	54 06.5	H = 12 34 56.4 h = 86 km MB = 5.2
				D = 123.82 Az = 331.5 (NEIC)
				h = 71 km
8.	ePKIKP	A	03 04 44	West Chile Rise 38.42 S 93.29 W
	e	A	04 50.5	H = 02 45 41.6 h = normal MB = 5.4
	LmH	B	59.5	D = 127.29 Az = 50.7 (NEIC)
	LmV	B	04 01.6	PKIKPV A 2.2s 81.5nm
				LmH B 20 0.3/um M = 5.0
				LmV B 18 0.4/um 5.1
8.	ePKIKP	A	19 46 16.5	Fiji Islands Region 20.14 S 178.66 W
	ePKHKP	A	46 23.5	H = 19 27 36.8 h = 613 km MB = 4.4
				D = 148.49 Az = 347.4 (NEIC)
8.	ePKP2	A	23 21 19.5	Tonga 24.75 S 175.59 W
				H = 23 01 05.0
				D = 153.55 Az = 350 (ISC)
				PKP2V A 1.4s 9.3nm
9.	eP	A	07 25 58.5	Southern Iran 28.17 N 56.03 E
	ePP	A	27 44	H = 07 18 23.7 h = normal MB=5.2 MS=5.0

Day	Phase		h m s	Remarks
cont. 9.	LmH	B	07 42.5	D = 40.21 Az = 316.4 (NEIC)
	LmV	B	47.0	PV A 1.0s 23.6nm M = 4.9
				LmH B 22.5 1.0 µm 4.6
				LmV B 18 0.51 µm 4.5
9.	eP	A	09 45 20	Kurile Islands 43.22 N 146.31 E
	e	A	45 23	H = 09 33 23.26 h = 58 km MB = 5.3
	e	A	45 32.5	D = 78.53 Az = 332.5 (NEIC)
	e	A	45 37	PV A 1.0s 21.6nm M = 5.1
9.	eP	A	09 51 58	Kurile Islands 43.12 N 146.42 E
				H = 09 39 58.7 h = 44 km MB = 4.9
				D = 78.65 Az = 332.6 (NEIC)
10.	eP	A	02 27 30.5	Dodecanese Islands 35.35 N 27.70 E
				H = 02 23 08.0 h = 50 km MB = 3.7
				D = 19.22 Az = 327.6 (NEIC)
10.	eP	A	04 35 53	Crete 35.18 N 25.44 E
				H = 04 31 40.2 h = 33 km MB = 4.8
				D = 18.43 Az = 331.2 (NEIC)
10.	e	A	12 31 07	South of Panama 7.27 N 79.99 W
				H = 12 18 24 h = 24 km MB = 4.6
				D = 85.76 Az = 40 (ISC)
10.	e	A	19 28 04.5	Crete 35.07 N 25.83 E
				H = 19 23 39.1 h = 41 km MB = 4.8
				D = 18.68 Az = 330.8 (NEIC)
				traces
10.	e	A	22 56 00	Chile-Bolivia Border Region
	ePP	A	59 41	20.10 S 68.73 W
	e	A	59 48	H = 22 42 18.52 h = 127 km MB = 5.5
	eSKS	C	23 06 20	D = 99.38 Az = 39.5 (NEIC)
	eS	B	07 10	LmH B 19s 0.6 µm
	eSP	C	08 40	LmV B 18 0.6 µm
	LmV	B	39.8	
	LmH	B	40.8	

Day	Phase		h m s	Remarks
11.	eP	A	00 41 21	Near East Coast of Kamchatka
	epP	A	41 48.5	53.63 N 159.17 E
	e	A	41 52	H = 00 30 04.5 h = 115 km MB = 5.0
				D = 72.57 Az = 339.0 (NEIC)
				PV A 1.6s 27.5nm M = 4.8
11.	ePKIKP	A	11 06 47	Fiji Islands Region 21.89 S 179.67 W
	iPKHKP	A	06 52.3	H = 10 48 09.2 h = 623 km MB = 5.1
	ePKP2	A	07 01	D = 149.96 Az = 345.6 (NEIC)
	e	A	09 11	PKHKPV A 1.0s 54.2nm
11.	eP	A	16 29 52.5	Near South Coast of Honshu
				35.78 N 139.85 E
				H = 16 17 41.5 h = 119 km
				D = 82.64 Az = 330 (ISC)
11.	ePKP	A	17 36 53.8	Loyalty Islands 20.56 S 168.52 E
	e	A	37 04.5	H = 17 17 14.3 h = 7 km
				D = 144.89 Az = 334.3 (NEIC)
12.	+iP	A	00 29 34	Northwestern Kashmir 35.92 N 73.33 E
				H = 00 21 14.2 h = 49 km MB = 5.2
				D = 45.95 Az = 308.7 (NEIC)
				PV A 1.7s 66.7nm
12.	eP	A	01 00 09.5	Kurile Islands 49.30 N 156.16 E
				H = 00 48 25.8 h = 50 km MB = 4.6
				D = 75.91 Az = 337.6 (NEIC)
12.	eP	A	18 03 47	Halmahera 1.81 N 127.39 E
	eX	A	08 00	H = 17 49 51.1 h = 108 km MB = 5.6
	ePP	AB	08 10	D = 104.63 Az = 323.7 (NEIC)
	eSKS	C	14 14	PV A 1.4s 16.3nm M = 5.8
	ePS	C	17 12	XV A 1.9 37.9nm
	eSS	C	23.3	PPV A 1.7 100.0nm 6.2
	LmH	B	56.7	PPV B 5 0.9 µm 6.7
	LmV	B	56.7	LmH B 17.5 0.5 µm
				LmV B 16 0.6 µm

Day	Phase		h m s	Remarks
13.	eP	AB	04 58 00.5	South of Alaska 52.83 N 163.06 W
	ePP	B	05 00 49	H = 04 46 11.0 h = 38 km MB = 5.9
	eS	B	07 47	D = 76.80 Az = 3.5 (NEIC)
	eSS	C	13 08	PV A 1.8s 507.0nm M = 6.2
	LmH	B	42.8	PV B 8 2.39/um 6.3
	LmV	B	42.8	LmH B 15 1.8/um 5.5
				LmV B 17 1.7/um 5.5
13.	eP	A	05 12 59	Northwestern Kashmir 35.84 N 73.31 E
				H = 05 04 36.7 h = normal MB = 5.2
				D = 45.99 Az = 308.8 (NEIC)
				PV A 1.1s 32.2nm M = 5.2
14.	-iP	A	00 11 23.5	Northwest of Kurile Islands
	e	A	11 35	48.41 N 148.70 E
	e	A	11 41	H = 00 00 24.5 h = 394 km MB = 5.6
	epP	A	12 56.5	D = 74.67 Az = 333.3 (NEIC)
				h = 429 km
				PV A 1.2s 32.5nm M = 4.9
14.	eP	A	03 41 25	Andreanof Islands, Aleutian Is.
	e	A	41 39	51.75 N 175.30 W
				H = 03 29 33.1 h = 62 km MB = 5.1
				D = 77.81 Az = 355.5 (NEIC)
14.	ePKP2	A	24 12 45.5	South of Kermadec Islands
	LmH	B	25 36.7	32.84 S 178.58 W
	LmV	B	46.9	H = 23 52 05.7 h = 33 km MB = 4.8 (NEIC)
				D = 160.68
				PKP2V A 1.6s 38.5nm
				LmH B 18 1.06/um M = 5.6
				LmV B 17 1.35/um 5.9
15.	ePKIKP	A	01 19 57	South of Kermadec Islands
	ePKP2	A	20 39	32.94 S 178.44 W
	LmH	B	02 43.3	H = 00 60 00.00 h = normal MB = 5.2
	LmV	B	54.2	D = 160.81 Az = 340.2 (NEIC)
				PKIKPV A 2.0s 34.2nm

Day	Phase		h m s	Remarks
cont. 15.				PKP2V A 1.6s 79.6nm
				LmH B 17 0.7µm M = 5.5
				LmV B 17 1.4µm 5.8
15.	ePKIKP	A	03 44 36	South of Kermadec Islands 33.1 S 178.4 W
				H = 03 24 38.5 h = 49 km
				D = 160.98 Az = 340 (ISC)
15.	eP	A	04 44 46	Kurile Islands 50.67 N 155.65 E
				H = 04 33 18.7 h = 118 km MB = 5.2
				D = 74.52 Az = 337.2 (NEIC)
				PV A 1.6s 27.5nm M = 4.8
15.	ePKP2	A	05 41 58	South of Kermadec Islands 32.90 S 178.35 W
	e	A	42 10.5	H = 05 21 18.7 h = 33 km MB = 4.5 (NEIC)
	LmH	B	07 11.0	D = 160.81
	LmV	B	15.5	PKP2V A 1.2s 30.4nm
				LmH B 18 0.5µm M = 5.3
				LmV B 17 0.6µm 5.5
15.	eP	A	07 57 43.5	South of Alaska 52.93 N 163.15 W
				H = 07 45 53.8 h = normal MB = 4.7
				D = 76.70 Az = 3.4 (NEIC)
				PV A 1.0s 13.8nm M = 4.9
15.	ePKIKP	A	10 47 38	Loyalty Islands Region 22.11 S 171.34 E
	+iPKHKP	AB	47 40.8	H = 10 28 08.3 h = 101 km MB = 5.6
	e	AB	11 01 20	D = 147.39 Az = 335.8 (NEIC)
	LmH	B	45.8	PKIKPV A 1.8s 40.5nm
	LmV	B	53.3	PKHKPV A 2.4 428.0nm
				PKHKPV B 8 1.7µm
				LmH B 24 0.7µm
				LmV B 22 0.4µm
15.	eP	A	14 56 12.5	Northwestern Kashmir 35.82 N 73.42 E
				H = 14 47 53.2 h = 63 km MB = 5.1
				D = 46.07 Az = 308.8 (NEIC)
				PV A 1.6s 27.4nm M = 4.9

Day	Phase	h m s	Remarks
15.	eP A	19 46 34.5	Andreanof Islands, Aleutian Is. 51.78 N 175.35 W H = 19 34 42.7 h = 62 km MB = 4.9 D = 77.78 Az = 355.5 (NEIC)
15.	eP A LmH B LmV B	22 06 11.5 11.5 14.1	Greece 38.06 N 21.10 E H = 22 02 55.7 h = 74 km MB = 4.7 D = 14.28 Az = 334.8 (NEIC) LmH B 13s 0.7/um LmV B 8 0.6/um
16.	eP A	03 33 27	Kurile Islands 44.09 N 148.41 E H = 03 21 29.0 h = 45 km MB = 5.0 D = 78.45 Az = 333.6 (NEIC) PV A 1.0s 15.8nm M = 5.0
16.	ePKP2 A	04 43 39	Kermadec Islands 30.03 S 177.74 W H = 04 23 13.7 h = 54 km MB = 4.4 (NEIC) D = 158.24 PKP2V A 1.5s 20.1nm
16.	ePKP2 A	09 03 37.5	South of Fiji 25.5 S 177.6 E H = 08 43 23.4 h = 0 km MB = 8.4 D = 152.68 Az = 340 (ISC) PKP2V A 1.5s 30.1nm
16.	ePKP2 A	14 31 40.5	South of Kermadec Islands 33.05 S 178.4 W H = 14 11 02 h = 32 km D = 160.92 Az = 340 (ISC)
16.	eP A	23 42 51	Greece 38.4 N 20.4 E H = 23 39 41 (BCIS) D = 13.79
17.	ePKP2 A LmH C LmV C	06 35 20 07 48.0 56.4	South of Kermadec Islands 33.04 S 178.37 W H = 06 14 40.4 h = 33 km MB = 4.3 (NEIC) D = 160.93 LmV C 18s 0.45/um M = 5.3

Day	Phase		h m s	Remarks
17.	eSg	A	18 21 48	Switzerland 47.3 N 7.5 E
				H = 18 19 33 h = 0 km (ISC)
				D = 4.3
17.	e	A	21 21 23	Lubin, Poland
				rockburst
17.	ePP	A	21 42 19	Sumba Island Region 10.30 S 120.50 E
	ePS	B	51 50	H = 21 23 16.2 h = 33 km MB = 5.7 (NEIC)
	ePPS	B	53 25	D = 109.86
	eSS	C	57 00	PPV A 1.8s 23.6nm M = 5.6
	LmH	B	22 35.0	LmH B 18.5 2.4/um 5.8
	LmV	B	35.0	LmV B 18 2.4/um 5.8
18.	+eP	A	02 00 35.3	Near East Coast of Honshu, Japan
				35.72 N 140.07 E
				H = 01 48 18.7 h = 76 km MB = 5.1
				D = 82.76 Az = 329.8 (NEIC)
				PV A 1.4s 30.2nm M = 5.0
18.	e	A	18 37 39.5	Timor 8.52 S 124.90 E
	ePP	AB	37 48	H = 18 18 25.1 h = normal MB = 5.7 (NEIC)
	eS diff	C	45 20	D = 111.25
	eSS	B	53 17	PPV A 1.6s 35.7nm M = 5.8
	e	C	19 04.0	LmH B 21 2.4/um 5.8
	LmH	B	23.3	LmV B 18 1.7/um 5.7
	LmV	B	36.1	
18.	eX	A	18 47 55	XV A 1.3s 17.5nm
18.	eP	A	23 04 01.5	Greece 38.23 N 20.52 E
	e	A	04 12	H = 23 00 50.7 h = 80 km MB = 4.4
	LmH	B	10.4	D = 13.94 Az = 335.8 (NEIC)
	LmV	B	10.4	PV A 1.1s 14.1nm M = 4.2
				LmV B 9 0.65/um

Day	Phase		h m s	Remarks
18.	ePKP2	A	23 27 12	South of Kermadec Islands 33.6 S 178.7 W H = 23 06 32 h = 36 km D = 161.37 Az = 339 (ISC) PKP2V A 1.3s 17.5nm
19.	e(Pn)	A	08 31 22	Czechoslovakia 40°07.3'N 13°52.5'E
	e	A	31 43	explosion yield 8.0 t (PRU)
	e(Sg)	A	31 53	H = 08 30.9 49.12 N 13.88 E (ISC) D = 2.10
19.	ePKP	A	09 47 45.5	Fiji Region 18.3 S 176.8 W H = 09 28 07.5 h = 0 km (ISC) D = 146.8
19.	ePKHKP	A	22 36 18.5	South of Australia 50.74 S 139.13 E
	ePKP2	A	36 21	H = 22 16 38.2 h = normal MB = 5.4 D = 147.33 Az = 290.7 (NEIC) PKHKPV A 1.3s 35.0nm PKP2V A 2.3 238.0nm
20.	ePn	A	04 35 39	D c. 2.4
	eSg	A	36 18	
20.	eP	AB	04 41 57	North Atlantic Ocean 20.57 N 29.67 W
	Pm	A	42 02.5	H = 04 33 48.9 h = normal MB = 5.7
	eS	B	48 26	D = 44.15 Az = 37.1 (NEIC)
	LmH	B	58.3	PV A 1.1s 113.0nm M = 5.6
	LmV	B	05 06.5	PV B 5 4.3 µm 6.4 PmV A 1.7 494.0nm 6.0 SH B 11 6.8 µm 6.3 LmH B 14.5 15.6 µm 6.1 LmV B 15 9.7 µm 5.9
20.	e(PKP)	A	06 59 07	South of Australia 50.34 S 139.57 E
	e	A	59 11	H = 06 39 17.4 h = normal MB = 5.3 D = 147.45 Az = 291.1 (NEIC) PKPV A 1.8s 43.9nm

Day	Phase		h m s	Remarks
20.	eP	A	08 31 00	Off Coast of Jalisco, Mexico
	ePP	AB	34 32	18.78 N 106.73 W
	eSKS	C	41 28	H = 08 17 48.6 h = 38 km MB = 5.7
	eSKKS	B	41 46	D = 92.27 Az = 34.1 (NEIC)
	eS	B	42 04	PV A 2.0s 55.5nm M = 5.6
	ePS	B	43 10	PPV A 1.8 47.3nm 5.6
	eSS	B	48 10	PPV B 10 1.3/um 6.3
	LmH	B	09 14.8	LmH B 16 14.0/um 6.5
	LmV	B	14.8	LmV B 15.5 17.1/um 6.6
21.	eP	A	04 28 38	Taiwan Region 22.30 N 121.05 E
	LmH	B	05 11.5	H = 04 16 05.1 h = 36 km MB = 4.9
	LmV	B	11.5	D = 84.57 Az = 322.9 (NEIC)
				PV A 1.2s 16.3nm M = 5.1
				LmH B 16 1.0/um 5.3
				LmV B 16 1.3/um 5.4
21.	ePKHKP	A	09 27 18	Tonga Islands 21.38 S 174.14 W
	ePKP2	A	27 23.5	H = 09 07 29.1 h = 39 km MB = 5.3
				D = 150.44 Az = 352.6 (NEIC)
				PKHKPV A 1.8s 54.0nm
21.	e(Sg)	A	15 19 07	
21.	eP	A	20 02 37	Central Alaska 63.15 N 151.06 W
	epP	A	03 09.5	H = 19 52 05.4 h = 132 km MB = 5.4
				D = 65.73 Az = 12.0 (NEIC)
				h = 136 km
				PV A 1.4s 18.6nm M = 4.8
22.	ePKP	A	06 27 34	Tonga Islands 15.77 S 173.67 W
	e	A	27 39	H = 06 08 08.2 h = 102 km MB = 5.3
	e	A	27 43.5	D = 144.95 Az = 354.1 (NEIC)
	e	A	28 06	
22.	ePKHKP	A	11 51 16.5	Fiji Islands Region 20.74 S 176.81 W
				H = 11 31 55.7 h = 248 km MB = 4.3
				D = 149.42 Az = 349.4 (NEIC)
				PKHKPV A 1.0s 15.8nm

Day	Phase		h m s	Remarks
22.	ePKP2	A	22 59 12	Kermadec Islands 30.36 S 178.11 W H = 22 38 48.9 h = 106 km MB = 4.1 (NEIC) D = 158.47
22.	e	A	23 15 03	Near East Coast of Honshu, Japan 38.64 N 142.95 E H = 23 02 39.3 h = 42 km MB = 4.4 D = 81.36 Az = 331.1 (NEIC)
23.	eP e	A A	10 00(20) 00 27	Dodecanese Islands 37.86 N 26.23 E H = 09 56 25.5 h = 18 km MB = 4.5 D = 16.48 Az = 325.5 (NEIC)
24.	LmV LmH	B B	03 56.8 57.2	LmH B 16.5s 2.7 μm LmV B 17 2.6 μm
24.	ePKP	A	07 50 20.5	Probably South of Mariana Islands 13.68 N 147.11 E H = 07 31 51.1 h = 15 km MB = 5.4 D = 105.07 Az = 332.5 (NEIC)
25.	eP	A	01 13 46.5	Off Coast of Oregon 43.44 N 127.73 W H = 01 01 41.0 h = normal MB=5.3 MS=5.2 D = 79.84 Az = 24.9 (NEIC) PV A 2.0s 25.6nm M = 4.9
25.	eP epP	A A	01 33 13 33 41	Alaska Peninsula 57.53 N 156.11 W H = 01 21 59.1 h = 93 km MB = 4.6 D = 71.72 Az = 8.2 (NEIC) h = 115 km
25.	LmH LmV	B B	01 50.7 50.8	Off Coast of Oregon 43.44 N 127.73 W H = 01 01 41.0 h = 33 km MB = 5.3 (NEIC) D = 79.84 LmH B 16.5s 0.7 μm M = 5.1 LmV B 18 0.8 μm 5.2

Day	Phase		h m s	Remarks
25.	e	A	02 13 00	
	e	A	13 05.5	
25.	+ePL-P	ABC	18 30 47.5	Jan Mayén Island Region 70.92 N 7.01 W
	eS	B	34 52	H = 18 25 55.4 h = normal MB = 5.3
	e	C	35 52	D = 22.08 Az = 147.3 (NEIC)
	e	C	36 10	PV A 1.0s 138.0nm M = 5.3
	LmH	B	41.1	PLV C 18 2.6/um
	LmV	B	42.5	SH B 13 3.7/um 5.5
				LmH B 13.5 5.9/um 5.2
				LmV B 13 5.6/um 5.3
25.	ePn	AB	21 57 40.5	Northern Italy 44.24 N 9.67 E
	ePg	A	58 04	H = 21 56 06.4 h = normal MB = 5.1
	e	A	58 30	D = 6.54 Az = 10.9 (NEIC)
	eSn	AB	58 48	PV A 0.8s 88.5nm M = 5.6
	LmH	B	22 00.5	LmH B 6 6.3/um 4.7
	LmV	B	00.5	LmV B 10 6.4/um
26.	eP	A	00 18 21	Jan Mayen Island Region 71.4 N 6.0 W
	e	A	18 55	H = 00 13 28 h = 38 km
				D = 22.35 Az = 150 (ISC)
26.	eP	A	03 33 07.5	Ionian Sea 37.54 N 20.21 E
	e	A	33 16	H = 03 29 40.0 h = 35 km MB = 4.1
	LmH	B	38.5	D = 14.47 Az = 337.6 (NEIC)
	LmV	B	40.5	LmH B 19.5s 1.4/um M = 4.1
				LmV B 15 0.5/um
26.	+iP	AB	17 17 33	Ryukyu Islands 27.44 N 128.54 E
	epP	B	17 48	H = 17 05 03.7 h = 47 km MB = 6.0
	ePP	BC	20 52	D = 84.42 Az = 325.2 (NEIC)
	eSKS	B	27 52	h = 56 km
	eS	C	28 18	PV A 1.2s 133.8nm M = 5.9
	eSS	C	34 08	PV B 7 1.2/um 6.1
	LmH	B	59.4	LmH B 17 3.5/um 5.8
	LmV	B	18 00.4	LmV B 14 3.4/um 5.9

Day	Phase		h m s	Remarks
26.	eP	A	17 27 33.5	<u>Kamchatka</u> 52.28 N 157.58 E
				H = 17 16 14.7 h = 144 km MB = 4.6
				D = 73.49 Az = 338.2 (NEIC)
				PV A 1.0s 6.5nm M = 4.4
26.	ePKIKP	A	23 07 40.5	<u>New Hebrides Islands</u> 14.32 S 167.23 E
	e	A	07 43	H = 22 48 34.4 h = 157 km MB = 5.4
	e	A	10 19	D = 138.73 Az = 336.5 (NEIC)
	ePP	A	11 07.5	
27.	+ePKIKP	AB	09 42 40	<u>New Hebrides Islands</u> 19.95 S 168.90 E
	-iPKHKP	A	42 42.0	H = 09 23 08.7 h = 39 km MB=5.7 MS=6.2
	PKP2m	A	42 44	D = 144.49 Az = 335.0 (NEIC)
	ePP	C	45 58	PKIKPV A 1.0s 39.4nm
	e	B	48 40	PKIKPV B 9 5.8 µm
	LmH	B	10 46.0	PKHKPV A 1.0 153.0nm
	LmV	B	55.2	PKP2mV A 1.0 228.0nm
				LmH B 22 4.9 µm M = 6.2
				LmV B 18 3.95 µm 6.2
27.	ePg	A	15 05 30.5	<u>Dorheim, Hessen, Fed. Rep. of Germany</u>
	eSg	A	05 51	50°57.87'N 9°12.12'E
				explosion
				H = 15 05 01.01 yield 5.6 t (HAN)
				D = 1.5
28.	ePP	A	01 19 47	<u>Halmahera</u> 2.56 N 128.67 E
	LmH	B	02 00.0	H = 01 01 20.4 h = 48 km MB = 5.4 (NEIC)
	LmV	B	06.6	D = 104.78
				LmH B 20s 1.6 µm M = 5.7
				LmV B 18 0.9 µm 5.4
28.	ePKIKP	A	02 46 09	<u>East New Guinea Region</u> 7.33 S 146.66 E
	eSKKS	B	54 44	H = 02 27 10.1 h = 2 km MB = 5.9
	ePKKP	A	56 02	D = 123.00 Az = 327.6 (NEIC)
	e	B	57 50	LmH B 23.5 7.1 µm M = 6.2
	eSS	B	03 04 40	LmV B 20 4.0 µm 6.1
	LmH	B	30.4	
	LmV	B	43.3	

Day	Phase		h m s	Remarks
28.	ePKP	A	03 35 59	<u>New Hebrides Islands</u> 15.28 S 167.47 E
	e	A	36 03	H = 03 16 52.0 h = 128 km MB = 5.8
	eX	AB	36 07	D = 139.69 Az = 336.3 (NEIC)
	ePP	AB	39 02.5	XV B 8s 2.2/um
	e	AB	39 32	PPV B 7 1.6/um M = 6.3
	eSS	C	57 40	LmH B 20 3.2/um
	LmH	B	04 28.4	LmV B 24 3.0/um
	LmV	B	30.5	
28.	eP	A	18 06 41	<u>South of Honshu, Japan</u> 33.31 N 140.12 E
				H = 17 54 15.8 h = 93 km MB = 4.8
				D = 84.86 Az = 330.0 (NEIC)
				PV A 1.0s 7.4nm M = 4.6
29.	eP	A	03 19 22	<u>Central Mid-Atlantic Ridge</u>
	LmH	C	33.5	7.65 N 36.73 W
	LmV	C	34.2	H = 03 09 24.5 h = normal MB = 4.9
				D = 58.58 Az = 33.9 (NEIC)
29.	e	A	05 08 49	<u>Peru-Ecuador Border Region</u>
				3.13 S 77.26 W
				H = 04 55 21.6 h = normal MB = 4.6
				D = 91.68 Az = 39.6 (NEIC)
				traces
29.	eP	A	07 32 33.5	<u>Near S. Coast of Honshu, Japan</u>
	epP	A	35 52.5	33.02 N 137.75 E
				H = 07 20 38.5 h = 331 km MB = 5.5
				D = 84.10 Az = 328.9 (NEIC)
				h = 341 km
				PV A 1.1s 26.4nm M = 5.0
29.	ePKHKP	A	18 01 22	<u>Kermadec Islands Region</u>
	ePKP2	A	01 47	31.40 S 178.12 W
				H = 17 41 17.3 h = 59 km MB = 5.3 (NEIC)
				D = 159.46

Day	Phase		h m s	Remarks
30.	ePKP	A	10 49 14	Loyalty Islands Region 21.08 S 170.21 E
	e	A	49 32.5	H = 10 29 51.2 h = 154 km MB = 5.0
	esP	A	50 07	D = 146.03 Az = 335.4 (NEIC)
				h = 172 km
30.	ePKIKP	A	14 02 30.5	Fiji Islands Region 18.43 S 177.91 W
	ePKHKP	A	02 33.5	H = 13 43 59.2 h = 645 km MB = 4.6
	ePKP2	A	03 36.5	D = 146.97 Az = 348.9 (NEIC)
				PKIKPV A traces
30.	eP	A	14 35 27	Greece 38.33 N 20.37 E
	LmV	B	41.9	H = 14 32 13.1 h = normal MB = 5.3
	LmH	B	42.0	D = 13.80 Az = 336.0 (NEIC)
				PV A 1.1s 101.0nm M = 5.5
				LmH B 12 5.0/um 4.8
				LmV B 12 5.7/um
30.	ePKIKP	A	17 07 08.5	Solomon Islands 6.33 S 154.80 E
	epP	A	07 20.5	H = 16 48 09.4 h = 50 km MB = 5.8
	ePP	B	09 04	D = 126.23 Az = 331.8 (NEIC)
	e	C	10 15	PKIKPV A 1.3s 35.0nm
	ePS	B	19 00	LmH B 22 3.3/um M = 6.0
	ePcPPKP	A	20 34	LmV B 19 2.6/um 5.9
	LmH	B	53.9	
	LmV	B	18 06.5	
31.	ePn	A	12 00 33	Czechoslovakia 49°35.7'N 13°40.2'E
	eSg	A	00 57	explosion yield 22 t (PRU)
				D = 1.66
31.	LmV	B	16 30.5	LmH B 15s 0.4/um
	LmH	B	30.9	LmV B 15 0.5/um
31.	eP	A	23 37 38	Sunda Strait 6.11 S 104.88 E
	ePP	A	41 32.5	H = 23 24 12.5 h = 73 km MB = 5.4
				D = 96.76 Az = 320.2 (NEIC)
				PPV A 1.2s 12.2nm M = 5.3

Day	Phase		h m s	Remarks
1.	eP	A	05 02 22.5	Jan Mayen Island-Region 71.91 N 0.27 W
				H = 04 57 29.0 h = normal MB = 4.4
				D = 22.00 Az = 159.5 (NEIC)
				PV A 1.6s 13.7nm M = 4.1
1.	eiPKHKP	AB	06 49 30	Tonga Islands 20.07 S 173.86 W
	LmV	B	08 04.8	H = 06 29 43.7 h = normal MB=5.1 MS=5.0
	LmH	B	05.7	D = 149.18 Az = 353.2 (NEIC)
				PKHKPV A 2.2s 180.0nm
				LmH B 17.5 0.4/um M = 5.2
				LmV B 20 0.6/um 5.4
1.	eP	A	15 36 27.5	Hokkaido, Japan Region 41.49 N 142.75 E
	e	A	36 37	H = 15 24 30.8 h = 76 km MB = 5.3
	e	A	36 47	D = 78.80 Az = 330.7 (NEIC)
	eX	A	36 53	PV A 1.8s 33.8nm M = 5.0
	e	A	38 40	XV A 2.0 85.5nm
	LmH	B	16 11.2	LmH B 14 0.5/um
	LmV	B	14.7	LmV B 16 0.4/um
1.	eP	A	16 51 48.5	Kurile Islands 43.37 N 146.29 E
				H = 16 39 50.7 h = 48 km MB = 4.8
				D = 78.39 Az = 332.5 (NEIC)
				PV A 1.1s 18.1nm M = 5.0
1.	eP	A	17 11 42	Norwegian Sea 72.42 N 3.79 E
	e	A	11 46	H = 17 06 47.5 h = normal MB = 4.8
				D = 22.13 Az = 166.7 (NEIC)
1.	ePKIKP	A	21 41 05.5	New Guinea 6.29 S 144.46 E
	epPKIKP	A	41 12.5	H = 21 22 15.4 h = 39 km MB = 5.4
	esPKIKP	A	41 18	D = 120.95 Az = 327.0 (NEIC)
	ePP	A	42 38.5	h = 30 km
				PKIKPV A 1.2s 16.2nm
				PPV A 1.9 37.9nm M = 5.6
1.	eP	A	22 04 29	Burma 26.42 N 96.32 E
	ipP	A	04 55.5	H = 21 53 46.0 h = 97 km MB = 5.3

Day	Phase		h m s	Remarks
cont. 1.	LmH	B	22 39.7	D = 66.80 Az = 316.4 (NEIC)
	LmV	B	40.0	h = 110 km
				PV A 0.8s 13.5nm M = 4.9
				pPV A 1.3 50.2nm
				LmH B 17 1.0 µm
				LmV B 17 0.7 µm
2.	+iP	AB	01 34 50	Eastern Kazakh SSR 49.91 N 78.84 E
	ePn	AB	36 20	H = 01 26 57.6 h = 0 km MB = 6.2
	ePP	AB	36 27	D = 41.61 Az = 297.9 (NEIC)
	eLg1	B	49 08	Underground explosion MB = 7.1 (UPP)
	LmH	B	52.3	PV A 1.2s 470.0nm M = 6.1
	LmV	B	52.6	PV B 1.5 0.8 µm 6.2
				PPV A 1.3 179.0nm 5.9
				Lg1: velocity = 3.48 km s^{-1}
2.	LmH	B	10 55.4	San Juan Province, Argentina
	LmV	B	11 10.6	31.73 S 69.31 W
				H = 09 57 26.2 h = 110 km MB = 4.1 (NEIC)
				D = 108.54
2.	eP	A	14 55 41	Republic of South Africa
				26.30 S 27.34 E
				H = 14 43 42.0 h = 20 km MB = 5.2
				D = 77.86 Az = 349.8 (NEIC)
2.	+iPKP	AB	20 14 55.3	Loyalty Islands 20.05 S 168.84 E
	-iPP	B	18 08	H = 19 55 20.1 h = 32 km MB=6.3 MS=7.0
	eSKSP	B	28 24	D = 144.56 Az = 334.8 (NEIC)
	LmH	B	21 28.2	PKPV A 1.2s 1240.0nm
	LmV	B	28.2	PKPV B 14 50.6 µm
				PPV B 18 19.5 µm M = 6.0
				LmH B 19 43.8 µm 7.2
				LmV B 19 51.2 µm 7.3
2.	+iX1	A	20 22 03.5	X1V A 1.2s 56.9nm
	+iX2	A	22 16.5	X2V A 1.0 96.4nm
	eX3	A	22 26	X3V A 1.8 115.0nm

Day	Phase		h m s	Remarks
2.	ePKP	A	21 28 12.5	New Hebrides Islands 20.32 S 168.99 E H = 21 08 37.3 h = 21 km MB = 5.0 D = 144.86 Az = 334.8 (NEIC) PV A 1.3s 37.2nm
2.	eP	A	21 40 29	Ascension Island Region 6.33 S 11.36 W H = 21 30 18.5 D = 60.12 Az = 17 (ISC) PV A 1.3s 15.3nm
2.	ePKP	A	23 29 27.5	Samoa Islands Region 16.51 S 172.69 W H = 23 09 50.3 h = normal MB = 5.0 D = 145.77 Az = 355.1 (NEIC) PKPV A 2.3s 61.0nm
3.	-ePKP ePP	A A	07 29 03.5 32 10	Loyalty Islands 20.07 S 168.63 E H = 07 09 28.7 h = 22 km MB = 4.8 D = 144.49 Az = 334.7 (NEIC) PKPV A 1.6s 27.5nm PPV A 1.6 35.7nm M = 5.3
3.	+ePKP epPKP	A A	08 17 44.5 17 54.5	New Hebrides Islands 20.16 S 169.06 E H = 07 58 12.3 h = 41 km MB = 4.8 D = 144.75 Az = 335.0 (NEIC) h = 37 km PKPV A 0.9s 27.2nm
3.	ePKP	A	15 26 01	New Hebrides Islands 19.79 S 169.19 E H = 15 06 21.8 h = 15 km MB = 5.0 D = 144.46 Az = 335.3 (NEIC) PV A 1.2s 36.6nm
3.	-iPKP LmH LmV	AB C B	16 34 37.5 17 15.5 17.7	Loyalty Islands 20.08 S 168.83 E H = 16 15 03.1 h = 26 km MB = 5.4 D = 144.58 Az = 334.8 (NEIC) PKPV A 1.3s 116.0nm LmH B 25 0.7/um M = 5.3

Day	Phase		h m s	Remarks
3.	ePKP	A	23 02 12	New Hebrides Islands 19.60 S 169.36 E
				H = 22 42 38.4 h = 22 km MB = 5.2
				D = 144.35 Az = 335.6 (NEIC)
				PKPV A 1.6s 87.9nm
3.	eP	A	24 06 14	Afghanistan 34.23 N 69.66 E
	ePP	A	07 58.5	H = 23 57 59.4 h = 20 km MB=5.4 MS=5.4
	e	A	08 22	D = 44.68 Az = 309.8 (NEIC)
	e	A	09 00	PV A 1.6s 63.1nm M = 5.3
	eS	C	12 56	LmH B 10 3.2/um 5.6
	eSS	C	16 10	LmV B 15 4.3/um 5.6
	LmV	B	28.6	
	LmH	B	31.2	
4.	e	A	00 15 20.5	
4.	ePKP	A	01 05 38.5	New Hebrides Islands 19.65 S 169.39 E
				H = 00 46 04.4 h = 18 km MB = 4.4
				D = 144.41 Az = 335.6 (NEIC)
4.	+iPKP	AB	04 09 28.5	New Hebrides Islands 20.07 S 169.13 E
	eipPKP	AB	09 38.5	H = 03 49 56.6 h = 53 km MB = 5.7
	ePP	C	12 50	D = 144.69 Az = 335.1 (NEIC)
	ePPP	C	15 50	h = 36 km
	eSKSP	C	23 10	PKPV A 1.3s 579.0nm
	ePPS	C	25 25	PKPV B 8 2.8/um
	LmH	B	05 22.9	pPKPV A 1.3 720.0nm
	LmV	B	23.2	PPV C 20 3.9/um M = 6.3
				LmH B 22 3.6/um
				LmV B 18 3.6/um
4.	ePg	A	07 45 22	Northern Italy 44.6 N 10.1 E
	eSb	A	46 23	H = 07 43 19 h = 0 km (ISC)
				D = 6.15
4.	eP	A	09 40 00	Southern Iran 28.54 N 57.00 E
				H = 09 32 22.0 h = normal MB = 4.9
				D = 40.54 Az = 315.8 (NEIC)

Day	Phase		h m s	Remarks
4.	ePKIKP	A	10 06 03	Near North Coast of New Guinea
	epPKIKP	A	06 25.5	4.79 S 144.56 E
	LmH	B	58.9	H = 09 47 19.9 h = 85 km MB = 5.6
	LmV	B	58.9	D = 119.76 Az = 327.5 (NEIC)
				h = 83 km
				PKIKPV A 1.3s 37.2nm
				LmH B 21.5 1.7/um
				LmV B 20 1.2/um
4.	ePKIKP	A	11 26 32	Fiji Islands Region 20.23 S 178.00 W
	ePKHKP	A	26 36	H = 11 07 49.7 h = 554 km MB = 4.7
	e	A	26 48	D = 148.71 Az = 348.2 (NEIC)
				PKHKPV A 1.3s 32.8nm
4.	ePKP	A	17 21 39	New Hebrides Islands 20.09 S 169.09 E
				H = 17 02 09.0 h = 45 km
				D = 144.69 Az = 335.0 (NEIC)
				PKPV A 1.0s 13.8nm
4.	eP	A	21 49 48	Java 8.19 S 112.34 E
	e	A	53 56	H = 21 36 01.7 h = 126 km MB = 6.0
	ePP	AB	54 00	D = 103.08 Az = 320.0 (NEIC)
	eSKS	C	22 00 14	PV A 1.3s 19.7nm M = 5.8
	eS	C	01 20	PPV A 2.2 206.5nm 6.4
	eSS	C	09 20	PPV B 6 1.0/um 6.6
	LmH	C	27.6	LmH C 40 1.1/um
4.	ePKP	AB	24 07 57.5	Loyalty Islands 20.16 S 168.88 E
	LmV	B	25 20.7	H = 23 48 22.3 h = 21 km MB=5.2 MS=6.1
	LmH	B	21.4	D = 144.68 Az = 334.8 (NEIC)
				PKPV A 1.2s 118.0nm
				PKPV B 10 2.9/um
				LmH B 17 2.8/um M = 6.1
				LmV B 19 2.6/um 6.0
5.	ePKP	A	00 24 26.5	Sawu Sea 9.68 S 122.24 E
	ePP	A	24 58	H = 00 05 50.2 h = normal MB=5.8 MS=5.9
	ePKKP	A	25 27.5	D = 110.48 Az = 320.5 (NEIC)
				PKPV A 2.0s 38.5nm

Day	Phase		h m s	Remarks
5.	e(P)	A	00 58 43.5	
5.	ePKP	A	01 19 07.5	Loyalty Islands 20.34 S 168.88 E
	e	A	19 18	H = 00 59 32.1 h = 21 km MB = 4.9
				D = 144.83 Az = 334.7 (NEIC)
				PKPV A 1.5s 75.4nm
5.	eP	A	05 32 53.5	Ryukyu Islands Region 28.36 N 131.56 E
				H = 05 20 18.5 h = normal MB = 4.6
				D = 85.16 Az = 326.4 (NEIC)
5.	eP	A	12 42 10	Hokkaido, Japan Region 42.01 N 142.47 E
				H = 12 30 15.6 h = 70 km MB = 4.6
				D = 78.24 Az = 330.5 (NEIC)
5.	epP	A	13 12 33.5	Eastern Caucasus 41.17 N 47.25 E
				H = 13 06 48.2 h = 61 km MB = 4.7
				D = 26.32 Az = 303.2 (NEIC)
5.	ePKHKP	A	13 51 00.5	South of Fiji Islands 22.16 S 179.77 W
				H = 13 32 18.4 h = 635 km MB = 4.9
				D = 150.20 Az = 345.3 (NEIC)
5.	eP1	A	19 29 54	Crete 35.06 N 24.85 E
	eP2	A	29 57.5	H = 19 25 42.4 h = normal MB=5.2 MS=5.4
	LmV	B	38.1	D = 18.30 Az = 332.3 (NEIC)
	LmH	B	38.2	P2V A 1.3s 30.6nm M = 4.3
				LmH B 13.5 1.7/um 4.5
				LmV B 15 2.0/um 4.7
5.	LmH	C	19 56.5	LmH C 45s 1.2/um
	LmV	C	57.0	LmV C 40 0.8/um
5.	LmV	C	20 04.5	LmV C 20s 0.7/um
	LmH	C	05.0	LmV C 18 0.9/um
5.	ePKIKP	A	20 26 29	East New Guinea Region 5.36 S 146.68 E
	e	A	27 30.5	H = 20 08 03.2 h = 230 km MB = 5.5

Day	Phase		h m s	Remarks
cont. 5.	e(pPP)	A	20 29 02	D = 121.36 Az = 328.2 (NEIC)
	LmH	B	21 10.3	LmH B 22.5s 2.4 µm
	LmV	B	19.6	LmV B 17 0.8 µm
5.	e	A	20 36 35	traces
5.	iPKP	A	22 26 37	Loyalty Islands 20.14 S 168.83 E
				H = 22 07 02.2 h = 25 km MB = 5.3
				D = 144.63 Az = 334.8 (NEIC)
				PKPV A 1.4s 137.0nm
6.	ePKP	A	07 28 54	Loyalty Islands 20.44 S 168.24 E
	epPKP	A	29 04	H = 07 09 21.3 h = 46 km MB = 4.8
	e	A	29 13.5	D = 144.67 Az = 334.1 (NEIC)
				h = 36 km
				PKPV A 1.3s 28.4nm
6.	e	A	11 06 40.5	Sikkim 26.96 N 88.71 E
	LmH	C	30.0	H = 10 56 08.8 h = normal MB = 4.8
				D = 61.66 Az = 315.2 (NEIC)
				LmH C 30s 0.5 µm M = 4.5
6.	eP	A	11 52 05.5	Honshu, Japan 36.10 N 139.71 E
				H = 11 39 49.7 h = 63 km MB = 5.1
				D = 82.28 Az = 329.6 (NEIC)
				PV A 1.3s 13.1nm
6.	eP	A	16 34 14.5	Kurile Islands 43.73 N 147.53 E
	epP	A	34 24	H = 16 22 16.0 h = 49 km MB = 4.5
				D = 78.48 Az = 333.1 (NEIC)
				h = 35 km
6.	ePKP	A	17 56 29.5	New Hebrides Islands 19.93 S 168.31 E
	e	A	57 01.5	H = 17 36 57.6 h = 45 km MB = 4.8
				D = 144.23 Az = 334.5 (NEIC)
7.	eP	A	06 53 06	Taiwan 22.75 N 120.77 E
	e	A	53 23	H = 06 40 35.8 h = 22 km MB = 5.4

Day	Phase		h m s	Remarks
cont. 7.	ePP	A	06 56 25	D = 84.05 Az = 322.8 (NEIC)
				PV A 1.4s 20.9nm M = 5.2
7.	eP	A	12 11 43	North Atlantic Ocean 49.02 N 39.37 W
	e	A	11 47	H = 12 05 13.8 h = normal MB = 5.2
	e	A	11 52	D = 32.40 Az = 67.4 (NEIC)
7.	eP	A	13 49 59	Bonin Islands Region 28.40 N 140.59 E
				H = 13 37 19.0 h = 155 km MB = 4.8
				D = 89.31 Az = 330.3 (NEIC)
7.	ePKP	A	15 34 24.5	Loyalty Islands Region 22.01 S 169.99 E
				H = 15 14 43.7 h = 32 km
				D = 146.77 Az = 334.6 (NEIC)
				PKPV A 1.8s 33.8nm
7.	e	A	16 47 32	
	e	A	47 40	
7.	e	A	18 56 18.5	
7.	ePKP	A	20 49 45.5	Loyalty Islands Region 22.02 S 169.73 E
				H = 20 30 07.4 h = 32 km
				D = 146.69 Az = 334.4 (NEIC)
				PKPV A 1.0s 11.8nm
8.	ePKHKP	A	13 08 22	Tonga Islands 20.08 S 174.07 W
	e	A	08 27	H = 12 48 36.9 h = normal MB = 4.6
				D = 149.17 Az = 352.9 (NEIC)
8.	eP	A	14 38 10.5	Taiwan 23.93 N 121.61 E
	ePP	A	41 28.5	H = 14 25 43.3 h = 27 km MB=5.5 MS=5.2
	e	A	41 38	D = 83.58 Az = 323.0 (NEIC)
	LmH	B	15 20.3	PV A 1.3s 26.2nm M = 5.2
	LmV	B	20.4	LmH B 15 3.5 μm 5.9
				LmV B 16 4.1 μm 5.9

Day	Phase		h m s	Remarks
8.	eP	A	21 30 40	South of Panama 6.94 N 82.23 W
	LmH	C	22 01.0	H = 21 17 54.5 h = normal MB = 5.0
	LmV	C	01.2	D = 87.12 Az = 39.5 (NEIC)
				LmH C 32s 0.7 µm M = 4.9
				LmV C 28 1.8 µm 5.4
9.	ePKHKP	A	03 07 51	Fiji Islands Region 21.39 S 179.05 W
				H = 02 49 03.5 h = 542 km MB = 4.7
				D = 149.62 Az = 346.5 (NEIC)
				PKHKPV A 1.2s 24.4nm
9.	ePKHKP	A	08 27 47	Fiji Islands Region 21.07 S 178.99 W
				H = 08 09 05.4 h = 621 km MB = 4.5
				D = 149.32 Az = 346.7 (NEIC)
				PKHKPV A 1.6s 22.0nm
9.	ePKIKP	A	08 28 20.5	Fiji Islands Region 20.94 S 179.01 W
	-iPKHKP	A	28 26	H = 08 09 44.3 h = 606 km MB = 5.7
	ePKP2	A	28 32.5	D = 149.19 Az = 346.8 (NEIC)
	e	A	30 58	PKIKPV A 1.9s 53.0nm
				PKHKPV A 1.5 176.0nm
				PKP2V A 1.0 106.0nm
9.	eP	A	11 49 39	Tyrrhenian Sea 40.3 N 13.3 E
				H = 11 47 16 h = 455 km
				D = 10.38 Az = 354 (ISC)
				PV A 0.9s 27.2nm
9.	+ePKP	A	15 59 11	New Hebrides Islands 20.07 S 169.16 E
				H = 15 39 41.6 h = 59 km MB = 5.0
				D = 144.70 Az = 335.1 (NEIC)
				PKPV A 1.2s 42.7nm
9.	-PKP	A	16 56 28	New Hebrides Islands 19.60 S 169.34 E
				H = 16 36 55.2 h = 25 km MB=5.6 MS=5.2
				D = 144.34 Az = 335.5 (NEIC)
				PKPV A 1.3s 98.2nm

Day	Phase		h m s	Remarks
9.	eP	AB	18 53 40.5	Taiwan 23.89 N 121.58 E
	ePP	C	56 56	H = 18 41 14.5 h = normal MB = 5.7
	eS	C	19 04 00	D = 83.60 Az = 323.0 (NEIC)
	eSS	C	09 24	PV A 3.6s 602.0nm M = 6.1
	e	C	13 25	PV B 12 2.6 um 6.2
	e	C	15 34	LmH B 15 21.5 um 6.7
	LmH	B	35.1	LmV B 15 26.2 um 6.8
	LmV	B	35.1	
10.	ePKIKP	A	00 42 29	Fiji Islands Region 20.89 S 178.19 W
	ePKHKP	A	42 33.5	H = 00 23 42.1 h = 516 km MB = 4.7
	ePKP2	A	42 41	D = 149.32 Az = 347.7 (NEIC)
				PKHKPV A 1.4s 27.9nm
10.	eX	A	04 52 56	Iran 30.25 N 57.66 E
				H = 04 45 09 h = 11 km MB = 4.6
				D = 39.75 Az = 314 (ISC)
				XV A 1.7s 24.2nm
10.	e	A	04 54 16	
10.	eSg	A	05 05 58	Poland 50.33 N 18.83 E
				H = 05 03 28.8 M = 3.1 (WAR)
				D = 4.65
10.	+iP	AB	07 52 41.3	Kurile Islands 44.53 N 147.40 E
	LmH	B	08 25.2	H = 07 40 55.8 h = 118 km MB = 5.6
	LmV	B	32.0	D = 77.73 Az = 333.0 (NEIC)
				PV A 1.8s 348.0nm M = 5.8
				LmH B 18 1.1 um
				LmV B 16 1.2 um
10.	+iP	AB	08 40 26.6	Near East Coast of Kamchatka
				54.64 N 160.45 E
				H = 08 29 13.4 h = 110 km MB = 5.2
				D = 71.90 Az = 339.7 (NEIC)
				PV A 1.5s 115.0nm M = 5.5

Day	Phase		h m s	Remarks
10.	LmH	C	15 55.5	Near Coast of Oaxaca, Mexico
	LmV	B	16 03	15.69 N 95.76 W
				H = 14 56 52.5 h = 37 km MB = 5.4 (NEIC)
				D = 88.62
				LmH C 16s 1.1/um M = 5.4
11.	eP	A	01 57 54.5	Burma 24.84 N 95.44 E
				H = 01 47 13.4 h = 143 km MB = 5.0
				D = 67.39 Az = 316.7 (NEIC)
				PV A 1.1s 22.2nm M = 4.9
11.	iPn	A	12 25 09.7	Explosion Czechoslovakia
	iSg	A	25 30.5	$50^{o}35.2'$N $14^{o}03.2'$E
				H = 12^{h} 24.7^{m}
				yield 10 t (PRU)
				D = 1.53
12.	eP	A	02 42 00	Lake Tanganyika Region 7.53 S 30.34 E
	e	A	42 10	H = 02 31 48.2
				D = 60.17 Az = 346
12.	eP	A	02 57 55	Near East Coast of Kamchatka
	e	A	58 01	55.55 N 162.47 E
	LmH	B	03 32.1	H = 02 46 33.7 h = 39 km MB=4.8 MS=4.7
	LmV	B	32.2	D = 71.43 Az = 340.9 (NEIC)
				LmH B 18s 1.3/um M = 5.2
				LmV B 20 1.9/um 5.4
12.	ePKP2	A	05 57 10	Loyalty Islands Region 22.49 S 171.67 E
				H = 05 37 24.1 h = 40 km MB = 4.7
				D = 147.86 Az = 335.9 (NEIC)
12.	ePKHKP	A	08 56 45.5	Tonga Islands 20.45 S 173.58 W
				H = 08 36 56.9 h = normal MB = 4.5
				D = 149.59 Az = 353.5 (NEIC)
				PKHKPV A 1.4s 20.9nm
12.	ePKHKP	A	12 13 38.5	Tonga Islands 21.37 S 174.44 W

Day	Phase		h m s	Remarks
cont. 12.	ePKP2	A	12 13 45.5	H = 11 53 48.7 h = normal MB = 4.9
				D = 150.39 Az = 352.2 (NEIC)
				PKHKPV A 1.6s 33.0nm
12.	-iP	AB	18 04 54.3	Tadzhik-Sinkiang Border Region
	+isP	AB	05 34	38.42 N 73.28 E
	ePP	B	06 38	H = 17 56 44.4 h = normal
	e	B	07 20	MB = 6.0 MS = 5.7 (NEIC)
	eS	BC	11 18	38.33 N 73.17 E
	esS	C	12 08	H = 17 56 52.9 h = 111 km MB = 5.9 (ISC)
	eSS	C	14 36	h = 120 km D = 44.39
	LmV	B	16.4	PV A 1.0s 254.0nm M = 5.9
	LmH	B	18.2	PV B 10 3.0/um 6.0
				sPV A 3 1448.0nm
				sPV B 5.5 4.4/um
				PPV B 14 3.2/um 6.1
				SH B 11.5 4.0/um 6.1
				LmH B 8 6.7/um 6.0
				LmV B 9.5 4.9/um 5.8
13.	eP	A	04 56 37	Near Coast of Oaxaca, Mexico
	eSKS	B	05 07 08	15.58 N 94.98 W
	ei	B	07 50	H = 04 43 45.2 h = normal MB=5.5 MS=6.5
	LmH	B	31.7	D = 88.25 Az = 37.6 (NEIC)
	LmV	B	38.1	PV A 2.5s 231.0nm M = 6.0
				LmH B 20 28.6/um 6.7
				LmV B 17 23.7/um 6.7
13.	ePKHKP	A	06 55 18.5	Fiji Islands Region 21.89 S 179.38 W
	ePKP2	A	55 27	H = 06 36 34.2 h = 601 km MB = 5.1
	e	A	55 32.5	D = 150.03 Az = 345.9 (NEIC)
13.	-iP	A	08 24 05.3	Bonin Islands Region 27.86 N 140.03 E
	ePP1	A	27 40.5	H = 08 11 48.8 h = 372 km MB = 5.5
	.iPP2	A	27 43.5	D = 89.53 Az = 330.1 (NEIC)
				PV A 1.3s 124.0nm M = 5.7
				PP1V A 1.3 70.0nm 5.6
				PP2V A 1.6 170.0nm 5.9

Day	Phase		h m s	Remarks
13.	e	A	23 27 11	Andaman Islands Region 12.2 N 95.3 E
				H = 23 15 12.3 h = 33 km
				D = 76.65 Az = 319 (ISC)
13.	eP	A	23 46 02	Andaman Islands Region 12.48 N 95.17 E
	e	A	46 09.5	H = 23 34 12.8 h = normal MB = 5.4
	eS	C	55 50	D = 76.39 Az = 319.4 (NEIC)
	eSSS	C	24 04 00	LmH B 19.5s 2.1 µm M = 5.5
	LmH	B	24.8	LmV B 16 1.5 µm 5.4
	LmV	B	28.8	
14.	+eP	A	04 36 39	Jan Mayen Island Region 71.05 N 7.91 W
	eS	B	40 42	H = 04 31 42.8 h = normal MB=5.1 MS=5.1
	LmH	B	47.3	D = 22.35 Az = 146.0 (NEIC)
	LmV	B	47.9	PV A 1.0s 70.9nm M = 5.1
				SH B 17 3.0 µm 5.4
				LmH B 13.5 5.5 µm 5.1
				LmV B 13.5 5.0 µm 5.2
14.	eP	A	11 40 10	Unimak Island Region 53.72 N 164.73 W
				H = 11 28 25.9 h = 50 km MB = 4.6
				D = 75.96 Az = 2.4 (NEIC)
14.	e	A	22 46 00.5	Molucca Sea 1.09 S 126.89 E
	e	C	53 15	H = 22 27 04.9 h = 45 km MB=5.4 MS=5.2
	ePPS	C	23 00 50	D = 106.64 Az = 323.1 (NEIC)
	eSS	C	04 50	LmH B 18s 0.5 µm M = 5.1
	eSSS	C	10 50	LmV B 17 0.4 µm 5.0
	LmH	B	43.0	
	LmV	B	44.9	
14.	ePKIKP	A	24 18 30	Fiji Region 1.1 S 177.5 W
				H = 23 58 47 h = 0 km
				D = 148.37 Az = 349 (ISC)
15.	e(PKP)	A	16 38 47	South of Australia 48.96 S 123.62 E
	eSS	C	59 55	H = 16 19 31.1 h = normal MB = 5.4
	eSSS	C	17 05 00	D = 137.49 Az = 299.1 (NEIC)

Day	Phase		h m s	Remarks
cont. 15.	LmV	B	17 47.3	PKPV A 2.0s 29.9nm
	LmH	B	52.5	LmH B 19 1.0 µm M = 5.5
				LmV B 20 1.0 µm 5.6
16.	+iP	AB	12 51 01.7	Hindu Kush Region 35.75 N 69.90 E
	ePcP	B	53 08	H = 12 43 05.5 h = 124 km MB = 5.6
	ePa	B	53 22	D = 43.88 Az = 308.6 (NEIC)
	eS	C	57 21	PV A 1.4s 198.0nm M = 5.6
	eSa	B	13 00 48	LmH C 46 1.2 µm
	LmH	C	05.0	LmV C 44 2.2 µm
	LmV	C	06.0	Pa: velocity = 7.9 km s^{-1}
				Sa: velocity = 4.6 km s^{-1}
17.	e	A	00 44 57	Loyalty Islands Region 22.58 S 171.28 E
				H = 00 25 04.8 h = 29 km MB = 5.0
				D = 147.80 Az = 335.5 (NEIC)
17.	eP1	A	02 46 01	Ionian Sea 37.39 N 20.32 E
	eP2	A	46 08.5	H = 02 42 35.5 h = 41 km MB = 4.3
	LmH	B	51.3	D = 14.65 Az = 337.6 (NEIC)
	LmV	B	54.2	P1V : traces
				P2V A 1.2s 20.4nm M = 4.5
				LmH B 16.5 1.3 µm 4.1
				LmV B 12 0.7 µm 4.2
17.	eP	A	09 16 51.5	Southern Iran 27.35 N 59.09 E
	epP	A	17 06.5	H = 09 09 00.1 h = 65 km MB = 5.4
	eS	C	23 12	D = 42.68 Az = 316.2 (NEIC)
	eSS	C	26 45	h = 71 km
	LmH	C	30.7	PV A 1.4s 18.6nm M = 4.7
				pPV A 1.7 54.5nm
				LmH C 45 1.1 µm
17.	iP	A	15 08 03	Mexico - Guatemala Border Region
	epP	A	08 57.5	15.61 N 91.46 W
				H = 14 55 45.2 h = 222 km MB = 5.2 (NEIC)
				h = 228 km
				D = 86.13 PV A : traces
				pPV A 1.6s 33.0nm

Day	Phase		h m s	Remarks
17.	eP	A	16 52 47	Southeastern Alaska 56.04 N 135.53 W
	e	A	52 50	H = 16 41 34.7 h = normal MB = 5.0
				D = 70.24 Az = 21.5 (NEIC)
18.	ePKP	A	03 04 13	Tonga Islands 16.21 S 174.27 W
				H = 02 44 55.1 h = 180 km MB = 4.7
				D = 145.32 Az = 353.4 (NEIC)
18.	eP	A	08 42 49.5	Off East Coast of Kamchatka
				52.68 N 160.22 E
				H = 08 31 15.9 h = normal MB = 4.6
				D = 73.69 Az = 339.8 (NEIC)
				PV A 0.8s 15.4nm M = 5.1
18.	ePKP	A	17 41 39	New Hebrides Islands 20.19 S 169.02 E
				H = 17 22 07.1 h = 50 km MB = 4.3 (NEIC)
				D = 144.76
				PV A 1.2s 24.4nm
19.	epP	A	04 48 18	El Salvador 12.97 N 88.52 W
				H = 04 35 24.1 h = 73 km MB = 5.6
				D = 86.40 Az = 38.9 (NEIC)
				h = 69 km
19.	eP	A	09 26 28	Mediterranean Sea 36.5 N 16.9 E
	eX	A	26 31	H = 09 22 58 h = 0 km MB = 4.1
				D = 14.63 Az = 347 (ISC)
				XV A 1.2s 22.4nm
19.	eP	A	17 42 21	Queen Elizabeth Islands 76.59 N 106.86 W
	eS	C	49 24	H = 17 33 47.9 h = normal MB=4.9 MS=5.3
	eSS	C	53 00	D = 47.24 Az = 49.7 (NEIC)
	LmH	B	18 04.6	PV A 2.0s 47.0nm M = 5.1
	LmV	B	05.4	LmH B 15.5 3.0 μm 5.4
				LmV B 15.5 2.8 μm 5.4
19.	+eP1	AB	20 17 01	North of Svalbard 80.37 N 2.54 W
	+iP2	AB	17 07.5	H = 20 10 52.3 h = normal MB=5.4 MS=5.2

Day	Phase		h m s	Remarks
cont. 19.	eS	B	20 22 00	D = 30.22 Az = 162.0 (NEIC)
	eSS	B	23 36	P1V A 2.0s 94.0nm M = 5.2
	LmH	B	27.5	P2V A 1.8 277.0nm 5.7
	LmV	B	29.5	LmH B 17 2.2/um 4.9
				LmV B 15 3.1/um 5.2
20.	eP	A	03 33 36	Greece 39.40 N 21.80 E
	e	A	33 49	H = 03 30 27.9 h = normal MB = 4.9
	LmH	B	38.3	D = 13.34 Az = 330.8 (NEIC)
	LmV	B	39.4	LmH B 14s 1.0/um M = 4.0
				LmV B 8 1.9/um
20.	ePKIKP	A	13 26 20.5	South of Fiji Islands 24.69 S 177.44 W
	ePKHKP	A	26 28.5	H = 13 06 47.5 h = 152 km MB = 5.2
	ePKP2	A	26 40	D = 153.16 Az = 347.2 (NEIC)
	epPKP	A	27 10	PKHKPV A 1.1s 36.0nm
21.	ePKIKP	A	01 42 23	Fiji Islands Region 17.58 S 178.97 W
	eiPKHKP	A	42 25	H = 01 23 46.6 h = 563 km MB = 5.1
				D = 145.94 Az = 347.9 (NEIC)
				PKHKPV A 1.3s 91.7nm
21.	eP	AB	02 59 43	Taiwan 23.83 N 121.55 E
	eS	C	03 10 06	H = 02 47 14.2 h = 14 km MB = 5.7
	eSS	C	15 39	D = 83.62 Az = 323.0 (NEIC)
	LmH	B	41.9	PV A 1.8s 50.6nm M = 5.4
	LmV	B	41.9	PV B 11.5 1.2/um 6.0
				LmH B 14 6.9/um 6.2
				LmV B 13 9.1/um 6.4
21.	ePKIKP	A	04 17 49.5	Solomon Islands 6.50 S 154.25 E
	e	A	18 14.5	H = 03 59 08.5 h = 34 km MB=5.4 MS=6.4
	e	A	20 19.5	D = 126.13 Az = 331.4 (NEIC)
	LmH	B	05 15.6	PKIKPV : traces
	LmV	B	21.4	LmH B 19s 9.0/um M = 6.5
				LmV B 17 8.4/um 6.5
21.				

Day	Phase		h m s	Remarks
21.	ePKIKP	A	07 51 12.5	<u>South Pacific Cordillera</u> 54.78 S 130.13 W H = 07 31 17.4 h = normal MB = 5.7 D = 156.67 Az = 84.3 (NEIC) PKIKPV A 1.6s 22.0nm
21.	eP	A	10 15 01.5	<u>Queen Elizabeth Islands</u> 76.55 N 106.69 W
	eX	A	15 08	H = 10 06 29.6 h = normal MB = 4.8
	LmH	B	37.4	D = 47.23 Az = 49.8 (NEIC)
	LmV	B	38.3	PV A : traces XV A 1.2s 30.5nm LmH B 14.5 3.7 µm M = 5.5 LmV B 15 4.1 µm 5.6
21.	+eP	A	17 13 39	<u>Near Islands, Aleutian Islands</u>
	LmH	B	44.6	52.45 N 173.61 E H = 17 01 55.3 h = 50 km MB = 5.5 D = 76.16 Az = 348.3 (NEIC) PV A 1.3s 78.6nm M = 5.5 LmH B 24 1.4 µm 5.2 The vertical component of the middle-period seismograph (type B) had been out of operation.
21.	eP	A	22 59 36	<u>West Pakistan</u> 32.16 N 69.81 E H = 22 51 15.4 h = 56 km MB = 4.4 D = 46.12 Az = 311.4 (NEIC) PV A 1.6s 27.5nm M = 4.9
22.	e	A	11 37 52	
22.	ePKP	A	14 10 37	<u>New Hebrides Islands</u> 15.43 S 167.18 E
	e	A	10 39	H = 13 51 10.4 h = 43 km MB=5.4 MS=5.7
	LmH	B	15 14.6	D = 139.71 Az = 336.0 (NEIC) PKPV A 1.2s 18.3nm LmH B 21 1.4 µm M = 5.6 The vertical component of the middle-period seismograph (type B) had been out of operation.

Day	Phase		h m s	Remarks
22.	e	A	15 02 44.5	
22.	+ePKP	A	18 13 23	<u>Tonga Islands</u> 15.50 S 172.99 W
	epPKP	A	13 41	H = 17 53 51.9 h = 57 km MB = 5.5
	esPKP	A	13 51	D = 144.75 Az = 354.9 (NEIC)
				h = 69 km
				PKPV A 1.3s 56.8nm
23.	eP	A	01 16 41.5	<u>North of Ascension Islands</u>
				0.26 S 15.91 W
				H = 01 07 04.9 h = normal MB = 4.9
				D = 55.87 Az = 20.8 (NEIC)
				PV A 1.3s 15.3nm
23.	ePKP	A	08 26 59.5	<u>Loyalty Islands</u> 20.18 S 168.70 E
	epPKP	A	27 11.5	H = 08 07 25.1 h = normal MB=4.8 MS=4.9
	esPKP	A	27 17	D = 144.62 Az = 334.7 (NEIC)
				h = 44 km
				PKPV A 1.2s 24.4nm
				pPKPV A 1.5 85.5nm
24.	eP	A	01 28 27.5	<u>Ionian Sea</u> 38.13 N 19.99 E
	e	A	29 05	H = 01 25 07.9 h = 0 km MB = 4.0
				D = 13.86 Az = 337 (ISC)
24.	eP	A	01 38 47	<u>Greece</u> 38.83 N 22.28 E
	e	A	38 55	H = 01 35 27.1 h = normal MB = 4.3
	LmH	B	44.6	D = 14.02 Az = 330.9 (NEIC)
	LmV	B	44.6	LmH B 13.5s 0.9 µm M = 4.0
				LmV B 12 1.0 µm
24.	eP1	A	03 51 3[illegible]	<u>Greece - Albania Border Region</u>
	iP2	A	51 39	39.51 N 20.34 E
	LmH	B	57.6	H = 03 48 35.1 h = 11 km MB = 5.3
	LmV	B	57.8	D = 12.71 Az = 334.0 (NEIC)
				P1V A 1.2s 32.5nm M = 5.4
				P2V A 0.8 88.5nm 6.0
				LmH B 12.5 3.8 µm 4.6
				LmV B 13 3.1 µm

Day	Phase		h m s	Remarks
24.	e	A	04 42 12.5	<u>Central Italy</u> 42.5 N 12.8 E
	eSn	A	43 31	H = 04 39 57 h = 68 km
				D = 8.23 Az = 355 (ISC)
24.	ePn	A	08 08 04.5	<u>Central Italy</u> 43.0 N 13.4 E
	ePg	A	08 35	H = 08 06 04 (BCIS)
	e	A	08 46	D = 7.78
	eSn	A	09 33	PnV A 0.8s 15.4nm
24.	ePn	A	08 30 57.5	<u>Central Italy</u> 43.0 N 13.4 E
	ePg	A	31 36	H = 08 29 02 (BCIS)
	e	A	31 45	D = 7.78
	eiSn	A	32 17	
24.	eP	A	09 05 24.5	<u>Western Russia</u> 52.78 N 51.07 E
				H = 09 00 08.0 h = normal MB = 4.7
				D = 24.33 Az = 280.9 (NEIC)
				PV A 1.0s 13.8nm M = 4.5
24.	eP	A	10 06 30	<u>Western Kazakh SSR</u> 51.84 N 64.15 E
	ePn	A	07 11	H = 09 59 57.8 h = 0 km MB = 5.2
	e	A	07 26	D = 32.32 Az = 289.1 (NEIC)
	e(PP)	A	07 39	PV A 0.7s 26.8nm M = 5.3
				Perhaps artificial (BCIS)
24.	eP	AB	13 30 26.5	<u>Bay of Bengal</u> 11.63 N 85.36 E
	eS	B	39 35	H = 13 19 12.4 h = normal MB = 5.4
	eSS	B	44 20	D = 70.79 Az = 319.7 (NEIC)
	LmH	B	14 07.4	PV A 1.4s 41.8nm M = 5.3
	LmV	B	07.8	LmH B 16 0.9 µm 5.1
				LmV B 16 1.2 µm 5.3
24.	ePg	A	15 05 20	<u>Hilders, Rhön / Fed. Rep. of Germany</u>
	iSg	A	05 36	50°32.53'N 10°02.28'E
				Explosion
				H = 15 05 00.83 yield 9.8 t (HAN)
				D = 1.0

Day	Phase		h m s	Remarks
24.	ePKP	A	20 26 27	North of New Zealand 16.45 S 175.4 W
				H = 20 06 47.8 h = 0 km MB = 5.2
				D = 145.43 Az = 352 (ISC)
				PKPV A 2.0s 38.4nm
25.	eP	AB	03 15 12.5	Near Coast of Oaxaca, Mexico
	eSKS	B	25 40	15.45 N 95.13 W
	eS	B	26 00	H = 03 02 20.2 h = normal MB=5.1 MS=5.3
	LmH	B	54.1	D = 88.44 (NEIC)
	LmV	B	58.1	PV A 1.5s 25.1nm M = 5.3
				SH B 12 1.2/um 5.9
				LmH B 18 1.3/um 5.4
				LmV B 16.5 1.4/um 5.4
25.	eP	A	13 52 36	Eastern Russia 56.29 N 123.46 E
	LmH	B	14 17.0	H = 13 42 34.4 h = 33 km MB = 5.1 (NEIC)
	LmV	B	20.6	D = 59.47
				LmH B 16s 0.9/um M = 5.0
				LmV B 16 0.5/um 4.8
25.	ePKP2	A	20 01 31	South of Kermadec Islands
				32.67 S 177.63 W
				H = 19 40 48.61 h = 33 km (NEIC)
				D = 160.79
25.	+eP	AB	20 09 37	North of Svalbard 80.28 N 2.39 W
	eS	B	14 40	H = 20 03 29.3 h = 33 km MB = 5.6 (NEIC)
	LmH	B	22.2	D = 30.13
	LmV	B	22.2	PV A 1.1s 105.0nm M = 5.5
				PV B 3.5 0.45/um 5.7
				SH B 11 0.7/um 5.1
				LmH B 14 1.0/um 4.6
				LmV B 15 0.9/um 4.7
25.	eP	A	22 50 52.5	Southern Iran 28.42 N 53.74 E
				H = 22 43 29.5 h = 41 km MB = 4.7 (NEIC)
				D = 38.64

Day	Phase		h m s	Remarks
26.	ePKIKP	A	04 41 31	South of Fiji Islands 22.10 S 179.54 W
	ePKHKP	A	41 37	H = 04 22 51.9 h = 599 km MB = 5.5 (NEIC)
	ePKP2	A	41 44.5	D = 150.20 h c. 600 km
	epPKP	A	43 53	PKIKPV A 1.3s 24.0nm
	eX	A	43 56	PKHKPV A 1.2 77.2nm
				XV A 1.9 113.6nm
26.	e	A	05 04 43	
26.	eP	A	08 03 50.5	Celebes Sea 4.20 N 124.44 E
				H = 07 50 35.9 h = 330 km MB = 5.3 (NEIC)
				D = 100.97
				PV A 0.8s 15.4nm M = 5.5
26.	eP	A	08 07 55	North Atlantic Ridge 35.40 N 36.40 W
	e	A	08 03.5	H = 08 00 40.8 h = 20 km MB = 5.0
	eS	B	13 52	D = 37.57 Az = 50.9 (NEIC)
	LmH	C	19.1	LmH C 31s 0.9 µm M = 4.4
	LmV	B	20.7	LmV B 19 1.1 µm 4.8
26.	ePKIKP	A	13 01 57	South of Kermadec Islands
	ePKP2	A	02 34.5	32.57 S 178.00 W
	LmV	B	25.5	H = 12 41 51.3 h = 5 km MB = 5.3 (NEIC)
	LmH	B	23.2	D = 160.59
				PKIKPV A 2.0s 34.2nm
				PKP2V A 1.3 41.5nm
				LmH B 18 0.5 µm M = 5.3
				LmV B 18.5 0.4 µm 5.3
26.	+iP	A	15 04 02.5	Near East Coast of Kamchatka
				52.10 N 158.80 E
				H = 14 52 31.1 h = 47 km MB = 5.2 (NEIC)
				D = 73.93
				PV A 1.0s 23.6nm M = 5.1
26.	ePKIKP	A	15 51 26.5	South of Tonga Islands 24.19 S 175.04 W
	ePKHKP	A	51 34	H = 15 31 38.0 h = normal MB = 5.5 (NEIC)

Day	Phase		h m s	Remarks
cont. 26.	LmV	B	17 04.6	D = 153.09
	LmH	B	05.0	LmH B 19s 0.3/um M = 5.0
				LmV B 19 0.4/um 5.3
26.	ePn	AB	16 05 03	<u>Central Italy</u> 42.95 N 13.40 E
	e	A	05 31	H = 16 03 11.9 h = normal MB = 4.9 (NEIC)
	ePg	A	05 41	D = 7.79
	eSn	A	06 28	PV A 0.9s 77.9nm M = 5.7
	iSn	A	06 29.5	LmH C 15.5 8.1/um 4.5
	eiSg	A	07 33	LmV B 10.5 2.7/um
	LmH	C	07.9	
	LmV	B	09.0	
27.	eP	A	02 58 23	<u>Philippine Islands Region</u>
	e	AB	58 36	20.17 N 121.50 E
	eSKS	B	03 08 49	H = 02 45 41.8 h = normal MB = 5.4 (NEIC)
	LmV	B	42.0	D = 86.51
	LmH	B	43.5	PV A 1.7s 42.4nm M = 5.4
				LmH B 17 1.5/um 5.5
				LmV B 17 1.8/um 5.6
27.	ePKP	A	03 44 28	<u>New Hebrides Islands</u> 13.83 S 167.21 E
				H = 03 25 34.2 h = 208 km MB = 4.1
				D = 138.28 Az = 336.7 (NEIC)
				traces
27.	ePKP2	A	04 48 07	<u>South of Kermadec Islands</u>
				32.21 S 178.14 W
				H = 04 27 18.9 h = 48 km (NEIC)
				D = 160.22
27.	e	A	07 00 43	<u>Molucca Passage</u> 0.57 N 126.24 E
	ePP	A	00 50	H = 06 42 19.6 h = 24 km MB = 5.7 (NEIC)
	e	A	01 18	D = 104.93
	LmV	B	47.9	LmH B 21.5s 1.3/um M = 5.4
	LmH	B	48.0	LmV B 24 1.8/um 5.6

Day	Phase		h m s	Remarks
27.	ePKIKP	A	15 35 21	Banda Sea 5.30 S 126.59 E
	ePP	A	36 09	H = 15 17 40.8 h = 425 MB = 5.8 (NEIC)
	epPP	B	37 40	D = 109.78
	esSKS	B	43 00	PPV A 2.0s 85.5nm M = 5.5
	eSP	B	44 34	LmH B 17.5 2.1 /um
	LmV	B	16 19.6	LmV B 22 1.6 /um
	LmH	B	20.0	
27.	eP	A	21 49 15.5	Off East Coast of Kamchatka
	e	A	49 38.5	53.43 N 161.25 E
				H = 21 37 46.6 h = 45 km MB = 4.7
				D = 73.20 Az = 340.4 (NEIC)
				PV A 0.8s 15.4nm M = 5.0
28.	eP	A	10 28 16	Arabian Sea 14.76 N 53.78 E
				H = 10 19 28.3 h = normal MB = 4.8
				D = 49.32 Az = 325.7 (NEIC)
28.	eP	A	13 30 55.5	Eastern Mediterranean Sea
				33.80 N 27.90 E
				H = 13 26 15.4 h = 33 km MB = 4.8 (NEIC)
				D = 20.63
				PV A 1.8s 40.5nm M = 4.5
28.	ePn	A	20 39 35	Central Italy 42.9 N 13.4 E
	eSn	A	40 48	H = 20 37 34 (BCIS)
	eSg	A	41 54	D = 7.87
29.	eP	A	02 02 32	Eastern Mediterranean Sea
				33.93 N 25.8 E
				H = 01 58 03 h = 44 km MB = 4.1
				D = 19.67 Az = 332 (ISC)
29.	eSg	A	02 12 53	Poland 50.27 N 18.87 E
				H = 02 10 19.4 M = 2.8 (WAR)
				D ≐ 4.65

Day	Phase		h m s	Remarks
29.	ePKP	A	02 33 26.5	Tonga Islands 17.47 S 173.85 W
	epPKP	A	33 47	H = 02 13 50.8 h = 80 km MB = 5.0
	esPKP	A	33 57	D = 146.61 Az = 353.7 (NEIC)
				h = 78 km
				PKPV A 1.8s 111.0nm
29.	ePKP	A	08 17 25	Samoa Islands Region 16.02 S 172.54 W
	e	A	17 38	H = 07 57 45.1 h = 13 km MB = 4.5
				D = 145.30 Az = 355.4 (NEIC)
				PKPV A 1.6s 22.0nm
30.	+iPKP	A	03 05 09	North of New Zealand 16.05 S 174.26 W
	e	A	05 17.5	H = 02 45 52.7 h = 201 km MB = 5.2
				D = 145.17 Az = 353 (ISC)
				PKPV A 1.5s 82.9nm
30.	eSg	A	08 02 05.5	Czechoslovakia 49°42.4'N 17°21.5'E
				Explosion yield 18 t (PRU)
				H = 08 00 02.0
				D = 3.86 Az = 286 (ISC)
30.	-eP1n	A	11 27 07.5	Adriatic Sea 44.11 N 13.04 E
	+iP2n	A	27 08.8	H = 11 25 31.6 h = normal MB = 4.6
	ePg	A	27 44	D = 6.62 Az = 352.1 (NEIC)
	eSn	A	28 19.5	P1nV A 0.6s 55.5nm M = 5.5
	iSn	A	28 21.5	P2nV A 0.8 113.0nm 5.7
	eSg	A	29 05	LmH B 9 1.5µm 3.9
	LmH	B	30.4	
	LmV	B	30.7	
30.	e(Pg)	A	13 40 27	
	e(Sg)	A	41 48	
30.	eP	A	22 04 14	Tadzhik-Sinkiang Border Region
				38.91 N 74.66 E
				H = 21 56 01.1 h = 59 km MB = 4.8
				D = 44.96 Az = 306.6 (NEIC)

Day	Phase		h m s	Remarks
1.	e	A	03 14 24	Near East Coast of Honshu, Japan
				35.64 N 140.57 E
				H = 03 02 10.3 h = 62 km MB = 4.7
				D = 83.03 Az = 330.1 (NEIC)
1.	ePn	A	07 26 56	Central Italy 42.74 N 13.25 E
	eSn	A	28 20	H = 07 24 52.9 h = normal MB = 3.9
	eSg	A	29 27	D = 7.99 Az = 352.5 (NEIC)
1.	eiP	A	11 46 07.5	Iran 35.42 N 57.91 E
	LmH	B	12 03.6	H = 11 39 03.6 h = 33 km MB = 5.4
	LmV	B	03.7	D = 36.47 Az = 309.3 (NEIC)
				PV A 1.7s 167.0nm M = 5.7
				LmH B 18.5 3.5/um 5.2
				LmV B 16 2.8/um 5.2
1.	eP	A	21 27 15	Near East Coast of Kamchatka
				54.91 N 162.01 E
				H = 21 15 51.8 h = normal MB = 4.8
				D = 71.94 Az = 340.7 (NEIC)
				PV A 1.8s 47.3nm M = 5.2
2.	eP1	AB	00 33 35.5	Mindanao, Philippine Islands
	iP2	A	33 38	6.47 N 126.60 E
	eiP3	AB	33 49	H = 00 19 47.2 h = normal MB = 6.3
	ePP	B	38 00	D = 100.42 Az = 324.1 (NEIC)
	eSKS	B	44 28	P1V A 2.4s 110.7nm M = 6.0
	eS	B	45 31	P2V A 2.2 392.0nm 6.6
	ePS	B	46 50	P3V A 1.5 482.0nm 6.8
	eSS	B	52 39	P3V B 13 18.1/um 7.4
	LmH	B	01 21.7	PPV B 15 34.6/um 7.6
	LmV	B	23.0	LmH B 20 481.0/um 8.0
				LmV B 23 754.0/um 8.1
2.	eP	A	01 54 32.5	Mindanao, Philippine Islands
				6.45 N 126.62 E
				H = 01 40 47.8 h = 54 km MB = 5.9
				D = 100.45 Az = 324.1 (NEIC)
				PV A 1.3s 56.8nm M = 6.0

Day	Phase		h m s	Remarks
2.	eX	A	02 36 14	Philippine Islands Region
				5.79 N 127.22 E
				H = 02 22 07.1 h = 51 km MB = 4.9
				D = 101.33 Az = 324.2 (NEIC)
				XV A 1.5s 30.2nm
2.	eP	A	04 10 42.5	Philippine Islands Region
	e	A	11 07.5	6.20 N 127.52 E
				H = 03 56 57.6 h = 65 km MB = 5.8
				D = 101.17 Az = 324.3 (NEIC)
				PV A 1.3s 28.4nm M = 5.7
2.	eP1	A	06 02 15	Philippine Islands Region
	eP2	A	02 28	5.62 N 127.31 E
	LmH	B	46.8	H = 05 48 19.5 h = normal MB = 5.4
	LmV	B	07 05.2	D = 101.52 Az = 324.2 (NEIC)
				P1V : traces
				P2V A 2.5s 84.5nm M = 5.9
				LmH B 17.5 3.4/um 5.9
				LmV B 20 4.1/um 6.0
2.	eX	A	13 31 59.5	XV A 1.4s 23.2nm
2.	eP	A	13 32 45	Dodecanese Islands 35.40 N 27.06 E
	LmH	B	39.9	H = 13 28 24.4 h = 55 km MB = 5.1
	LmV	B	41.4	D = 18.90 Az = 328.4 (NEIC)
				PV A 1.7s 109.0nm M = 4.8
				LmH B 13 4.1/um 4.9
				LmV B 12.5 2.6/um 4.9
2.	LmH	B	23 10.3	Mid - Indian Rise 36.89 S 78.41 E
	LmV	B	14.3	H = 21 58 49.8 h = 33 km MB = 5.2 (NEIC)
				D = 105.06
				LmH B 16s 1.5/um M = 5.6
				LmV B 17 1.7/um 5.7
3.	e(pP)	A	01 15 43.5	Mindanao, Philippine Islands
				5.72 N 126.92 E
				H = 01 01 46.9 h = 62 km MB = 5.1

Day	Phase		h m s	Remarks
cont. 3.				D = 101.22 Az = 324.1 (NEIC) h = (41) km pPV A 1.4s 14.0nm
3.	LmV	B	04 10.2	Philippine Islands Region
	LmH	B	10.6	5.97 N 127.22 E
				H = 03 04 33.1 h = 65 km MB = 5.0 (NEIC) D = 101.19 LmH B 18s 0.6 µm LmV B 17 1.0 µm
3.	e(P)	A	07 58 46	Mindanao, Philippine Islands 6.54 N 126.79 E H = 07 45 02.6 h = 82 km MB = 5.3 D = 100.47 Az = 324.1 (NEIC) (P)V A 1.5s 20.1nm M = (5.5)
3.	eP	A	09 03 02	Southern Sinkiang Prov., China
	e	A	03 23.5	39.37 N 75.26 E
	LmH	B	22.0	H = 08 54 47.2 h = normal MB = 5.0
	LmV	B	23.4	D = 45.07 Az = 306.3 (NEIC) PV A 1.4s 27.9nm M = 5.0 LmH B 18 0.7 µm 4.6 LmV B 12 1.0 µm 5.0
3.	ePKP	A	09 13 12	Fiji Islands Region 17.84 S 178.52 W H = 08 54 36.1 h = 604 km MB = 4.6 D = 146.28 Az = 348.4 (NEIC)
3.	ePKP2	A	10 52 59	South of Fiji Islands 23.98 S 175.98 W H = 10 32 49.9 h = normal MB = 4.9 D = 152.74 Az = 349.4 (NEIC)
3.	eP	A	13 24 39	Near East Coast of Kamchatka 53.81 N 160.59 E H = 13 13 12.8 h = normal MB = 4.4 D = 72.70 Az = 339.9 (NEIC)

Day	Phase		h m s	Remarks
3.	eP	A	18 11 53	South Indian Ocean 11.32 S 87.50 E
				H = 17 58 55.9 h = normal MB = 5.3
				D = 89.89 Az = 321.9 (NEIC)
				PV A 1.2s 16.3nm M = 5.2
3.	eP	A	19 39 20	Philippine Islands Region
	esP	A	39 49	5.80 N 127.27 E
	e	A	53 08	H = 19 25 32.4 h = 58 km MB = 5.3
	LmH	B	20 21.1	D = 101.35 Az = 324.2 (NEIC)
	LmV	B	42.0	h = 75 km
				PV A 1.4s 20.9nm M = 5.6
				LmH B 21 3.1 μm
				LmV B 17 2.6 μm
3.	e	A	20 49 39	Philippine Islands Region
				5.47 N 127.41 E
				H = 20 35 46.3 h = 66 km MB = 5.2
				D = 101.70 Az = 324.2 (NEIC)
3.	eP	A	22 20 53	Kyushu, Japan 31.94 N 131.54E
	LmH	B	23 01.4	H = 22 08 32.0 h = 13 km MB = 5.3
	LmV	B	01.6	D = 82.18 Az = 326.2 (NEIC)
				LmH B 15.5s 2.2 μm M = 5.6
				LmV B 17 2.6 μm 5.7
3.	e(Sg)	A	23 52 49	Poland 50.35 N 18.82 E
				H = 23 50 19.7 M = 3.0 (WAR)
				D = 4.6
4.	eP	A	03 29 19	Dodecanese Islands 35.23 N 27.18 E
	e	A	29 26.5	H = 03 24 55.2 h = 39 km MB = 4.3 (NEIC)
	e	A	29 45	D = 19.10
4.	+eiP	AB	10 28 41	South of Honshu, Japan 33.33 N 140.68 E
	Pm	AB	28 55	H = 10 16 12.0 h = 66 km MB = 6.6 (NEIC)
	ePP	B	32 07	D = 85.08
	eS	B	38 52	PV A 1.7s 1120.0nm M = 6.8
	eSP	A	40 06	PmV A 1.8 4090.0nm 7.4
	eSS	B	44 52	PmV B 12 55.2 μm 6.6

Day	Phase		h m s	Remarks
cont. 4.	eSSS	B	10 47 51	SH B 20s 18.6 µm M = 6.8
	eSSSS	B	51 20	SPV A 2.6 935.0nm
	eP'P'	A	54 45	LmH B 19 276.0 µm 7.7
	LmH	B	11 14.4	LmV B 18.5 323.0 µm 7.8
	LmV	B	14.5	
	eP'P'P'	A	15 32	
4.	eP	A	11 34 26.5	<u>South of Honshu, Japan</u> 33.31 N 140.71 E H = 11 21 58.0 h = 79 km MB = 4.8 D = 85.11 Az = 330.3 (NEIC)
4.	eP1	A	12 03 09.5	<u>South of Honshu, Japan</u> 33.16 N 140.71 E
	eP2	A	03 12	H = 11 50 38.3 h = 64 km MB = 5.3
	ePP	A	06 33	D = 85.24 Az = 330.3 (NEIC) P2V A 1.4s 32.6nm M = 5.2
4.	e(pP)	A	13 43 35	<u>South of Honshu, Japan</u> 33.17 N 140.50 E H = 13 30 53.4 h = 88 km MB = 4.4 D = 85.14 Az = 330.2 (NEIC) traces
4.	eP	A	14 50 23.5	<u>Fox Islands, Aleutian Islands</u> 52.89 N 168.31 W H = 14 38 34.5 h = 49 km MB = 4.7 D = 76.84 Az = 0.1 (NEIC)
4.	-eP	A	15 03 28	<u>South of Honshu, Japan</u> 33.22 N 140.63 E H = 14 50 56.8 h = 59 km MB = 5.3 D = 85.15 Az = 330.2 (NEIC) PV A 1.4s 60.5nm M = 5.2
4.	eP	A	15 12 53.5	<u>South of Honshu, Japan</u> 33.13 N 140.55 E
	epP	A	13 09.5	H = 15 00 22.5 h = 61 km MB = 5.3 (NEIC) D = 85.19 h = 63 km PV A 1.4s 27.9nm M = 5.1
4.	ePKP2	A	16 02 16	<u>South of Fiji Islands</u> 23.73 S 176.17 W H = 15 42 23.1 h = 132 km MB = 4.7 D = 152.45 Az = 349.3 (NEIC)

Day	Phase		h m s	Remarks
4.	e	A	18 06 04.5	
	e	A	06 17.5	
4.	ePP	A	18 10 39	West New Guinea Region 1.52 S 136.68 E
	e	A	10 51	H = 17 51 22.4 h = 33 km MB = 6.0 (NEIC)
				D = 112.7
				PPV A 1.8s 33.8nm M = 5.8
4.	eP	A	18 25 41	South of Honshu, Japan 33.37 N 140.51 E
	epP	A	26 00	H = 18 13 16.2 h = 94 km MB = 4.6
	LmH	B	52.2	D = 84.97 Az = 330.2 (NEIC)
	LmV	B	19 03.5	h = 69 km
				LmH B 20.5s 3.8/um
				LmV B 20 3.0/um
4.	ePKHKP	A	21 57 30.5	South of Fiji Islands 22.07 S 179.77 W
	ePKP2	A	57 39	H = 21 38 46.3 h = 611 km MB = 4.7
				D = 150.12 Az = 345.4 (NEIC)
				PKHKPV A 1.3s 30.6nm
				PKP2V A 1.4 28.0nm
5.	LmV	B	03 51.3	West New Guinea Region 1.62 S 136.62 E
	LmH	B	51.8	H = 02 46 20.7 h = 33 km MB = 5.4 (NEIC)
				D = 112.75
				LmH B 18s 3.3/um M = 6.0
				LmV B 20 3.1/um 5.9
5.	e(P)	A	04 23 44.5	South of Honshu, Japan 33.33 N 140.79 E
				H = 04 11 07.0 h = 53 km MB = 4.7
				D = 85.13 Az = 330.3 (NEIC)
				(P)V A 1.7s 24.2nm M = (5.1)
5.	eP	A	09 14 04.5	Chagos Archipelago Region
				5.32 S 68.55 E
				H = 09 02 30.0 h = normal MB = 4.7
				D = 74.04 Az = 326.3 (NEIC)
				PV A 1.8s 13.5nm M = 4.6

Day	Phase		h m s	Remarks
5.	eP	A	10 15 24	South of Honshu, Japan 33.31 N 140.27 E
				H = 10 02 53.3 h = 60 km MB = 4.8
				D = 84.93 Az = 330.1 (NEIC)
				PV A 1.9s 45.5nm M = 5.2
5.	eP	A	12 03 41.5	Aegean Sea 38.96 N 23.90 E
	e	A	03 14.5	H = 12 00 16.3 h = 62 km MB = 4.7
				D = 14.55 Az = 327.4 (NEIC)
5.	eP	A	14 16 10	Off East Coast of Honshu, Japan
	Pm	A	16 15	33.49 N 141.20 E
	epP	A	16 24	H = 14 03 38.3 h = 55 km MB = 5.2
	LmH	B	57.8	D = 85.16 Az = 330.5 (NEIC)
	LmV	B	57.9	h = 50 km
				PmV A 1.4s 39.6nm M = 5.4
				LmH B 17 3.8 µm
				LmV B 17 3.2 µm
5.	ePKHKP	A	19 51 23	Tonga Islands Region 23.60 S 175.87 W
	ePKP2	A	51 34.5	H = 19 31 29.2 h = normal MB = 5.3
	e	A	51 47	D = 152.38 Az = 349.7 (NEIC)
				PKHKPV A 1.5s 35.2nm
5.	LmH	B	23 01.0	LmH B 18s 3.4 µm
	LmV	B	20.7	LmV B 16 2.4 µm
6.	ePKIKP	A	02 02 08	West New Guinea 2.99 S 139.14 E
	ePP	A	03 17.5	H = 01 43 24.8 h = 33 km MB = 5.4 (NEIC)
	e	A	03 32	D = 115.30
	e	A	03 46.5	PKIKPV: traces
	LmH	B	55.2	PPV A 1.6s 38.5nm M = 5.9
	LmV	B	56.0	LmH B 18.5 6.8 µm 6.3
				LmV B 18.5 14.3 µm 6.6
6.	eP	A	04 16 06	South of Honshu, Japan 33.25 N 140.46 E
				H = 04 03 35.2 h = 62 km MB = 4.9
				D = 85.06 Az = 330.1 (NEIC)

Day	Phase		h m s	Remarks
6.	eP	AB	05 43 33	<u>Mid-Indian Rise</u> 9.23 S 67.30 E H = 05 31 44.8 h = normal MB=5.6 MS=5.6 D = 76.62 Az = 327.3 (NEIC) PV A 1.8s 162.0nm M = 5.7
6.	ePKHKP	A	09 38 25	<u>South of Fiji Islands</u> 23.84 S 179.05 E H = 09 19 32.2 h = 553 km MB = 4.7 D = 151.51 Az = 343.1 (NEIC)
6.	eP	A	10 46 09	<u>South of Honshu, Japan</u> 33.35 N 140.49 E H = 10 33 41.1 h = 65 km MB = 4.9 D = 84.98 Az = 330.2 (NEIC)
6.	e	A	14 08 21	
6.	eP	A	16 19 09	<u>South of Honshu, Japan</u> 32.56 N 141.29 E H = 16 06 31.5 h = 49 km MB = 4.9 D = 86.00 Az = 330.6 (NEIC) PV A 1.0s 11.8nm M = 5.1
6.	eP ePP	A A	20 24 39 28 11	<u>South of Honshu, Japan</u> 33.23 N 140.62 E H = 20 12 08.1 h = 58 km MB = 5.3 D = 85.14 Az = 330.2 (NEIC) PV A 1.2s 20.3nm M = 5.1
6.	eP LmH	A B	23 53 40.5 24 28.5	<u>Off East Coast of Honshu, Japan</u> 40.25 N 143.73 E H = 23 41 29.2 h = 17 km MB = 5.3 D = 80.25 Az = 331.4 (NEIC) PV A 1.2s 26.4nm M = 5.1 LmH B 17 2.0 µm 5.5
7.	eP	A	01 06 56.5	<u>South of Honshu, Japan</u> 33.27 N 140.80 E H = 00 54 24.8 h = 61 km MB = 4.8 D = 85.18 Az = 330.3 (NEIC) PV A 1.4s 23.3nm M = 5.1

Day	Phase		h m s	Remarks
7.	e(P)	A	02 14 27	Off East Coast of Honshu, Japan 34.18 N 141.43 E H = 02 01 54.9 h = 63 km MB = 4.9 D = 84.65 Az = 330.6 (NEIC)
7.	ePKHKP ePKP2	A A	02 33 34 33 39	Fiji Islands Region 20.40 S 178.42 W H = 02 14 49.1 h = 572 km MB = 4.5 D = 148.79 Az = 347.6 (NEIC)
7.	ePKHKP	A	08 10 02.5	Tonga Islands Region 23.64 S 175.92 W H = 07 50 08.7 h = normal MB = 4.7 D = 152.41 Az = 349.6 (NEIC) traces
7.	LmV LmH	B B	12 52.7 53.0	Near N. Coast of West New Guinea 2.93 S 139.11 E H = 11 41 24.1 h = 20 km MB = 5.6 (NEIC) D = 115.22
7.	LmH LmV	B B	15 56.0 56.0	North of Halmahera 3.75 N 128.50 E H = 14 50 15.0 h = 53 km MB = 5.4 (NEIC) D = 103.72
7.	e(P) e	A A	17 37 04.5 31 11.5	North of Svalbard 81.80 N 5.48 W H = 17 30 38.2 h = normal MB = 4.8 D = 31.73 Az = 159.1 (NEIC)
7.	eP e	A A	19 30 33.5 30 41.5	Near Islands, Aleutian Is. 53.12 N 171.10 E H = 19 18 53.0 h = normal MB = 5.5 D = 75.16 Az = 346.6 (NEIC) PV A 1.4s 60.5nm M = 5.4
7.	eP	A	23 50 13.5	Near East Coast of Honshu, Japan 35.58 N 139.96 E H = 23 37 56.0 h = 79 km MB = 5.0 D = 82.83 Az = 329.8 (NEIC) traces

Day	Phase		h m s	Remarks
8.	ePKP	A	01 53 27	New Hebrides Islands 20.67 S 169.52 E
	epPKP	A	53 39	H = 01 33 52.2 h = 42 km MB = 4.5
	esPKP	A	53 45	D = 145.38 Az = 335.1 (NEIC)
				h = 47 km
				PKPV A 1.1s · 36.3nm
8.	pP	A	04 16 54.5	Off East Coast of Honshu, Japan
				33.18 N 141.41 E
				H = 04 04 06.7 h = 46 km MB = 4.8
				D = 85.51 Az = 330.6 (NEIC)
				h = (50) km
8.	eP	A	04 24 22	South of Honshu, Japan 32.51 N 141.66 E
	epP	A	24 34.5	H = 04 11 42.7 h = 38 km MB = 5.0
	e	A	25 48	D = 86.19 Az = 330.8 (NEIC)
	ePP	A	27 42	h = 42 km
8.	eP	A	04 47 10.5	North of Severnaya Zemlya
				85.27 N 91.18 E
				H = 04 39 46.5 h = 32 km MB = 4.5
				D = 38.92 Az = 265.4 (NEIC)
8.	ePKP2	A	05 10 39	South of Tonga Islands 24.02 S 175.93 W
				H = 04 50 34.5 h = normal MB = 5.1
				D = 152.79 Az = 349.5 (NEIC)
8.	ePKHKP	A	06 37 20.5	South of Fiji Islands 23.42 S 175.97 W
	ePKP2	A	37 32.5	H = 06 17 26.4 h = normal MB = 4.5
				D = 152.19 Az = 349.6 (NEIC)
8.	ePKHKP	A	07 01 27.5	Tonga Islands Region 23.71 S 175.70 W
	ePKP2	A	01 37.5	H = 06 41 32.2 h = normal MB = 5.0
				D = 152.51 Az = 349.9 (NEIC)
8.	eP	A	08 01 41	Southern Sumatra 1.88 S 99.66 E
	e	A	01 49.5	H = 07 48 42.0 h = normal MB = 5.5
	LmV	B	48.8	D = 90.20 Az = 320.5 (NEIC)

Day	Phase		h m s	Remarks
cont. 8.	LmH	B	08 49.5	PV A 1.3s 39.3nm M = 5.5
				LmH B 18.5 1.5 /um 5.5
				LmV B 20 1.8 /um 5.5
8.	eP	A	09 01 14	Southern Sumatra 1.92 S 99.72 E
	e	A	01 25.5	H = 08 48 14.0 h = normal MB = 5.3
				D = 90.27 Az = 320.5 (NEIC)
8.	e(P)	A	10 32 05.5	Philippine Islands Region
	LmH	B	11 24.1	5.69 N 127.36 E
	LmV	B	34.5	H = 10 18 12.8 h = 64 km MB = 5.0
				D = 101.49 Az = 324.2 (NEIC)
				LmH B 16s 1.1 /um
				LmV B 18 1.3 /um
8.	eP	A	17 09 23	Philippine Islands Region
	LmH	B	57.6	7.24 N 127.12 E
	LmV	B	58.3	H = 16 55 40.4 h = 57 km MB = 5.4
				D = 100.10 Az = 324.3 (NEIC)
				PV A 2.0s 42.8nm M = 5.7
				LmH B 15.5 1.5 /um
				LmV B 17 1.9 /um
8.	ePKIKP	A	18 21 01.8	Fiji Islands Region 21.42 S 179.15 W
	ePKHKP	AB	21 07	H = 18 02 23.5 h = 605 km MB = 5.4
	ePKP2	A	21 15	D = 149.63 Az = 346.4 (NEIC)
				PKIKPV A 1.8s 40.5nm
				PKHKPV A 1.4 172.0nm
				PKP2V A 1.4 69.7nm
8.	eP	A	18 26 04	Southern Sumatra 1.86 S 99.78 E
	e	A	26 12.5	H = 18 13 03.0 h = 20 km MB=5.8 MS=5.9
	e	A	26 19.5	D = 90.26 Az = 320.5 (NEIC)
	eS	B	37 00	PV A 1.8s 77.7nm M = 5.7
	e	B	50 16	LmH B 20 2.6 /um 5.7
	LmH	B	19 08.8	LmV B 20 2.7 /um 5.7
	LmV	B	12.7	

Day	Phase		h m s	Remarks
8.	e(P)	A	20 31 05.5	Philippine Islands Region
	e	A	31 13.5	5.67 N 127.30 E
	e	A	31 19	H = 20 17 12.9 h = 63 km MB = 5.4
				D = 101.48 Az = 324.2 (NEIC)
				PV A 1.2s 16.3nm M = 5.6
8.	e(P)	A	20 57 08	Philippine Islands Region
	e	A	57 21	5.69 N 127.37 E
	ePP	A	21 01 16	H = 20 43 16.0 h = 65 km MB = 5.3
	LmH	B	49.8	D = 101.50 Az = 324.2 (NEIC)
	LmV	B	52.6	PV A 1.5s 17.6nm M = 5.5
				PPV A 1.4 18.6nm 6.5
				LmH B 19.5 2.1 /um
				LmV B 20 2.1 /um
9.	eP	A	00 21 32.5	Philippine Islands Region
	epP	A	21 46.5	5.64 N 127.45 E
	LmH	B	01 13.9	H = 00 07 45.8 h = 68 km MB = 5.1
	LmV	B	17.1	D = 101.59 Az = 324.2 (NEIC)
				h = 50 km
				LmH B 18s 0.9 /um
				LmV B 18 1.1 /um
9.	eP	A	06 18 41	South of Honshu, Japan 33.12 N 140.80 E
	e	A	18 46.5	H = 06 06 07.6 h = 51 km MB = 5.3
	epP	A	18 54.5	D = 85.31 Az = 330.3 (NEIC)
	esP	A	19 04	h = 55 km
				PV A 1.2s 30.5nm M = 5.3
9.	+eP1	A	06 54 27.5	North Atlantic Ridge 15.21 N 45.16.W
	eP2	AB	54 32.5	H = 06 44 40.4 h = normal MB=5.7 MS=5.7
	eS	B	07 02 28	D = 57.47 Az = 39.2 (NEIC)
	LmH	B	21.2	P1V A 1.6s 52.2nm M = 5.3
	LmV	B	23.2	P2V A 1.9 114.0nm 5.6
				SH B 15.5 2.8 /um 5.8
				LmH B 16 4.1 /um 5.6
				LmV B 16 5.0 /um 5.8

Day	Phase		h m s	Remarks
9.	LmH	B	08 30.0	Mindanao, Philippine Islands
	LmV	B	30.3	7.2 N 126.8 E
				H = 07 27 49.0 h = 93 km MB = 5.1 (NEIC)
				D = 99.99
9.	eP	A	08 55 36.5	Greece 38.35 N 20.20 E
	eX	A	59 15	H = 08 52 22.0 h = 21 km MB = 4.7
				D = 13.73 Az = 336.4 (NEIC)
				PV A 0.7s 17.2nm M = 5.0
				XV A 1.4 20.9nm
9.	LmH	B	13 58.0	Mindanao, Philippine Islands
	LmV	B	58.0	7.4 N 126.9 E
				H = 12 55 47.5 h = 66 km MB = 5.2 (NEIC)
				D = 99.85
				LmH B 18s 0.9/um
				LmV B 18 1.3/um
10.	eP	A	00 43 48	South of Honshu, Japan 33.21 N 140.78 E
				H = 00 31 15.9 h = 53 km MB = 4.3
				D = 85.23 Az = 330.3 (NEIC)
10.	eP	A	03 47 23.5	Hokkaido, Japan Region 43.11 N 141.54 E
	e	A	47 26.5	H = 03 35 48.2 h = 172 km MB = 5.1
				D = 76.95 Az = 329.9 (NEIC)
				PV A 1.1s 32.2nm M = 5.0
10.	+iP	AB	04 34 46.2	Eastern Kazakh SSR 49.85 N 78.10 E
	ePn	A	36 19	H = 04 26 57.7 h = 0 km MB = 5.7
				D = 41.22 Az = 297.6 (NEIC)
				underground explosion MB = 5.9 (UPP)
				PV A 0.8s 240.0nm M = 6.0
10.	+iP	AB	04 34 59.3	Eastern Kazakh SSR 50.11 N 78.81 E
	ePn	A	36 32.5	H = 04 27 08.4 h = 0 km (NEIC)
	LmH	B	51.6	underground explosion MB = 7.0 (UPP)
	LmV	B	52.9	D = 41.50
				PV A 1.0s 266.0nm M = 5.9

Day	Phase		h	m	s	Remarks
10.	-iP	A	04	45	55.5	Chiapas, Mexico 17.46 N 93.61 W
						H = 04 33 37.2 h = 211 km MB = 5.2
						D = 85.96 Az = 38.0 (NEIC)
						PV A 0.8s 32.7nm M = 5.2
10.	ePKIKP	A	08	18	30.5	Tonga Islands Region 23.93 S 175.80 W
	e	A		18	44.5	H = 07 58 35.5 h = normal MB = 5.1
	ePKP2	A		18	52.5	D = 152.71 Az = 349.7 (NEIC)
10.	e(PKP)	A	11	15	04:5	New Hebrides Islands 20.65 S 169.29 E
						H = 10 55 31.0 h = 33 km
						D = 145.27 Az = 334.9 (NEIC)
10.	eP	A	12	16	57	Southern Greece 37.46 N 21.25 E
	e	A		17	17	H = 12 13 29.2 h = 68 km MB = 4.3
						D = 14.87 Az = 335.5 (NEIC)
10.	ePKP	A	13	20	17	Fiji Region 18.54 S 177.8 W
						H = 13 00 38.5 h = 0 km MB = 4.8
						D = 147.10 Az = 349 (ISC)
10.	+iP1	AB	18	38	07.6	Kurile Islands 44.77 N 149.39 E
	-iP2	A		38	09.0	H = 18 26 07.1 h = 13 km MB = 6.0
	i	AB		38	15	D = 78.14 Az = 334.1 (NEIC)
	i	A		38	24	P1V A 1.0s 74.8nm M = 5.7
	LmV	B	19	16.7		P1V B 8.5 2.7 µm 6.4
	LmH	B		16.8		P2V A 1.2 439.0nm 6.4
						P2H A 1.2 210.0nm 6.4
						LmH B 17 7.7 µm 6.1
						LmV B 17 7.6 µm 6.1
10.	PKHKP	A	22	27	24.5	Fiji Islands Region 18.38 S 177.88 W
						H = 22 08 48.4 h = 609 km MB = 4.6
						D = 146.93 Az = 348.9 (NEIC)
						PKHKPV A 1.2s 16.3nm

Day	Phase		h m s	Remarks
11.	eP	A	00 34 18.5	Kurile Islands 44.45 N 149.45 E
	e	A	34 45	H = 00 22 19 h = 28 km MB = 4.9
				D = 78.45 Az = 33.4 (ISC)
				PV A 0.9s 23.4nm M = 5.2
11.	eP	A	01 47 49	Southern Sumatra 1.90 S 99.58 E
	e	A	47 58	H = 01 34 49.7 h = normal MB=5.1 MS=5.4
	e	A	49 24	D = 90.16 Az = 320.5 (NEIC)
	e	A	49 37	PV A 1.5s 25.2nm M = 5.3
	e	A	49 45	XV A 1.6 35.8nm
11.	eP	A	06 12 28	South of Honshu, Japan 33.45 N 140.94 E
	e	A	12 30.5	H = 05 59 56.9 h = 55 km MB = 4.7
	e	A	12 49.5	D = 85.08 Az = 330.4 (NEIC)
11.	ePKP	A	08 59 22	Fiji Islands Region 16.19 S 177.50 W
				H = 08 39 40.3 h = 20 km MB = 4.8
				D = 144.85 Az = 349.9 (NEIC)
11.	ePKHKP	A	13 36 43.5	Fiji Islands Region 20.32 S 176.15 W
				H = 13 17 12.7 h = 148 km MB = 4.7
				D = 149.12 Az = 350.3 (NEIC)
				PKHKPV A 1.0s 25.6nm
11.	eP	A	17 41 53	South of Honshu, Japan 33.21 N 140.44 E
				H = 17 29 21.7 h = 67 km MB = 4.9
				D = 85.08 Az = 330.1 (NEIC)
				PV A 1.4s 16.3nm M = 4.9
11.	eP	A	19 44 48.5	Kurile Islands 44.72 N 149.33 E
				H = 19 32 52.3 h = 45 km MB = 4.8
				D = 78.18 Az = 334.0 (NEIC)
12.	eP	A	00 47 32.5	Near East Coast of Kamchatka
				53.15 N 160.05 E
				H = 00 36 04.1 h = 40 km MB = 4.7
				D = 73.22 Az = 339.6 (NEIC)

Day	Phase		h m s	Remarks
12.	e	A	03 30 00	Zambia 16.84 S 27.91 E H = 03 18 45.0 h = 10 km MB = 4.8 D = 68.70 Az = 348.9 (NEIC)
12.	ePKHKP	A	05 59 07.5	South of Fiji Islands 22.88 S 176.28 W H = 05 39 20.4 h = 70 km MB = 5.0 D = 151.61 Az = 349.4 (NEIC) PKHKPV A 0.8s 19.2nm
12.	e	A	07 47 24	Norwegian Sea 71.42 N 11.30 E H = 07 42 26.5 h = normal MB = 4.6 D = 20.85 Az = 179.4 (NEIC)
12.	eP epP ePP	A A A	09 15 59.5 16 10 18 51	Kurile Islands Region 44.12 N 151.01 E H = 09 03 56.5 h = 38 km MB = 5.7 D = 76.33 Az = 21.4 (NEIC) h = 39 km PV A 1.2s 34.5nm M = 5.2
12.	eP	A	11 02 54	South of Honshu, Japan 33.25 N 140.68 E H = 10 50 26.1 h = 83 km MB = 5.2 D = 85.14 Az = 330.3 (NEIC) PV A 1.3s 21.8nm M = 5.0
12.	eP	A	16 58 43	Southern Sumatra 1.87 S 99.70 E H = 16 45 50.5 h = 92 km MB = 4.6 D = 90.21 Az = 320.5 (NEIC)
12.	eX	A	17 52 32.5	North Atlantic Ridge 26.0 N 43.6 W H = 17 43 41 h = 0 km MB = 4.9 D = 48.47 Az = 44 (ISC) XV A 1.0s 13.8nm
12.	eP ePP LmH LmV	A A C B	18 00 44 04 49 41.4 51.9	Mindanao, Philippine Islands 6.47 N 126.49 E H = 17 46 47.8 h = 3 km MB = 5.3 D = 100.36 Az = 324.0 (NEIC) PV A 2.0s 64.1nm M = 5.8 PPV A 2.0 68.4nm 5.7

Day	Phase		h m s	Remarks
12.	ePKHKP	A	20 07 25.5	South of Fiji Islands 23.38 S 179.91 W
	ePKP2	A	07 37	H = 19 48 29.1 h = 491 km MB = 4.6
	epPKP	A	09 33.5	D = 151.34 Az = 344.6 (NEIC)
				PKHKPV A 1.2s 16.2nm
				pPKPV A 1.5 25.1nm
13.	e	A	03 01 34	Greece - Bulgaria Border Region
	e(Sg)	A	05 35.5	41.62 N 24.02 E
				H = 02 58 51.6 h = 29 km MB = 4.5
				D = 12.46 Az = 320.6 (NEIC)
13.	eP	A	16 14 00.5	Rat Islands, Aleutian Is.
				51.61 N 175.30 E
				H = 16 02 08.3 h = normal MB = 5.1
				D = 77.19 Az = 349.4 (NEIC)
				PV A 1.5s 22.6nm M = 5.0
14.	eP	A	05 12 14	North Atlantic Ridge 28.45 N 43.71 W
				H = 05 03 45.5 h = normal MB = 4.9
				D = 46.81 Az = 45.9 (NEIC)
				PV A 2.0s 29.9nm M = 4.9
14.	eP	A	08 54 23.5	South of Honshu, Japan 33.46 N 140.38 E
				H = 08 41 55.3 h = 68 km MB = 5.2
				D = 84.84 Az = 330.1 (NEIC)
				PV A (1.7)s 30.3nm M = 5.1
14.	ePKHKP	A	08 57 05.5	Fiji Islands Region 19.72 S 178.08 W
	ePKP2	A	57 10	H = 08 38 09.1 h = 448 km MB = 4.8
				D = 148.20 Az = 348.3 (NEIC)
14.	eP	A	17 53 39	Greece 38.17 N 20.17 E
	e	A	53 44	H = 17 50 21.7 h = 10 km MB = 4.7
	e	A	53 49.5	D = 13.87 Az = 336.7 (NEIC)
	e	A	56 36	
14.	ePKP	A	20 27 35	New Hebrides Islands 19.19 S 168.66 E
	epPKP	A	27 47.5	H = 20 08 06.5 h = 42 km MB = 4.8
				D = 143.71 Az = 335.2 (NEIC) h = 45 km

Day	Phase		h m s	Remarks
14.	eP	A	21 02 00	South Indian Ocean 1.31 S 89.33 E
	eX	A	02 08.5	H = 20 49 34.8 h = normal MB=5.4 MS=5.2
	LmV	C	35.8	D = 83.23 Az = 321.2 (NEIC)
	LmH	C	42.2	PV A 1.2s 26.4nm M = 5.2
				XV A 1.3 83.0nm
				LmH C 20 1.1/um 5.2
				LmV C 29 1.5/um 5.2
15.	eP	A	02 04 32	South of Honshu, Japan 33.21 N 140.83 E
	epP	A	04 46.5	H = 01 51 58.3 h = 54 km MB = 5.1
				D = 85.24 Az = 330.3 (NEIC)
				h = 53 km
				PV A 1.6s 22.0nm M = 5.0
15.	eP	A	06 27 55	Off East Coast of Honshu, Japan
	ePP	A	31 13	33.31 N 141.30 E
	LmH	B	07 09.5	H = 06 15 20.1 h = 35 km MB = 5.2
	LmV	B	09.5	D = 85.35 Az = 330.6 (NEIC)
				PV A 1.6s 44.0nm M = 5.4
				PPV A 1.6 16.5nm 5.2
				LmH B 16 1.0/um 5.3
				LmV B 18 1.0/um 5.3
15.	eP	A	14 53 29.5	Kurile Islands 44.65 N 149.41 E
				H = 14 41 31.9 h = 40 km MB = 4.7
				D = 78.26 Az = 334.1 (NEIC)
				PV A 1.4s 46.5nm M = 5.3
15.	eP	A	18 00 07.5	Dodecanese Islands 35.23 N 27.24 E
				H = 17 55 46.5 h = normal MB = 4.7
				D = 19.13 Az = 328.4 (NEIC)
15.	eP	A	20 01 17.5	North of Svalbard 82.66 N 4.79 W
	e	A	01 20	H = 19 54 48.2 h = normal MB = 5.0
				D = 32.52 Az = 160.5 (NEIC)
				PV A 1.0s 27.6nm M = 5.1

Day	Phase		h m s	Remarks
16.	eP	A	02 35 09	South of Honshu, Japan 32.56 N 141.35 E H = 02 22 29.8 h = normal MB = 5.0 D = 86.03 Az = 330.6 (NEIC) PV A 1.2s 16.3nm M = 5.1
16.	ePKP	A	04 24 31	Fiji Islands Region 17.83 S 178.58 W H = 04 05 49.2 h = 557 km MB = 5.0 D = 146.26 Az = 348.3 (NEIC)
16.	epP	A	05 01 04	Rat Islands, Aleutian Is. 51.60 N 175.29 E H = 04 49 03.2 h = 41 km MB = 4.9 D = 77.19 Az = 349.4 (NEIC) h = (37) km
16.	ePKIKP e LmH	A A C	11 49 15.5 49 29 12 36.5	Solomon Islands 6.97 S 155.76 E H = 11 30 18.6 h = 79 km MB = 5.8 D = 127.25 Az = 332.1 (NEIC) PKIKPV A 1.2s 77.2nm LmH C 28 0.8 µm
16.	e	A	12 02 34.5	
16.	ePKP e	A A	16 43 32.5 43 41.5	Samoa Islands Region 16.66 S 171.69 W H = 16 23 58.2 h = 63 km MB = 5.0 D = 145.99 Az = 356.2 (NEIC) PKPV A 1.3s 41.5nm
16.	ePKHKP	A	20 48 31.5	Tonga Islands Region 23.76 S 175.73 W H = 20 28 39.4 h = 59 km MB = 4.8 D = 152.56 Az = 349.8 (NEIC)
16.	eP	A	23 07 20	North of Severnaya Zemlya 85.61 N 85.12 E H = 22 60 00.0 h = normal MB = 4.4 D = 38.48 Az = 258.8 (NEIC)

Day	Phase		h m s	Remarks
17.	eP	A	00 25 58	<u>Philippine Islands Region</u> 6.11 N 127.40 E H = 00 12 10.5 h = 56 km MB = 5.4 D = 101.18 Az = 324.2 (NEIC) traces
17.	+iP	AB	00 30 29.8	<u>Kurile Islands</u> 44.70 N 149.20 E
	epP	AB	30 47	H = 00 18 33.75 h = 50 km MB=5.7 MS=5.8
	esP	A	30 55	D = 78.15 Az = 334.0 (NEIC)
	eS	C	40 35	h = 68 km
	eSS	C	45 28	PV A 1.9s 269.0nm M = 5.9
	eSSS	C	48 35	PV B 5 2.4 μm 6.4
	LmH	B	01 09.2	LmH B 18 9.9 μm 6.2
	LmV	B	09.3	LmV B 18 10.1 μm 6.2
17.	eP	A	00 44 21.5	<u>Kurile Islands</u> 44.68 N 149.34 E H = 00 32 22.8 h = normal MB = 4.8 D = 78.21 Az = 334.1 (NEIC)
17.	eP	A	00 53 24	<u>Kurile Islands</u> 44.8 N 149.2 E H = 00 41 26.8 h = 30 km MB = 4.1 D = 78.04 Az = 334 (ISC)
17.	eP	A	01 30 14.5	<u>Kurile Islands</u> 44.77 N 149.22 N H = 01 18 16.6 h = normal MB = 4.7 D = 78.09 Az = 334.0 (NEIC) PV A 1.2s 12.2nm M = 4.8
17.	eP	A	01 38 02	<u>Kurile Islands</u> 44.68 N 149.41 E H = 01 26 04.3 h = normal MB = 4.8 D = 78.24 Az = 334.1 (NEIC) PV A 1.6s 19.2nm M = 4.9
17.	e	A	03 58 18	<u>South of Honshu, Japan</u> 33.35 N 139.51 E H = 03 45 51.5 h = 153 km MB = 4.3 D = 84.57 Az = 329.7 (NEIC) traces

Day	Phase		h m s	Remarks
17.	eP	A	05 57 56.5	Kurile Islands 44.65 N 149.28 E H = 05 45 57.7 h = normal MB = 4.6 D = 78.22 Az = 334.0 (NEIC) PV A 1.0s 11.8nm M = 4.9
17.	eP	A	06 36 51.5	Kurile Islands 44.65 N 149.36 E H = 06 24 51.9 h = normal MB = 4.9 D = 78.25 Az = 334.1 (NEIC) PV A 1.0s 25.6nm M = 5.2
17.	eP	A	07 42 42	North of Severnaya Zemlya 85.8 N 75.3 E H = 07 35 22 h = 0 km MB = 9.3 D = 37.84 Az = 249 (ISC)
17.	eP	A	12 48 58	Crete 34.26 N 26.19 E
	e	A	49 02	H = 12 44 30.3 h = 35 km MB = 4.7
	e	A	49 32	D = 19.54 Az = 331.4 (NEIC)
	LmH	B	58.5	PV A 2.2s 54.6nm M = 4.4
				LmV B 14 1.1/um 4.7
17.	ePKP	A	14 31 18	Loyalty Islands Region 22.36 S 171.61 E H = 14 11 34.2 h = 37 km MB = 4.9 D = 147.7 Az = 335.9 (NEIC)
17.	ePKIKP	A	18 19 48	New Ireland Region 5.49 S 153.21 E
	e	A	19 55	H = 18 00 50.2 h = 44 km MB = 5.5 D = 124.74 Az = 331.2 (NEIC)
17.	LmH	B	19 18.2	Probably to North of Ascension Island
	LmV	B	18.2	1.23 S 16.13 W H = 18 46 10.0 h = 33 km MB = 4.9 (NEIC) D = 56.85
17.	eP	A	21 02 48	North of Ascension Island 1.42 S 15.99 W H = 20 53 02.9 h = normal MB = 5.1 D = 56.98 Az = 20.6 (NEIC) PV A 1.4s 23.2nm M = 5.0

Day	Phase		h m s	Remarks
17.	e	A	24 05 44	Off East Coast of Honshu, Japan
	LmH	B	47.5	34.15 N 141.43 E
	LmV	B	51	H = 23 53 01.3 h = 42 km MB = 4.6
				D = 84.67 Az = 330.6 (NEIC)
18.	e	A	01 25 37.5	Off East Coast of Honshu, Japan
				33.9 N 141.4 E
				H = 01 12 44.3 h = 0 km MB = 4.1 (ISC)
				D = 85.0
18.	eP	AB	01 30 00.5	Zambia 16.61 S 28.11 E
	LmH	B	02 02.4	H = 01 18 54.2 h = 2 km MB=5.2 MS=5.7
	LmV	B	04.4	D = 68.51 Az = 348.8 (NEIC)
				PV A 1.4s 32.6nm M = 5.4
				LmH B 18.5 1.8 µm 5.3
				LmV B 18 1.3 µm 5.2
18.	ePKIKP	A	07 12 20	South of Fiji Islands 22.25 S 179.83 W
	+iPKHKP	AB	12 26.5	H = 06 53 39.6 h = 591 km MB = 5.4
	-PKP2	AB	12 35.5	D = 150.27 Az = 345.2 (NEIC)
	e	A	14 43	PKIKPV A 1.4s 37.2nm
				PKHKPV A 1.2 191.0nm
				PKP2V A 1.3 166.0nm
18.	eP	A	11 31 33.5	Burma 21.23 N 94.17 E
	e	A	31 52	H = 11 20 31.4 h = 72 km MB = 5.3
				D = 69.24 Az = 317.5 (NEIC)
18.	eP	AB	14 10 03	Southern Sumatra 1.83 S 99.65 E
	eX	A	10 16	H = 13 57 03.8 h = 33 km MB=5.6 MS=5.6
	eS	B	20 56	D = 90.14 Az = 320.5 (NEIC)
	LmV	B	57.0	PV A 1.7s 84.9nm M = 5.7
	LmH	B	57.8	XV A 1.3 52.4nm
				LmH B 19 2.0 µm 5.6
				LmV B 19 1.8 µm 5.6
18.	eP	A	23 39 41	South of Honshu, Japan 33.29 N 140.83 E
				H = 23 27 02.6 h = 8 km MB = 4.6
				D = 85.17 Az = 330.3 (NEIC)

Day	Phase		h m s	Remarks
19.	ePKIKP	A	00 31 05	New Britain Region 6.16 S 148.85 E
	epPKIKP	A	31 15	H = 00 12 11.7 h = 64 km MB = 5.0
	ePP	A	32(54)	D = 123.16 Az = 328.9 (NEIC)
	epPP	A	33 06.5	LmH B 18.5s 0.8/um
	LmH	B	01 25.5	LmV B 20 0.7/um
	LmV	B	25.7	
19.	eP	A	10 45 43	Fox Islands, Aleutian Is.
	LmH	B	11 30.8	52.04 N 169.65 W
	LmV	B	30.4	H = 10 33 49.0 h = 44 km MB = 5.0
				D = 77.69 Az = 359.2 (NEIC)
				PV A 1.5s 30.2nm M = 5.1
				LmH B 20 0.5/um 4.9
				LmV B 16 0.8/um 5.2
19.	e	A	15 04 33	
19.	e	A	15 36 19	Mindanao, Philippine Islands
	LmH	B	16 21.0	6.11 N 126.62 E
	LmV	B	33.4	H = 15 22 22.2 h = 14 km MB = 5.0
				D = 100.72 Az = 324.0 (NEIC)
				eV : traces
				LmH B 19.5s 2.1/um M = 5.7
				LmV B 18 1.4/um 5.5
19.	ePKHKP	A	16 07 17	South of Fiji Islands 22.18 S 179.76 W
	ePKP2	A	07 23.5	H = 15 48 34.8 h = 570 km MB = 4.7
				D = 150.22 Az = 345.3 (NEIC)
19.	eP	A	19 38 52.5	Dodecanese Islands 35.48 N 27.81 E
	e	A	39 01	H = 19 34 31.4 h = 54 km MB = 4.6
	LmH	B	45.6	D = 19.16 Az = 327.3 (NEIC)
	LmV	B	48.6	PV A 1.9s 60.6nm M = 4.5
				LmH B 15.5 1.8/um 4.5
				LmV B 16 1.6/um 4.6

Day	Phase		h m s	Remarks
19.	eP	A	23 16 36	Fox Islands, Aleutian Is. 52.18 N 169.60 W H = 23 04 38.3 h = 14 km MB = 4.5 D = 77.55 Az = 359.2 (NEIC) PV: traces
19.	eP	A	23 24 55.5	South of Honshu, Japan 33.22 N 140.64 E H = 23 12 23.2 h = 42 km MB = 4.5 D = 85.15 Az = 330.2 (NEIC)
19.	eP	A	23 27 12	South of Honshu, Japan 33.38 N 140.53 E H = 23 14 42.4 h = 61 km MB = 4.6 D = 84.97 Az = 330.2 (NEIC) PV: traces
20.	e(P)	A	00 09 30	Off East Coast of Honshu, Japan 34.19 N 141.54 E H = 23 56 51.0 h = normal MB = 4.6 D = 84.68 Az = 330.6 (NEIC) (P)V A 1.8s 20.3nm M = (5.0)
20.	LmH LmV	B B	08 32.8 38.8	Mariana Islands 20.82 N 144.49 E H = 07 33 56.8 h = 43 km MB = 5.0 (NEIC) D = 97.63 LmH B 15s 0.7 µm M = 5.2 LmV B 15.5 0.7 µm 5.3
20.	ePKIKP +iPKHKP +iPKP2 epPKHKP	A AB A A	18 11 57 12 03 12 11 14 22.5	Fiji Islands Region 21.75 S 179.63 W H = 17 53 19.4 h = 612 km MB = 5.4 D = 149.84 Az = 345.7 (NEIC) h = 629 km PKIKPV A 1.0s 19.7nm PKHKPV A 0.9 121.0nm PKP2V A 1.1 54.4nm
20.	ePKHKP	A	19 16 14	Fiji Islands Region 21.80 S 179.76 W H = 18 57 32.3 h = 645 km MB = 4.9

Day	Phase		h m s	Remarks
cont. 20.	ePKP2	A	19 16 22	D = 149.86 Az = 345.5 (NEIC) PKHKPV A 1.0s 19.7nm PKP2V A 1.5 22.6nm
20.	ePKHKP	A	21 31 09.5	<u>South of Fiji Islands</u> 22.39 S 179.44 E H = 21 12 22.1 h = 567 km MB = 4.5 D = 150.23 Az = 344.3 (NEIC) PKHKPV A 1.3s 41.5nm
20.	eP	A	22 56 25.5	<u>Mindanao, Philippine Islands</u> 8.56 N 126.02 E H = 22 42 50.8 h = 47 km MB = 5.3 D = 98.40 Az = 324.1 (NEIC) PV A 1.3s 15.3nm M = 5.4
21.	e	A	00 34 21.5	<u>South Indian Ocean</u> 24.98 S 69.79 E H = 00 21 07.0 h = normal MB = 4.6 D = 91.10 Az = 327.2 (NEIC)
21.	ePKHKP ePKP2	A A	01 25 08 25 16	<u>Fiji Islands Region</u> 21.79 S 178.91 W H = 01 06 22.6 h = 587 km MB = 4.7 D = 150.03 Az = 346.5 (NEIC) PKHKPV A 1.2s 16.3nm
21.	e	A	08 44 28.5	
21.	eP	A	10 05 08	<u>Negros, Philippine Islands</u> 9.20 N 123.95 E H = 09 52 36.8 h = 565 km MB = 5.2 D = 96.68 Az = 323.6 (NEIC) PV A 1.1s 20.2nm M = 5.3
21.	eP	A	20 27 20	<u>Southern Nevada</u> 37.14 N 116.08 W H = 20 15 00.2 h = 0 km MB = 5.0 D = 81.23 Az = 30.6 (NEIC) Nuclear explosion FLAX (AEC)

Day	Phase		h m s	Remarks
22.	ePn	A	00 36 28	Central Italy 42.82 N 13.01 E
	eSn	A	37 52	H = 00 34 29.1 h = 0 km
	e(Sg)	A	38 48	D = 7.89 Az = 354 (ISC)
22.	ePKHKP	A	01 08 52	Fiji Islands Region 20.30 S 178.40 W
				H = 00 50 11.2 h = 623 km MB = 4.7
				D = 148.69 Az = 347.7 (NEIC)
22.	eX	A	01 13 20.5	XV A 0.8s 11.5nm
22.	eP	A	01 17 23	Taiwan 23.17 N 120.72 E
	LmH	C	52.0	H = 01 04 56.4 h = normal MB = 4.9
	LmV	C	57.2	D = 83.69 Az = 322.8 (NEIC)
				PV A 2.0s 34.2nm M = 5.1
				LmH C 16 1.7/um 5.5
22.	ePKHKP	A	08 40 55	Tonga Islands 18.19 S 175.53 W
				H = 08 21 39.7 h = 250 km MB = 4.4
				D = 147.12 Az = 351.6 (NEIC)
22.	eP	A	08 50 52.5	South of Honshu, Japan 33.46 N 140.86 E
	e(PP)	A	54 25	H = 08 38 22.9 h = 64 km MB = 4.9
				D = 85.04 Az = 330.3 (NEIC)
22.	ePKP	A	09 13 26	Tonga Islands Region 17.36 S 172.61 W
	e	A	13 40	H = 08 53 46.8 h = normal MB = 4.9
	eX	A	13 53.5	D = 146.82 Az = 355.1 (NEIC)
				PKPV A 1.5s 20.1nm
				XV A 1.6 71.4nm
22.	eP	A	10 59 28	South of Alaska 54.57 N 159.51 W
				H = 10 47 47.8 h = normal MB = 4.7
				D = 74.90 Az = 5.8 (NEIC)
22.	e(Sg)	A	12 06 13	Czechoslovakia 49°01'N 13°59.9'E
				H = 12 04.7
				Explosion yield 6 t (PRU)
				D = 2.24 Az = 318 (ISC)

Day	Phase		h m s	Remarks
22.	eX	A	12 34 22.5	XV A 1.9s 22.7nm
22.	ePKP	A	12 37 17	South Sandwich Islands Region
	e	A	37 22	55.40 S 28.17 W
	ePP	AB	38 03.5	H = 12 19 00.1 h = normal MB=5.8 MS=6.4
	eSKS	B	44 10	D = 110.75 Az = 25.8 (NEIC)
	eSKKS	C	45 10	PKPV A 1.6s 30.2nm
	eSdiff.	B	45 52	LmH B 21.5 12.6 /um M = 6.5
	ePS	B	47 40	LmV B 22 13.8 /um 6.5
	ePPS	B	48 45	
	eSS	B	53 40	
	LmV	B	13 18.3	
	LmH	B	18.4	
22.	ePKP	AB	16 41 44	Samoa Islands Region 15.98 S 172.42 W
	LmH	C	17 38.0	H = 16 22 08.9 h = normal MB=5.7 MS=6.0
				D = 145.27 Az = 355.5 (NEIC)
				PKPV A 2.3s 329.0nm
22.	+ePKP	A	19 00 06.5	Samoa Islands Region 16.41 S 172.44 W
				H = 18 40 29.7 h = normal MB = 5.1
				D = 145.69 Az = 355.4 (NEIC)
				PKPV A 1.4s 46.5nm
22.	ePKP	A	20 13 39	Samoa Islands Region 16.13 S 172.60 W
	e	A	13 44	H = 19 54 02.5 h = normal MB = 5.0
				D = 145.40 Az = 355.3 (NEIC)
23.	eP	A	02 29 35.5	Rat Islands, Aleutian Is.
	e	A	29 53	51.27 N 179.12 E
	LmH	C	03 08.0	H = 02 17 40.9 h = 45 km MB=5.2 MS=4.9
	LmV	C	11.0	D = 77.91 Az = 351.9 (NEIC)
23.	eP1	A	06 42 22	Nicaragua 12.36 N 86.12 W
	eP2	A	42 24.5	H = 06 29 42.5 h = 5 km MB=5.6 MS=6.2
	eS	C	52 45	D = 85.40 Az = 39.3 (NEIC)
	LmH	B	18.7	P1V A 1.5s 27.6nm M = 5.3

Day	Phase		h m s	Remarks
cont. 23.	LmV	B	06 18.7	P2V A 1.6s 90.5nm M = 5.8 LmH B 18 9.9/um 6.2 LmV B 19 13.3/um 6.4
23.	eP	A	07 32 30	<u>Nicaragua</u> 12.00 N 86.19 W H = 07 19 49.0 h = 5 km MB = 5.2 D = 85.72 Az = 39.2 (NEIC)
23.	ePKHKP	A	17 31 25	<u>South of Fiji Islands</u> 24.27 S 179.93 E H = 17 12 28.3 h = 510 km MB = 4.0 D = 152.15 Az = 344.0 (NEIC) traces
24.	e(P)	A	03 44 00	<u>Dodekanese Islands</u> 36.19 N 27.77 E H = 03 39 39.6 h = 35 km MB = 4.3 D = 18.56 Az = 326 (ISC)
24.	ePg eSn e(Sg)	A A A	08 14 04.5 14 42 15 33	<u>Northern Italy</u> 44.6 N 10.1 E H = 08 12 06 (BCIS) D = 6.20
24.	ePg e(Sg)	A A	17 37 00 38 16	<u>Northern Italy</u> 44.6 N 10.1 E H = 17 34 54 (BCIS) D = 6.20
24.	ePKP2	A	20 51 41	<u>Macquarie Islands Region</u> 52.37 S 160.5 E H = 20 31 00.1 h = 33 km MB = 5.5 D = 160.62 Az = 277 (NEIC) PKP2V A 1.5s 22.6nm
24.	ePKP	A	23 18 15.5	<u>Loyalty Islands</u> 20.70 S 168.73 E H = 22 58 40.6 h = normal D = 145.09 Az = 334.4 (NEIC)
25.	ePKP ePP	A A	03 52 05.5 55 05	<u>New Hebrides Islands</u> 14.54 S 167.83 E H = 03 32 45.0 h = 26 km MB = 5.6 D = 139.16 Az = 336.9 (NEIC) PKPV A 1.2s 16.3nm

Day	Phase		h m s	Remarks
25.	eP	A	12 42 25	Off East Coast of Kamchatka 53.12 N 162.90 E H = 12 30 52.60 h = 40 km MB = 4.8 D = 73.81 Az = 341.4 (NEIC)
25.	eP	A	12 55 54.5	Rumania 45.80 N 26.73 E H = 12 53 16.4 h = 132 km MB = 4.3 D = 11.18 Az = 301.1 (NEIC)
25.	eP	A	13 07 45	Kurile Islands 45.50 N 149.82 E H = 12 55 52.3 h = 50 km MB = 4.7 D = 77.62 Az = 334.3 (NEIC) traces
25.	ePKP2	A	14 10 47.5	Kermadec Islands Region 30.98 S 179.85 E H = 13 50 59.3 h = 401 km MB = 4.6 (NEIC) D = 158.48
25.	+iP	AB	19 07 25.5	Near East Coast of Kamchatka
	ei	A	07 47	53.01 N 159.38 E
	ePS	C	17 30	H = 18 55 58.1 h = 52 km MB = 5.7
	LmV	B	39.0	D = 73.20 Az = 339.2 (NEIC)
	LmH	B	41.7	PV A 1.2s 246.0nm M = 6.0 LmH B 16.5 4.0 /um LmV B 24 5.6 /um
25.	ePKP	A	22 38 45	Samoa Islands Region 16.59 S 172.66 W
	e	A	38 52	H = 22 19 06.8 h = normal MB = 5.2
	e	A	38 56	D = 145.85 Az = 355.1 (NEIC) PKPV A 1.5s 35.2nm
26.	eP	A	18 42 41.5	Southern Iran 28.40 N 52.73 E H = 18 35 22.9 h = normal MB = 4.6 D = 38.05 Az = 317.2 (NEIC)
26.	eP	A	22 15 34.5	Andreanof Islands, Aleutian Is.
	e	A	15 45	51.67 N 176.28 W
	ePP	A	18 32	H = 22 03 42.2 h = 57 km MB = 5.5

Day	Phase		h m s	Remarks
cont. 26.	ePPS	C	22 26 30	D = 77.84 Az = 354.9 (NEIC)
	eSS	C	30 40	PV A 1.4s 23.2nm M = 5.0
	LmH	C	44.9	LmH C 27 2.1/um
	LmV	C	44.9	LmV C 26.5 2.4/um
27.	-eiP	A	14 17 15	Sea of Okhotsk 46.29 N 144.02 E
	epP	A	18 33	H = 14 06 09.0 h = 342 MB = 5.4
	e	A	18 56	D = 75.05 Az = 330.9 (NEIC)
	e	A	19 46.5	h = 341 km
				PV A 1.5s 100.6nm M = 5.3
27.	ePKP	A	18 18 07.5	Samoa Islands Region 16.16 S 172.62 W
				H = 17 58 29.3 h = normal MB = 4.9
				D = 145.43 Az = 355.2 (NEIC)
27.	eP	A	23 08 01	Queen Elizabeth Islands 76.79 N 107.04 W
	eS	C	14 50	H = 22 59 29.7 h = normal MB=5.0 MS=6.0
	eSS	B	18 25	D = 47.14 Az = 49.7 (NEIC)
	LmH	B	28.6	LmH B 15s 12.9/um M = 6.0
	LmV	B	29.9	LmV B 16 10.6/um 6.0
28.	-eiPKIKP	A	03 00 32.5	Loyalty Islands Region 21.54 S 169.90 E
	+iPKHKP	AB	00 33.8	H = 02 40 53.0 h = 28 km MB=5.6 MS=5.6
	LmH	C	04 09.5	D = 146.32 Az = 334.9 (NEIC)
	LmV	C	19.0	PKHKPV A 1.8s 236.0nm
				PKHKPV B 10 2.3/um
				LmH C 19 1.1/um M = 5.6
28.	e(PKP)	A	04 24 07.5	New Hebrides Islands 17.86 S 167.47 E
				H = 04 04 23.8 h = 5 km MB = 5.1
				D = 142.04 Az = 335.0 (NEIC)
28.	eP	A	06 07 59.5	Philippine Islands Region
	LmV	B	57.7	5.58 N 127.13 E
	LmH	B	07 00.5	H = 05 54 10.6 h = 70 km MB = 5.8
				D = 101.45 Az = 324.1 (NEIC)
				PV A (2.0s) 115.0nm M =(6.2)
				LmH B 17.5 3.8/um
				LmV B 17 4.6/um

Day	Phase		h m s	Remarks
28.	-iPKP	AB	09 19 30	New Hebrides Islands 20.54 S 169.36 E
	eipPKP	AB	19 39.5	H = 08 59 55.4 h = 47 km MB = 5.5
				D = 145.21 Az = 335.0 (NEIC)
				PKPV A 1.2s 364.0nm
28.	ePKP	A	10 56 21	Fiji Region 19.6 S 178.6 W
				H = 10 36 40.9
				D = 147.95 Az = 348 (ISC)
				PKPV A 1.6s 27.5nm
28.	ePKP	A	11 05 28	Loyalty Islands Region 21.71 S 169.88 E
				H = 10 45 50.2 h = 42 km
				D = 146.46 Az = 334.7 (NEIC)
28.	eP	A	14 44 37	Queen Elizabeth Islands 76.78 N 106.30 W
	LmH	B	15 06.1	H = 14 36 07.3 h = normal MB=4.8 MS=5.2
	LmV	B	07.7	D = 47.01 Az = 50.3 (NEIC)
				PV A (2.0)s 85.5nm M = (5.4)
				LmH B 13.5 3.6 µm 5.5
				LmV B 15 2.7 µm 5.4
28.	+eP	AB	17 05 55.5	Afghanistan 34.68 N 70.37 E
	eScP	A	11 23	H = 16 57 45.0 h = 63 km MB = 5.6
	eS	C	12 36	D = 44.85 Az = 309.5 (NEIC)
	eSS	C	15 32	PV A 1.7s 78.3nm M = 5.3
	LmH	C	22.1	ScPV A 1.5 40.2nm
	LmV	C	28.3	LmH C 32 2.8 µm
				LmV C 17 2.6 µm
29.	eP	A	02 39 16.5	Atlantic-Indian Rise 30.93 S 59.09 E
				H = 02 26 09.7 h = normal MB = 4.7
				D = 91.41 Az = 332.0 (NEIC)
29.	eP	A	03 48 55	Crete 34.86 N 23.59 E
				H = 03 44 43.9 h = 35 km MB = 4.3
				D = 18.03 Az = 334.7 (NEIC)

Day	Phase		h m s	Remarks
29.	esPKP	A	05 09 49.5	Near Coast of Central Chile
	LmV	B	53.3	30.56 S 71.02 W
	LmH	B	54.4	H = 04 51 01.6 h = 60 km MB = 6.0
				D = 108.64 Az = 41.8 (NEIC)
				sPKPV A 1.4s 18.6nm
				LmH B 18 0.7 μm
				LmV B 18 0.8 μm
29.	ePKIKP	A	09 06 00	New Britain Region 4.94 S 152.42 E
				H = 08 47 07.1 h = 64 km MB = 5.5
				D = 123.89 Az = 331.0 (NEIC)
				PKIKPV A 1.5s 25.1nm
29.	+ePKP	A	18 38 31	New Hebrides Islands 19.76 S 168.92 E
	epPKP	A	38 38.5	H = 18 18 58.0 h = 31 km MB = 4.7
				D = 144.33 Az = 335.1 (NEIC)
				h = 27 km
				PKPV A 1.6s 44.0nm
29.	ePKIKP	A	19 25 01	Loyalty Islands Region 21.59 S 169.74 E
	ePKP2	A	25 05	H = 19 05 23.4 h = normal MB = 5.4
	LmH	C	20 24.5	D = 146.30 Az = 334.7 (NEIC)
	LmV	B	33.8	PKIKPV A 1.5s 37.7nm
				PKP2V A 1.4 53.5nm
				LmH C 28 0.45 μm M = 5.0
29.	e	A	19 35 26	
29.	e	A	21 26 20.5	
29.	eP	A	23 43 20	North of Svalbard 85.00 N 5.15 E
				H = 23 36 32.6 h = normal MB = 4.7
				D = 34.55 Az = 172.7 (NEIC)
				PV A 1.8s 33.8nm M = 5.0
30.	eSg	A	05 35 49	Netherlands 51.8 N 6.1 E
				H = 05 33 51 (BCIS)
				D = 3.60

Day	Phase		h m s	Remarks
30.	ePn	A	05 49 26	<u>Netherlands</u> 51.61 N 5.66 E
	e	A	49 52	H = 05 48 28.4 h = normal
	e(Sn)	A	50 09	D = 3.88 Az = 102.0 (NEIC)
	eiSg	A	50 23	H = 05 48 26 51.8 N 6.1 E (BCIS)
30.	ePKP	A	06 17 22.5	<u>Samoa Islands Region</u> 16.09 S 172.63 W
				H = 05 57 45.9 h = normal MB = 4.9
				D = 145.36 Az = 355.2 (NEIC)
30.	eP	A	06 27 35.5	<u>North Atlantic Ridge</u> 47.24 N 27.36 W
				H = 06 22 03.0 h = normal MB = 4.7
				D = 25.62 Az = 67.8 (NEIC)
30.	e(P)	A	07 25 05	<u>Philippine Islands Region</u>
				5.72 N 127.11 E
				H = 07 11 16.9 h = 81 km MB = 4.7
				D = 101.33 Az = 324.1 (NEIC)
				PV: traces
30.	eP	A	13 39 16	<u>Tibet</u> 33.67 N 87.57 E
				H = 13 29 35.5 h = normal MB = 4.5
				D = 56.33 Az = 312.1 (NEIC)
30.	eP	A	24 03 48	<u>Tibet</u> 33.60 N 87.72 E
				H = 23 54 05.9 h = normal MB = 4.9
				D = 56.48 Az = 312.2 (NEIC)
31.	eSg	A	07 22 49	<u>France</u> 43°25' N 0°38' W
				H = 07 16 56 (BCIS)
				D = 11.05
31.	ePn	A	07 41 35	<u>Northern Italy</u> 44.8 N 12.9 E
	e	A	41 51.5	H = 07 40 07 (BCIS)
	iSn	A	42 48	D = 5.95
	e(Sg)	A	43 18	

Day	Phase		h m s	Remarks
31.	eSg	A	08 02 17	Netherlands 51.8 N 6.1 E H = 08 00 20 (BCIS) D = 3.60
31.	e	A	10 18 42.5	Kodiak Island Region 58.30 N 151.70 W H = 10 07 19.1 h = 21 km MB = 4.6 D = 70.56 Az = 112 (NEIC)
31.	eP	A	16 13 59	Jan Mayen Islands Region 71.1 N 9.4 W H = 16 08 57.8 h = 0 km MB = 4.2 D = 22.68 Az = 144 (ISC) PV A 1.6s 30.2nm M = 4.6
31.	e	A	18 09 57	

Bei Fragen zur Produktsicherheit wenden Sie sich bitte an:
If you have any questions regarding product safety,
please contact:

Walter de Gruyter GmbH
Genthiner Straße 13
10785 Berlin
productsafety@degruyterbrill.com